T0094261

Introduction to Computing Applications in Forestry and Natural Resource Management

Introduction to Computing Applications in Forestry and Natural Resource Management

Jingxin Wang

CRC Press is an imprint of the
Taylor & Francis Group, an **informa** business

CRC Press
Taylor & Francis Group
6000 Broken Sound Parkway NW, Suite 300
Boca Raton, FL 33487-2742

© 2018 by Taylor & Francis Group, LLC
CRC Press is an imprint of Taylor & Francis Group, an Informa business

No claim to original U.S. Government works

Printed on acid-free paper

International Standard Book Number-13: 978-1-138-62630-0 (Hardback)

This book contains information obtained from authentic and highly regarded sources. Reasonable efforts have been made to publish reliable data and information, but the author and publisher cannot assume responsibility for the validity of all materials or the consequences of their use. The authors and publishers have attempted to trace the copyright holders of all material reproduced in this publication and apologize to copyright holders if permission to publish in this form has not been obtained. If any copyright material has not been acknowledged, please write and let us know so we may rectify in any future reprint.

Except as permitted under U.S. Copyright Law, no part of this book may be reprinted, reproduced, transmitted, or utilized in any form by any electronic, mechanical, or other means, now known or hereafter invented, including photocopying, microfilming, and recording, or in any information storage or retrieval system, without written permission from the publishers.

For permission to photocopy or use material electronically from this work, please access www.copyright.com (http://www.copyright.com/) or contact the Copyright Clearance Center, Inc. (CCC), 222 Rosewood Drive, Danvers, MA 01923, 978-750-8400. CCC is a not-for-profit organization that provides licenses and registration for a variety of users. For organizations that have been granted a photocopy license by the CCC, a separate system of payment has been arranged.

Trademark Notice: Product or corporate names may be trademarks or registered trademarks, and are used only for identification and explanation without intent to infringe.

Library of Congress Cataloging-in-Publication Data

Names: Wang, Jingxin, 1963-
Title: Introduction to computing applications in forestry and natural resource management / Jingxin Wang.
Description: Boca Raton : CRC Press, 2017. | Includes bibliographical references.
Identifiers: LCCN 2017011246 | ISBN 9781138626300 (hardback : alk. paper)
Subjects: LCSH: Forest management--Data processing. | Forest management--Computer programs | Forest management--Problems, exercises, etc. | Natural resources--Management--Data processing. | Natural resources--Management--Problems, exercises, etc.
Classification: LCC SD381.5 .W36 2017 | DDC 634.9/20285--dc23
LC record available at https://lccn.loc.gov/2017011246

Visit the Taylor & Francis Website at
http://www.taylorandfrancis.com

and the CRC Press Website at
http://www.crcpress.com

Printed and bound in the United States of America by Sheridan

To my dear wife, Xiaoming Liang, and our dear son, Jerry Wang.

Contents

Section II Data Manipulation and Analysis

Section V Visual Basic .NET Programming

Section VI Web-Based Applications

Preface

Due to the complexity of operational forestry problems, computing applications are becoming pervasive in all aspects of forest and natural resource management. We therefore wanted to provide a comprehensive introduction to computing and applications in forest and natural resource management that would be designed for both undergraduate and graduate students of these topics. This book introduces state-of-the-art applications for several of the most important computing technologies in terms of data acquisition, data manipulations, basic programming techniques, and other related computer and Internet concepts and applications.

We organized the information contained in this book in six parts. Section I introduces basic computing concepts and software engineering design. Section II addresses data manipulation, basic statistical analysis, and mathematical programming. Section III deals with database management such as entity relationships and structural query language. Section IV focuses on spatial technology, handheld programming, and applications in natural resources. Section V introduces object-oriented programming using Visual Basic .NET, while Section VI addresses web-based programming and applications.

Specifically, this text covers

- Computer operating systems and resources
- Elementary data manipulations, statistical computing, and mathematical programming using MS Excel
- Basic database management with MS Access
- Relational database management
- Data acquisition using handheld computers and their applications in natural resources
- GPS and GIS applications in forest resource management
- Object-oriented programming with Visual Basic .NET
- HTML and web-based programming
- Application examples in forest resource management

We discuss in detail several major computing application examples in the lessons of this book, which are the products of our previous research projects; and they address

- Databases—databases built for optimal bucking, forest cruising, and inventory

- VB.NET applications—forest harvesting simulation and log inventory
- VB.NET for mobile devices—time study of timber harvesting operations and log inventory
- VBA—forest and biomass harvesting productivity and cost analysis
- ASP.NET and HTML—logging safety initiatives and online forest health management systems
- Optimization and spatial analysis applications—3D log bucking and sawing, identifying and mapping forest vegetation phenology, and biomass harvesting and scheduling

Since 2002, the materials covered in this book have been used in required courses for both undergraduate and graduate students in the Division of Forestry and Natural Resources at West Virginia University: FOR 240—Introduction to Computing in Natural Resources, and WDSC 555—Computer Applications in Forest Resource Management. For these courses, we designed class exercises that are also provided in this text for classroom use. We also included a comprehensive project in which students develop an application for natural resources in three distinct modules that each focus on a specific application topic:

Module 1: Data Storage—Design and build a simple database to store field data using MS Access.

Module 2: Data Acquisition—Collect field data using a handheld PC.

Module 3: Data Manipulation—Design an interface and implement business functions to transfer field data from a handheld to a desktop PC and manipulate/analyze the data to generate desired reports.

These modules are cumulative; each module or component must be integrated to accomplish a subsequent module. The project should be programmed based on the principles of software engineering.

I thank many individuals who inspired me to accomplish this book. I am grateful to Drs. Bojan Cukic, John Atkins, James Mooney, and Camille Hayhurst for their lectures and notes while I took their classes as an MS student in computer science at West Virginia University. I also thank Drs. Amy Falcon, Damon Hartley, Wenshu Lin, Weiguo Liu, Benktesh Sharma, Jinzhuo Wu, and Zhen Yu, Yuxi Wang, Changle Jiang, and John Vance for their contributions to the book while they assisted me in teaching. Special thanks go to Sarah Owen for her great and professional efforts to review and improve the book.

Jingxin Wang
Morgantown, West Virginia

Author

Dr. Jingxin Wang is professor and associate director of research and the director of the Renewable Materials and Bioenergy Research Center in the Division of Forestry and Natural Resources at West Virginia University, Morgantown, West Virginia. He received his BS in forest/mechanical engineering from Jilin Forestry College, China, MS and PhD in forest/mechanical engineering from Northeast Forestry University, China. He received an MS in computer science from West Virginia University and a PhD in forest resource management from the University of Georgia, USA. He has taught undergraduate and graduate students for more than 20 years. His research interests include biomass energy and bioproducts, forest carbon sequestration and optimization, computer simulation and system modeling, and forest ecosystem management and climate change. Dr. Wang has authored or coauthored 150 refereed papers and 12 books or book chapters. He has served as an editorial board member and associate editor for four international journals and as adjunct professor for four Chinese universities/institutions. He is an active member of six international professional societies.

Section I

Foundations

1

Fundamental Computing Concepts

1.1 Computer Operating Systems

An **operating system** (OS) is a software program that acts as an intermediary between the user of a computer system and the computer hardware (Silberschatz et al. 2002, 2013). "The purpose of an operating system is to provide an environment in which a user can execute programs in a convenient and efficient manner" as Silberschatz et al. (2013) indicated. The OS makes the computer more convenient to use, allows users to execute application programs in an environment without the necessity of communicating directly with hardware devices, and helps to ensure the efficient use of computer resources. An OS is the first program loaded into memory when a computer system is booted. It is a program that manages the computer hardware and controls interactions between application programs and the hardware.

A **computer system** can be divided roughly into four components (Silberschatz et al. 2002, 2013):

1. *Computer hardware* provides the basic computing resources including a processor or central processing unit (CPU), computer memory, and input/output (I/O) devices such as monitors, printers, disk drives, USB storage devices, and others.
2. *Operating systems* control and coordinate the use of the hardware among application programs for users.
3. *System and application programs* include system or custom-written software products (like compilers, database management systems, and Office programs) for solving computing problems.
4. *Users* include operators, machines, and other computers that interact with a computer system to accomplish some specific tasks.

1.2 Brief History of Operating Systems

Computer OSs are complicated programs that have evolved with the development of hardware and the increased demand of computing services. The long history of computer OSs can be roughly categorized into the following five phases (Hayhurst 2002, Silberschatz et al. 2002, 2013):

1. 1940s–1950s: First Generation, early computers—In the early days of computing, there were no OSs. Early computers were large and expensive. Typically one user ran one program at a time. One of the major deficiencies of early computers was their inefficient use of CPU time, which was also the primary driver for the change to OSs.
2. 1950s–1960s: Second Generation, beginning of batch OSs—Due to the high cost and inefficiency of computers, programmers began "batching" or running the programs with similar resources through the computer together as a group to improve a computer's performance. The solution was generally adopted and became known as the **batch system** (Tanenbaum 2001). During this time period, significant changes also occurred in computing (Hayhurst 2002): (a) advances were made in hardware, including magnetic tape, line printers, and magnetic disks; (b) assemblers were first created to allow programs to be written in something other than machine language; (c) device drivers became available to communicate directly with specific I/O devices; and (d) compilers for higher-level languages such as FORTRAN and others were developed.
3. 1970s–1980s: Third Generation, interactive multitasking systems—Multitasking allows users to share the computer and run multiple tasks (processes) at the same time. The processor executes multiple tasks by switching among them, but the switches occur so frequently that the users can interact with each program while it is running (Silberschatz et al. 2002). In multitasking systems, as Hayhurst (2002) and Silberschatz et al. (2002) stated, "Several jobs must be kept in memory at the same time; this requires memory management and protection."
4. 1980s–Present: Fourth Generation, personal computers—IBM started to design personal computers (PCs) in the 1970s and early 1980s with the introduction of a new OS. The Disk Operating System (DOS) was first launched by Seattle Computer Products. A revised system was renamed Microsoft Disk Operating System (MS-DOS) and quickly came to dominate the IBM PC market (Tanenbaum 2001). MS-DOS was later widely used on the 80386 and 80486. Although the initial version of MS-DOS was fairly primitive, subsequent versions included more advanced features, including many taken from Unix.

In the 1960s, Doug Engelbart at Stanford Research Institute had invented the graphical user interface (GUI) with windows, icons, menus, and a mouse. These ideas were adopted by researchers at Xerox PARC and incorporated into machines they built. In the early 1980s, Steve Jobs started building Apple Macintosh with a GUI. Microsoft then produced a GUI-based system called Windows, which originally ran on top of MS-DOS. The Windows OS has continued to evolve into the current version.

5. 1990s–Present: Fifth Generation, handheld and tablet computers—The early version of the handheld PC (HPC) was designed by Hewlett Packard around 1990. Since then, many products have become available from different companies. Most of these HPCs were run under the Microsoft Windows CE OS. Some models of HPCs were discontinued in the early 2010s as the market shifted to tablet computers and smartphones. However, HPCs are still available as data loggers for field data collection using the most current mobile OSs such as Microsoft Windows Mobile or Windows 10 or a later version.

   Tablet computers were introduced in 2010 and have only grown in popularity since then. Many tablet products are available such as iPad, Surface, and Android tablets. The OSs they use include iOS, Android, and MS Windows 10 (or later versions).

1.3 Types of Operating Systems

In today's market, there are several types of OSs available, such as Windows, MacOS, Linux, and Unix, and each of these has evolved through various versions. MS Windows, for example, has gone through Windows 3.x, Windows 95, 98, Windows ME, Windows NT, Windows 2000, Windows XP, Vista, 7, 8, 10, and so on. One way of categorizing OSs depends on the computers they control and the applications they support (Silberschatz et al. 2002, 2013):

1. Desktop/laptop systems: Desktop/laptop systems are the most commonly used OS on PCs. Microsoft Windows, MacOS, and Linux are three well-known examples of this type of OS. In the 1970s, during the first decade of PCs, PCs lacked the features needed to protect an OS from user programs (Silberschatz et al. 2002); therefore, they were neither multiuser nor multitasking. Modern desktop/laptop systems are both multitasking and multithreading. Multithreading is an extension of multitasking, where a single application or task can be divided into threads and each of the threads can be executed in parallel.

2. Multiprocessor systems: A multiprocessing OS (also called a parallel system) uses two or more processors or CPUs on one computer, sharing main memory and peripherals. Multiprocessor systems have three main advantages: increased throughput, economy of scale, and increased reliability (Silberschatz et al. 2002). Examples of these systems include Linux, Unix, and Windows. There are two types of relationships among these processors: symmetric multiprocessing, in which all processors are peers, and asymmetric multiprocessing, where one processor controls the others (Silberschatz et al. 2013).
3. Distributed systems: A distributed OS considers the users like an ordinary centralized OS but runs on multiple, independent, and networked CPUs (Tanenbaum 1993). Distributed systems depend on networking for their functionality (Silberschatz et al. 2002). "With the introduction of the Web in the mid-1990s, network connectivity became an essential component of a computer system" (Silberschatz et al. 2002).

   Typical examples of distributed systems include Unix and Mac OS. Distributed systems provide these advantages: sharing of computer resources, reliability, and computing efficiency.
4. Real-time systems: Real-time OSs are used to control machinery, scientific instruments, and industrial systems. In general, the user does not have much control over the functions performed by this type of OS. Real-time OSs must guarantee a response within a specified time and data flow.
5. Handheld systems: Handheld systems, also known as mobile OSs, are designed to run on mobile devices such as personal digital assistants, smartphones, tablet computers, and other handheld devices. Handheld devices typically have a limited amount of memory, slower processors, and smaller display screens. Commonly used systems include Android, iOS, and Windows Mobile such as Windows 10 or a later version.

1.4 Major Operating System Responsibilities

Operating systems tackle many responsibilities in order to run computers conveniently, efficiently, and safely. These responsibilities range from simple tasks such as I/O device management to complicated tasks such as memory and process management. Here are a few major responsibilities of most OSs (Lane and Mooney 2001).

1.4.1 User Interface

An OS is responsible for the interaction between users and their computer programs and hardware. This interaction is usually carried out through input commands such as typing lines of text, selecting menu items or graphic icons, or speaking phrases. Input commands should be consistent and user friendly.

The user interacts with the computer system through an interface mechanism that is called the **application programming interface (API).** In Microsoft's version, called a dynamic link library (DLL), APIs are centralized in a binary file that has a specialized executable format that Windows can read (Bock 2000). The Windows API is a core set of application programming interfaces in the Windows OSs, including kernel32.dll, user.dll, and gdi32.dll.

1.4.2 Device Management

Computers can connect to a variety of input, output, and storage devices that must be controlled by the OS, including monitors, printers, hard drives, USB drives, and other devices (Lane and Mooney 2001). **Device management** encompasses all aspects of controlling these devices: starting operation, requesting and waiting for data transfers, tracking these devices, and responding to errors that may occur all by use of device drivers (Lane and Mooney 2001). A **device driver** is a program that is written to communicate with a specific type of I/O device.

1.4.3 Time Management

Time management controls the time and sequence of computing events. A special category of I/O device is the **timer**, whose role is to measure time and cause events to occur at specific times.

1.4.4 Memory Management

Memory management is the process of controlling and coordinating the use of computer memory. It refers to the management of computer main memory, a **critical resource** in any computer system. Early forms of memory management were concerned primarily with allocating portions of main memory to each process as it began, while newer strategies allow additional areas of memory to be allocated and deallocated as desired using swapping (Lane and Mooney 2001). **Swapping** is a mechanism by which a process can be moved temporarily out of main memory to secondary storage, making that memory available to other processes. This strategy makes the memory space available for more immediate needs. An evolved form of the swapping technique, **virtual memory** systems have now become common

(Lane and Mooney 2001). Virtual memory can extend the use of physical memory and provide memory protection.

1.4.5 File Management

Data stored on computers are always organized into files. **File management** is a process of naming, storing, and handling files. The OS controls file operations such as writing, reading, and security protection.

1.5 Computer Resources and File Systems

1.5.1 Computer Resources

The resources provided by a computing system may be grouped into two major categories (Lane and Mooney 2001): (1) **Physical resources** (also called hardware resources) and (2) **Logical resources** (also known as software resources). As the name implies, physical resources are the permanent physical components of a computer system. Logical resources are collections of information, such as data or programs. Logical resources must be stored within physical resources (e.g., within main or secondary memory). The two principal objectives of an OS's resource management are convenient use and controlled sharing of these physical and logical computer resources (Lane and Mooney 2001). The resources that must be managed by a typical computer OS are summarized in Table 1.1.

TABLE 1.1

Computer Resources Managed by an OS

Physical Resources	Logical Resources
Processor (a critical resource)	Applications and sessions
Main memory (a critical resource)	Processes and tasks
I/O devices and controllers	Files
Secondary storage (disks and memory cards)	Shared programs and data
Timers and clocks	Procedures that perform useful services

Source: Modified based on Lane, M. and Mooney, J., *A Practical Approach to Operating Systems* (Lecture Notes), Lane Department of Computer Science and Electrical Engineering, West Virginia University, Morgantown, WV, 2001.

1.5.2 File Systems

A collection of information maintained for a set of users in long-term storage is called a **file system.** "The file system consists of two distinct parts: a collection of files, each storing related data, and a **directory** structure, which organizes and provides information about all the files in the system" (Silberschatz et al. 2002, 2013).

A **file** is a named collection of related information that is recorded on secondary storage (Silberschatz et al. 2002). As Silberschatz et al. (2013) described, "From a user's perspective, a file is the smallest allotment of logical secondary storage." The information in a file is defined by its creator. Many different types of information may be stored in a file, such as source programs, object programs, executable programs, numerical data, text, graphic images, and audio/video recordings (Table 1.2).

TABLE 1.2

Common File Types

File Type	Extension	Function
Executable	`exe, com`	Machine-language program
Source code	`c, cpp, java, vb`	Source code in various languages such as C, C++, Java, and Visual Basic
Text	`txt, doc, docx`	ASCII text data, documents
Presentation, spreadsheet	`ppt, pptx, xls, xlsx, xlsm`	PowerPoint or Excel file formats
Library	`lib, dll`	Libraries of routines for programmers
Archive	`zip, tar`	Compressed files for archiving storage
Multimedia	`jpg, jpeg, mpg, avi, wmv`	Binary file containing image, audio/video
Web pages	`html, htm, php, asp`	Web-based source code

Sources: Modified based on Silberschatz, A. et al., *Operating System Concepts*, 6th edn., John Wiley & Sons, Inc., New York, 2002; Silberschatz, A. et al., *Operating System Concepts*, 9th edn., John Wiley & Sons, Inc., New York, 2013.

1.5.2.1 File Attributes or Properties

A file has certain properties, which vary from one OS to another (Silberschatz et al. 2013). Typical properties of a file include name, type of file, location, size, protection, and dates of creation and modification. File size is typically measured in bytes. A **byte** is equal to 8 bits, and a bit is the basic computer storage unit. A kilobyte is 1024 bytes, a megabyte is 1024^2 bytes, a gigabyte is 1024^3 bytes, and a terabyte is 1024^4 bytes.

1.5.2.2 File Operations

The computer OS is responsible for managing the file system that provides users and programs with a suitable set of operations to manage files. Some of the typical operations could be related to creating files, reading files, modifying files, or deleting files (Silberschatz et al. 2013).

1.5.2.3 File Types and Naming

A common way to implement a file type is to include the type as part of the file name (Silberschatz et al. 2013). A file name is usually a string of characters, such as "FOR240.doc" or "FOR240.vb." In some systems, a file name is case sensitive. "The name is typically split into two parts: a name and an extension, usually separated by a period" (Silberschatz et al. 2013). Typical file types in MS-DOS, Windows, or UNIX are summarized in Table 1.2. Application programs either automatically put file types as extensions or allow users to select specific file types as they save the files.

For efficient and convenient file management, we should always use meaningful file names and structure. What follows are some examples of complete, structured file names:

```
A:\BOOK\CHAPTERS\FILEMGT.TXT (MS-DOS)
C:\Book\Chapters\File Management\FileMgt.txt (Windows)
/usr/jwang/book/chapters/filemgt.txt (Unix)
```

In the above examples, the main file name is the same: `FileMgt`. In a Windows system, we can easily perform the above operations on a directory. However, in MS-DOS, the related commands need to be used to perform the file operations.

`cd`—change directory

`dir`—list files in a directory

`del`—delete a file

`rename`—rename the file

`copy`—copy files

`format`—format the disks

1.5.2.4 Directory Structure

A directory contains files and subdirectories. A typical directory structure is the **tree-structured directory,** which allows users to create their own subdirectories and organize their files accordingly. The MS-DOS/Windows system, for example, is structured as a tree. In fact, a tree is the most common directory structure. The tree has a **root directory.** Every file in the system has a unique **path name**. A path name of a file is the path from the root, through all the subdirectories, to that specified file.

Class Exercises

1. Operating systems:
   a. What is an operating system?
   b. What are the major responsibilities of an operating system?
   c. List the types of operating systems.
   d. Compare and contrast multiprogramming vs. multitasking.
   e. What are the basic components of a computer system?
2. Computer resources and file systems:
   a. Compare and contrast physical and logical resources.
   b. Define file, file attributes, and file systems.
   c. Compare and contrast directory and tree-structured directory.

References

Bock, J. 2000. *Visual Basic 6 Win32 API Tutorial*. Wrox Press, Birmingham, U.K., 368pp.

Hayhurst, C. 2002. *Semantics of Programming Language* (Lecture Notes). Lane Department of Computer Science and Electrical Engineering, West Virginia University, Morgantown, WV.

Lane, M. and J. Mooney. 2001. *A Practical Approach to Operating Systems* (Lecture Notes). Lane Department of Computer Science and Electrical Engineering, West Virginia University, Morgantown, WV.

Silberschatz, A., P.B. Galvin, and G. Gagne. 2002. *Operating System Concepts* (6th Edition). John Wiley & Sons, Inc., New York.

Silberschatz, A., P.B. Galvin, and G. Gagne. 2013. *Operating System Concepts* (9th Edition). John Wiley & Sons, Inc., New York.

Tanenbaum, A. 1993. Distributed operating systems anno 1992. What have we learned so far? *Distributed Systems Engineering* 1(1): 3–10.

Tanenbaum, A. 2001. *Modern Operating Systems* (2nd Edition). Prentice Hall, Upper Saddle River, NJ.

2

Programming Languages and Software Engineering

2.1 Programming Languages

Programming languages were invented to make the computer easier to use and are designed to be both higher level and general purpose (Sethi 1996). A higher-level language such as C++, VC, VB, or Java is independent of the underlying machine, while a general-purpose language can be used in a wide range of applications. Existing programming languages can be classified into the following four families based on their models of computation (Sethi 1996, Hayhurst 2002):

1. **Imperative programming**
   a. Includes action-oriented languages such as Pascal, C, and FORTRAN
   b. Focuses on how the computer should perform its task
   c. Views computation as a sequence of actions
   d. Views instructions as performing actions on data stored in memory
   e. Approaches a program as a series of steps, each of which performs a calculation, retrieves input, or produces output
   f. Uses languages that encapsulate procedural abstraction, assignments, loops, sequences, and conditional statements
2. **Functional programming**
   a. Employs a computational model based on the recursive definition of functions (originated with Lisp)
   b. Considered a program as a function from inputs to outputs, defined in terms of simpler functions through a process of refinement

   c. Views a program as a collection of mathematical functions each with an input (domain) and a result (range)
   d. Employs functions that interact and combine with each other using functional composition, conditionals, and recursion: Lisp, Scheme
3. **Logic programming**
   a. Takes inspiration from prepositional logic
   b. Performs computation as an attempt to find values that satisfy certain specified relationships using a goal-directed search through a list of logical rules
   c. Attempts to use logical reasoning to answer queries
   d. Views a program as a collection of logical declarations about what outcome a function should accomplish rather than how that outcome should be accomplished
   e. Executes a program by applying these declarations to achieve a series of possible solutions to a problem, such as Prolog
4. **Object-oriented programming**
   a. Is relatively recent and closely related to imperative language
   b. Has much more structure and a distributed model of both memory and computation
   c. Implements computation as interactions among semi-independent objects, each of which has both its own internal state and executable functions to manage that state
   d. Views a program as a collection of objects that interact with one another by passing messages that transform an object's state
   e. Uses object modeling, classification, inheritance, and information hiding as fundamental building blocks for object-oriented programming (OOP) languages, such as C++, Java, and VB.NET

There are two basic approaches to implementing a program in a higher-level language: compilation and interpretation.

- In **compilation**, the language is brought down or converted to the level of the machine using a translator called a compiler. In the compilation process, a source program in a high-level language such as C++ is first compiled to an executable/target program. We can then call/execute the target program with input to generate output (Figure 2.1a). Compilation is usually more efficient than interpretation.
- In **interpretation**, the machine is brought up to the level of the language, building a higher-level machine (called virtual machine) that

can run the language directly. For the interpretation process, we first develop a source program in a higher-level language such as VB or Python (Figure 2.1b). The program is then executed or interpreted together with input under the language environment, and there is no compilation phase involved. Interpretation can provide more functionality, such as debugging during execution.

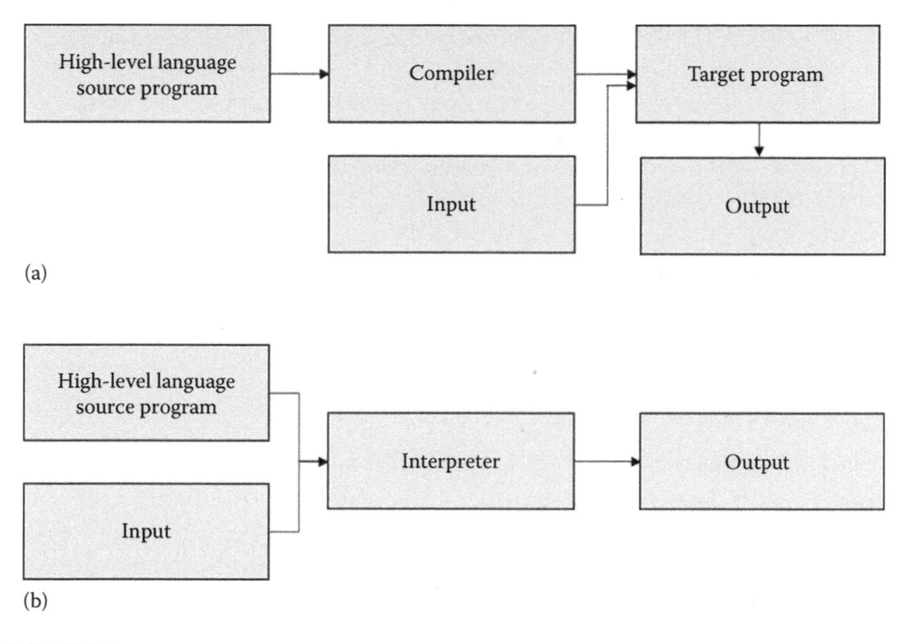

FIGURE 2.1
Program implementation—compilation (a) and interpretation (b).

2.2 Object-Oriented Programming

An object-oriented approach produces a modular solution of programming that is a collection of objects that interact with each other, and are not simply a sequence of actions. The fundamental idea behind object-oriented languages is to integrate or combine both data and functions (or methods) into a single unit that is called an **object**. Objects are members of a **class** that is a grouping of similar objects. **Object-oriented programming** embodies three fundamental principles (Lafore 1998):

1. Encapsulation: Objects combine data and operations. An alarm clock object, for example, contains both time and operations such as "set the time." A tree object in forest simulation could contain data on

tree parameters such as diameter and height, and functions on tree growing and related silvicultural activities.

2. Inheritance: The idea of classes leads to the idea of inheritance. We understand the biological classes of animals and trees. Similarly, an OOP class can be divided into subclasses. Classes can inherit properties from other classes.
3. Polymorphism: Using operators or functions in different ways depending on what they are operating on is called polymorphism. Polymorphism means "many forms." Objects can determine the appropriate operations at execution time. For example, "+" (which is an operator in C++ and VB) when with numeriç values, addition occurs; when with string variables, concatenation will occur. This means the compiler determines the correct meaning of "+" at execution time, and we say "+" is overloaded.

Simula was the first OOP language. Java, Python, C++, and Visual Basic .NET are the most popular OOP languages today.

2.2.1 C and C++

C was created in 1972 by Dennis Ritchie as an implementation language for software associated with a Unix operating system at AT&T Bell Lab. C++ is derived from the C language. It was first introduced in the early 1990s. The most important elements added to C to create C++ were concerned with classes, objects, and OOP (Lafore 1998). However, C++ has many other features such as improved input/output.

2.2.2 Visual Basic .NET

Before Visual Basic (VB) 1.0/Windows 3.X was introduced to the world in 1991, developers had to be well versed in C++ programming as well as the rudimentary building blocks (Windows API) of the Windows system itself. VB changed the face of Windows programming by removing the complex burden of writing code for the user interface.

When Microsoft introduced VB 3.0 in 1993, the programming world was changed again. VB 3.0 did include user-friendly interfaces, but also introduced database applications with data access objects (DAOs). VB 3.0 included version 1.1 of the Microsoft Jet Database Engine.

VB 4.0 and 5.0 were released in 1995 and 1997 specifically for Windows 95. In addition to DAOs, VB 4.0 and 5.0 introduced a new data object, remote data object (RDO), for client/server applications. The program could be used for both 16-bit and 32-bit versions of Windows.

VB 6.0 was released in 1998 for Windows 1995 and 2000. It contained DAOs, RDOs, and ActiveX Data Objects (ADOs) for database applications.

It first introduced web-based applications. VB 6.0 has been considered the most successful VB programming packet.

VB.NET is the successor to VB 6.0 and is part of the Microsoft .NET platform for Windows XP, 2000, 7, 8, 10, or later versions. Since 2003, several versions of VB.NET have been released together with Microsoft Visual Studio. VB.NET enhanced the programming with ADO.NET and more functions on web-based applications. In the past, VB has been criticized as a "toy" language, compared to C++/Java. However, VB.NET has become a great choice for programmers of all levels using OOP techniques.

2.3 Software Engineering

Computer software has become a driving force. It is the engine that drives business decision-making and serves as the basis for modern scientific investigation and engineering problem-solving (Pressman 2001). Computer software is the product that software engineers design and build. It includes programs that execute, within a computer of any size and architecture, documents that encompass hard copy and virtual forms, and data that combine numbers and text (Pressman 2001).

Software engineering (SE) is the establishment and use of sound engineering principles in order to obtain economical software that is reliable and works efficiently on real machines (Pressman 2001). It is a branch of computer science that provides techniques to facilitate the development of computer programming. SE originated in the 1960s due to the software crisis and is not synonymous with programming. SE encompasses all of the stages involved in the development of a large software product. Its goal is to produce or provide "high quality, well-designed, and well-engineered" software. The software products developed using SE principles usually have the following characteristics (Kochut 1996):

- Perform precisely under all circumstances
- Reliable (bug free)
- Maintainable (easy to modify)
- Use appropriate user-friendly interfaces
- Highly efficient and cost effective

When we start to develop a software product, we always ask ourselves if we need to use SE for the process. The answer could be "We should always use SE." However, in some cases, it also depends on the size and budget of the software program. For example, while writing 100 lines of a program in either C++ or VB may not be a tough task, writing a program of several hundred thousand lines of code will take dramatically longer to accomplish. An exponential growth relationship exists between software size and its development time and cost. Well-designed software would enable programmers to efficiently use their time to develop and test the product. Therefore, using SE to design a software product will help reduce its development time and cost.

To solve actual problems, a software engineer or a team of engineers must incorporate a development strategy that encompasses the process, methods, tool layers, and generic phases (Pressman 2001). This strategy is often referred to as a **process model** or a **software engineering paradigm**. There are several software process models, such as the waterfall model, incremental model, V-model, and iterative model. Each model follows a particular life cycle of software in order to ensure its success in the process of software development. A common process model is established by defining a small number of framework activities that are applicable to all software projects, regardless of their size or complexity (as illustrated in the example of a waterfall model) (Figure 2.2).

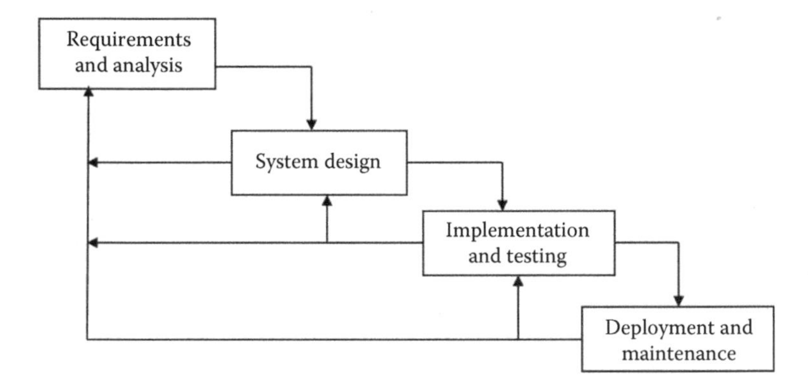

FIGURE 2.2
Waterfall model of software development process.

The **software life cycle** is a process that ensures a robust software product is developed using the SE principles. A software life cycle can be depicted as phases or segments of the software life on a wheel (Carrano 1995). Each phase produces deliverables required by the next phase. We typically start by specifying the requirements of the software, then move sequentially to the phases of design, verification, etc. However, these phases are interrelated. For example, testing a program can suggest changes to either the program specification or solution design (Carrano 1995).

In a typical SE process, there are eight phases surrounding the documentation core. Documentation is not a separate phase, as you might expect. It is integrated into all eight phases of the software life cycle (Carrano 1995):

1. *Phase 1—Specification*: In this phase, we need to identify the initial statements of the software's purpose and also specify clearly all

aspects of the program requirements and potential problems. Some issues that we must address as we write the specifications for the software include (1) data input and validity, (2) potential software users and types of user interfaces, (3) error detection and handling, (4) special cases that need to be considered, (5) output, (6) necessary documentation, and (7) future enhancements of the software product.

2. *Phase 2—Design*: Once we have completed specification, we need to design solutions to the problems associated with the software program. The best way to simplify the problem-solving process is to divide a large problem into small, manageable parts (Carrano 1995). The resulting program will contain modules, which are self-contained units of code. A module could be a single function or several functions and other blocks of code. System design helps in specifying hardware and system requirements and also helps in defining overall system architecture.
3. *Phase 3—Verification*: In this phase, we need to use available methods to prove that the algorithms implemented in the program are correct. This process could be time-consuming.
4. *Phase 4—Implementing/Coding*: In this phase, we divide the program into components and modules, then start coding. This could be the longest phase of a software life cycle. There are two main ways to implement coding: bottom-up implementation and top-down implementation.
5. *Phase 5—Testing*: After coding is complete, we perform tests to ensure the software addresses the required needs, and to remove any errors we may encounter. There are a few ways to test software, including unit testing, integration testing, system testing, and acceptance testing.
6. *Phase 6—Refining*: In previous phases, we make some simplifying assumptions. We should always leave room to improve and refine solutions to the problems of the software product. For example, input or output formats might need to be refined.
7. *Phase 7—Production or Deployment*: After successful testing and refinements, the software is deployed to its intended users, installed on their computers, and used.
8. *Phase 8—Maintenance*: Real-world users of the software always seem to discover some unexpected problems that will need to be fixed. Developers may also need to add new features to enhance the software product after its release.

Time and cost necessary for each phase in the life cycle of a software product vary from system to system. For example, more than 40% of time or cost may be needed for the design phase in scientific and business types of systems. However, there may be more than 50% of time or cost needed for the testing phase of an operating system.

2.4 Example of Forest Harvesting Simulator Design

The Forest Harvesting Simulator is a more convenient and cost-effective way to perform simulations of forest operations (Wang and Greene 1999). This interactive simulator allows you to make applications that make full use of the graphical user interface (GUI). Here are the few major phases in developing this forest harvesting simulator.

2.4.1 System Design

Object-oriented modeling techniques (OMTs) were employed in the system design. A hierarchical structure among different modules is very useful while modifying the program with OMT (Rumbaugh et al. 1991). A schematic hierarchy of the forest harvesting simulation system was demonstrated with the following major layers/components (Wang and LeDoux 2003, Figure 2.3):

- GUI application layer: This layer is composed of multiple subapplication type modules that deal with GUI support of different functions like browsing files, performing operations, analyzing, viewing, and reproducing outputs. This layer talks to the underlying class layer following the system hierarchy.
- Module layer: This is the major part of the system. Objects, controls, and object-oriented data models are implemented here. The modules contain form, class, and standard modules.
- Data store layer: The module layer talks to this layer to obtain the persistent data support.

Interactive simulation is a major part of this harvesting simulator, which was described and published in detail in a previous paper (Wang and Greene 1999). The modularity or the internal organization of this simulation program serves primarily as the functional tool for project management. Usually, an event is executed by clicking a corresponding command button or item. The main event procedures in the Multiple Document Interface form of this harvesting simulator are `FILE`, `EDIT`, `RUN`, `ANALYSIS`, `VIEW`, `OUTPUT`, and `HELP` (Figure 2.4). Other forms and procedures under this form can be invoked at this point.

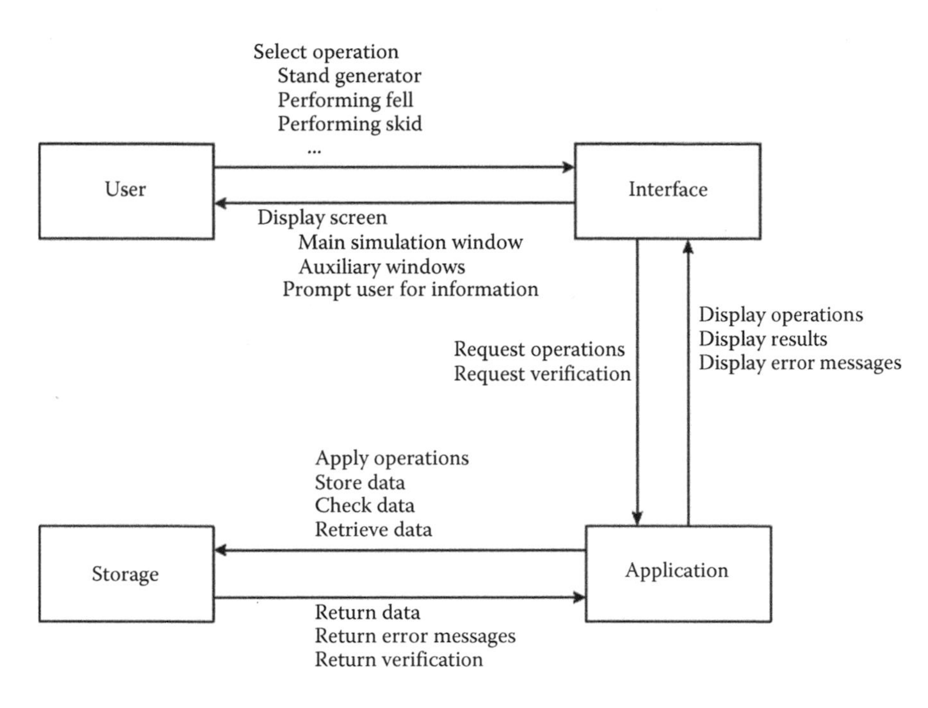

FIGURE 2.3
Event flow diagram of the system.

Communication among different modules in the system uses Dynamic Data Exchange, Dynamic-link Libraries, and Windows API. The front end of the harvesting simulation system could be Windows platform, while the relational database is used in the back end.

The whole system is event-driven. Event-driven applications execute codes in response to an event. Each object (such as a form or a control) in the system has a predefined set of events. If one of these events occurs, the system invokes the codes in the associated event procedure. Objects in the system using VB automatically recognize a predefined set of events if you invoke an event.

2.4.2 Functional Requirements

Through GUI, the user can browse or search the files under a specified directory and drive. This allows the user to find files created while performing simulations. Stand generation, felling, and skidding or forwarding simulations are performed under the specific event procedures incorporated with other objects. The simulation can be analyzed statistically and economically. The simulation results can be retrieved and viewed again. The outputs can also be reproduced. The event procedures or functional commands are

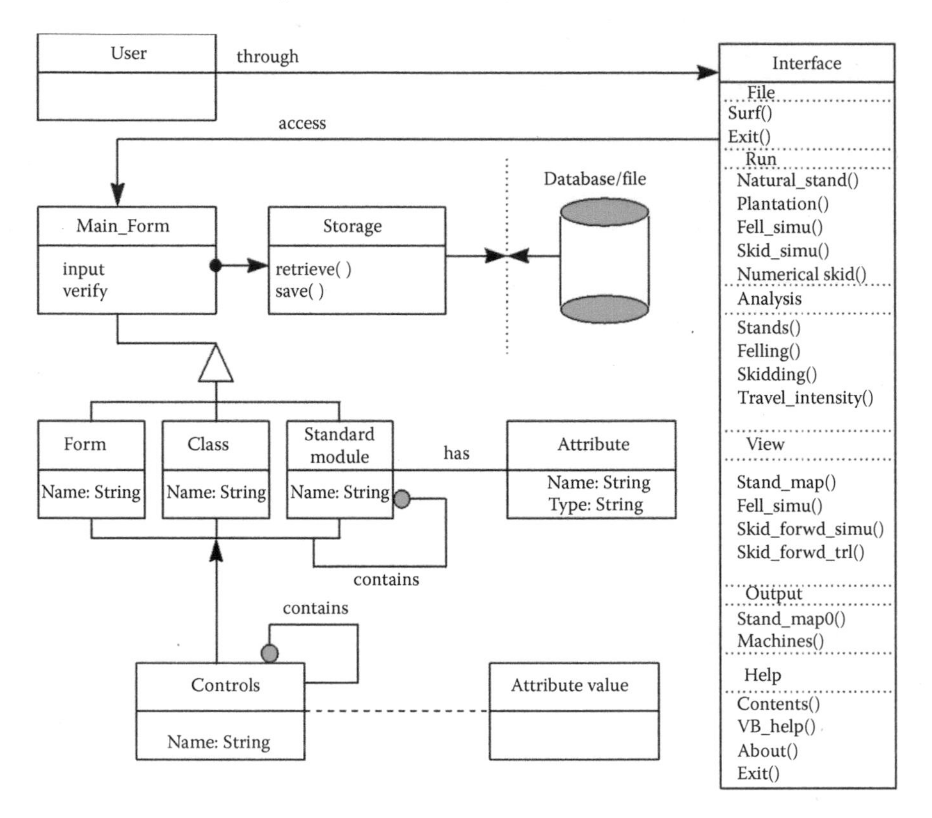

FIGURE 2.4
Architecture of the system.

organized in a hierarchical manner. The following is a list of the major functions or event procedures in the system:

- `File()`—to browse or find files, close, and exit the application
- `Edit()`—to edit a new map or an existing stand map
- `Run()`—to perform stand generation simulations, interactive simulations, and numerical felling and extraction simulations
- `Analysis()`—to analyze the simulation results
- `View()`—to retrieve and view the simulation performed earlier
- `Output()`—to reproduce the simulation results on paper
- `Help()`—to present the online help about the simulator

2.4.2.1 Specification Document

The specifications are a blueprint for the application in the sense that a database developer and a programmer should be able to build the application directly from the specifications. At minimum, the specifications should contain the following sections:

- An introduction that contains a description of the problem space, a brief history of the problem, and benefits that will accrue from the application being developed.
- A description of all the tables, including all structural, performance, and behavioral characteristics. This includes the table names, attribute names, attribute data types, referential and semantic integrity, high-level language interfaces, and additional tools required.
- Graphical and written descriptions of the forms/interfaces and relationships between the forms including any behavioral characteristics that are relative to the forms/interfaces.
- A list of the reports that will be generated via the forms/interfaces.
- The menus that will constitute the interface to the application.
- Any security considerations if this is a multiuser application.

Class Exercises

1. Describe the programming language families and their typical languages.
2. What is object-oriented programming?
3. What are the three fundamental principles of object-oriented programming?
4. What is software engineering?
5. What are the key phases in the software life cycle?

References

Carrano, F.M. 1995. *Data Abstraction and Problem Solving with C++: Walls and Mirrors.* Addison-Wesley Publishing Company, Menlo Park, CA.

Hayhurst, C. 2002. *Semantics of Programming Language* (Lecture Notes). Lane Department of Computer Science and Electrical Engineering, West Virginia University, Morgantown, WV.

Kochut, K. 1996. *Software Engineering* (Lecture Notes). The University of Georgia, Athens, GA.

Lafore, R. 1998. *Object-Oriented Programming in C++* (3rd Edition). SAMS, Indianapolis, IN, 925pp.

Pressman, R.S. 2001. *Software Engineering: A Practitioner's Approach*. McGraw Hill, New York.

Rumbaugh, J., M. Blaha, W. Premerlani, F. Eddy, and W. Lorensen. 1991. *Object-Oriented Modeling and Design*. Prentice Hall, Englewood Cliffs, NJ, 500pp.

Sethi, R. 1996. *Programming Languages: Concepts & Constructions* (2nd Edition). Addison-Wesley, Reading, MA, 640pp.

Wang, J. and W. Greene. 1999. An interactive simulation system for modeling stands, harvests, and machines. *Journal of Forest Engineering* 10(1): 81–99.

Wang, J. and C. LeDoux. 2003. Estimating and validating ground-based timber harvesting production through computer simulation. *Forest Science* 49(1): 64–76.

Section II

Data Manipulation and Analysis

3

Elementary Data Manipulation Using Excel

3.1 Excel Formulas

MS Excel is a spreadsheet program that includes features of data manipulation, graphing, data analysis, and Visual Basic for Applications (VBA). Since it was first introduced in 1987, Excel has evolved through more than 10 versions. Depending on the version, an Excel worksheet can consist of 1,048,576 rows and 16,384 columns, and the number of sheets in a workbook is also limited by the available memory of the computer (with a default of one sheet).

Excel's formulas are what make its spreadsheets more useful. Without formulas, a spreadsheet would be little more than a word document file with some table features. To add a formula to a worksheet, you enter it into a cell. You can delete, move, and copy formulas just like any other items of data. Formulas use arithmetic operators to work with values, text, worksheet functions, and other formulas to calculate a value for the cell (Walkenbach 1999). A **formula** entered into a cell can consist of any of the following elements:

- Operators such as + (for addition) and * (for multiplication)
- Cell references (including named cells and ranges)
- Values or text
- Excel functions (such as SUM, COUNT, or AVERAGE)

Here are a few examples of formulas:

=125^(1/3)	Returns the cube root of 125.
=100*.06	Multiplies 100 by 0.06.
=B2+B4	Adds the values in cells B2 and B4.
=B2&B4	Concatenates the contents of cells B2 and B4.
=Revenue – Costs	Subtracts the cell named "Costs" from the cell named "Revenue."
=SUM(A1:A10)	Adds the values in the range A1:A10.
=B1=C14	Compares cell B1 with cell C14. If they are identical, the formula returns "TRUE"; otherwise, it returns "FALSE."
=B1>=C14	Returns "TRUE" if the value in cell B1 is greater than or equal to the value in cell C14; otherwise, it returns "FALSE."

3.1.1 Operators in Formulas

We can use a variety of operators in Excel formulas (Table 3.1). Excel also has many built-in functions such as SUM, SIN, COS that enable us to perform more operations in formulas. Table 3.1 lists Excel's commonly used operators and their precedence. As stated by Walkenbach (1999), "Exponentiation has the highest precedence (that is, it's performed first), and logical comparisons have the lowest precedence." We can use parentheses to override Excel's built-in order of precedence. We can also nest parentheses in formulas, which means putting parentheses inside of parentheses. If you do so, Excel evaluates the most deeply nested expressions first and works its way out (Walkenbach 1999).

3.1.2 Entering Formulas

A formula must begin with an equal sign to inform Excel that the cell contains a formula rather than text. There are two ways that we enter a formula into a cell: enter it manually or enter it by pointing to cell references.

If entering a formula manually, you simply type an equal sign (=), followed by the formula. As you type, the characters appear in the cell and in the formula bar. You can edit the formula as needed.

If entering a formula by pointing to cell references, you will still need to do some manual typing. For example, to enter the formula `=B2+B3` into cell B4, you can simply follow these steps:

a. Move the cell pointer to cell B4 and then left click the mouse.
b. Type an equal sign "=" to begin the formula.
c. Use the mouse to point to cell B2 and then left click the mouse, and the cell B2 reference appears in cell B4 and in the formula bar.
d. Type a plus sign "+."
e. Repeat step (c) for cell B3.
f. Press "Enter" to finish the formula.

TABLE 3.1

Operators and Their Precedence in Excel Formulas

Arithmetic	Name	Logical Comparison	Name
^	Exponentiation	=	Equal to
*	Multiplication	<>	Inequality
/	Division	<	Less than
+	Addition	>	Greater than
-	Subtraction	<=	Less than or equal to
&	Concatenation	>=	Greater than or equal to

Source: Based on Walkenbach, J., *Microsoft Excel 2000 Bible*, IDG Books Worldwide, Inc., Foster City, CA, 1999.

3.1.3 Referencing Cells Outside the Worksheet

Most of the time, we work on one spreadsheet within a workbook. However, formulas can refer to cells in other worksheets in the same workbook or to worksheets in different workbooks. Excel uses a special type of notation to handle these types of references. To reference a cell in another worksheet in the same workbook, we use the following format:

```
=SheetName!CellAddress
```

In this format, we need to put the worksheet name and the cell address, separated by an exclamation point. Here is an example of a formula that uses a cell on the *Sheet2* worksheet or a worksheet named "Forest Sale":

```
=A1*Sheet2!A1 or =A1*'Forest Sale'!A1
```

This formula multiplies the value in cell A1 on the current worksheet by the value in cell A1 on *Sheet2* or the *Forest Sale* sheet.

Here is the format to reference a cell in a different workbook:

```
=[WorkbookName]SheetName!CellAddress
```

In this case, the workbook name (in square brackets), the worksheet name, and an exclamation point precede the cell address (Walkenbach 1999). The following is an example of a formula that uses a cell reference in the *Sheet1* worksheet in a workbook named "Stumpage":

```
=[Stumpage.xlsx]Sheet1!A1
```

If the workbook name or the Excel file in the reference includes multiple words with white spaces, we have to enclose it (and the sheet name) in single quotation marks. For example, if we want to multiply the cell A1 of the current sheet by the value of cell A1 on *Sheet1* in a workbook named *Stumpage* in 2014:

```
=A1*'[Stumpage in 2014]Sheet1'!A1
```

When a formula refers to cells in a different workbook, the other workbook does not need to be opened. If the workbook is not opened, you must add the complete path to the reference. Here's an example:

```
=A1*'C:\YourAppFolder\[Stumpage in 2014]Sheet1'!A1
```

3.1.4 Relative versus Absolute References

There are two types of cell references: relative and absolute. By default, all Excel cell references are relative references in formulas except when the formula includes cells in different worksheets or workbooks (Walkenbach 1999). The difference between these two types of references can be easily understood while we work and copy a formula from one cell to another.

3.1.4.1 Relative Reference

For example, we would like to create a simple formula to calculate the total sale values of Christmas trees by species at a fixed rate of sales tax. This could be formulated as $SaleValueSpecies_i$ = *(# of trees sold for species i)*(unit price of species i)*(1+sales tax rate)*. Figure 3.1 shows a worksheet with this formula in cell D3. The formula, which uses the default relative references, is as follows:

```
=B3*C3*(1+B7)
```

You should notice that we only use the B7 cell to hold the sales tax rate. When you copy this formula to the two cells below it (cells D4 and D5),

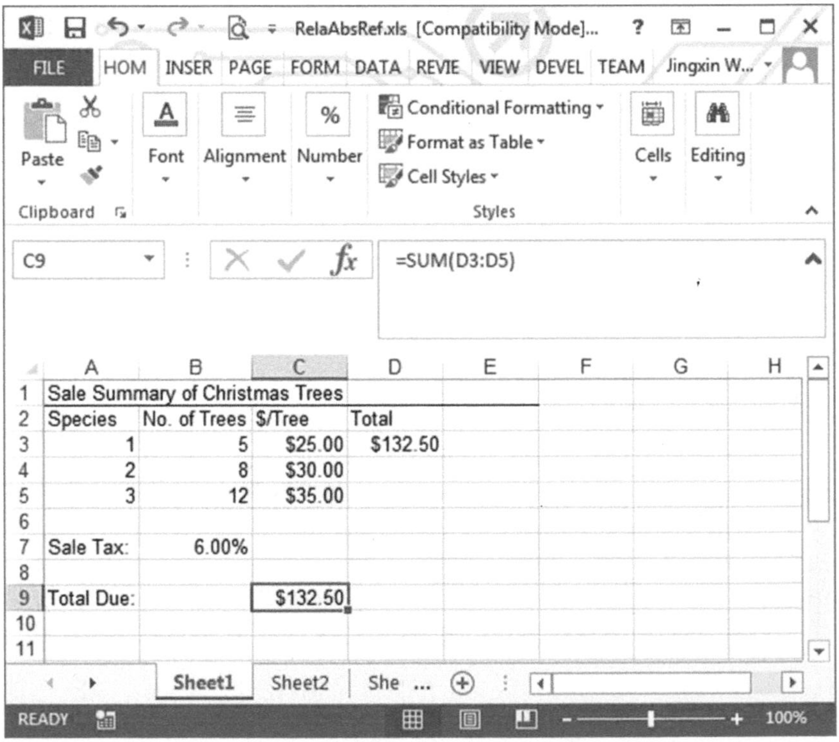

	A	B	C	D
1	Sale Summary of Christmas Trees			
2	Species	No. of Trees	$/Tree	Total
3	1	5	$25.00	$132.50
4	2	8	$30.00	
5	3	12	$35.00	
6				
7	Sale Tax:	6.00%		
8				
9	Total Due:		$132.50	

FIGURE 3.1
Copying Excel formula using relative references. (Based on Walkenbach, J., *Microsoft Excel 2000 Bible*, IDG Books Worldwide, Inc., Foster City, CA, 1999.)

Excel will not produce an exact copy of the formula. However, it will create these formulas:

- Cell D4: `=B4*C4*(1+B8)`
- Cell D5: `=B5*C5*(1+B9)`

In that case, Excel adjusts the cell references to refer to the cells that are relative to the new formula. It simply means that a formula that contains a relative cell reference changes as you copy it from one cell to another. As you can imagine, these two copied formulas will not work for us since the B8 and B9 cells do not reference the sales tax rate. To solve this problem, we need to use an absolute reference.

3.1.4.2 Absolute Reference

Sometimes we want to maintain the original cell reference when we copy a formula. In the above example, we need to use the same sales tax rate for each of the three species. We therefore need an absolute reference in the formula (Figure 3.2).

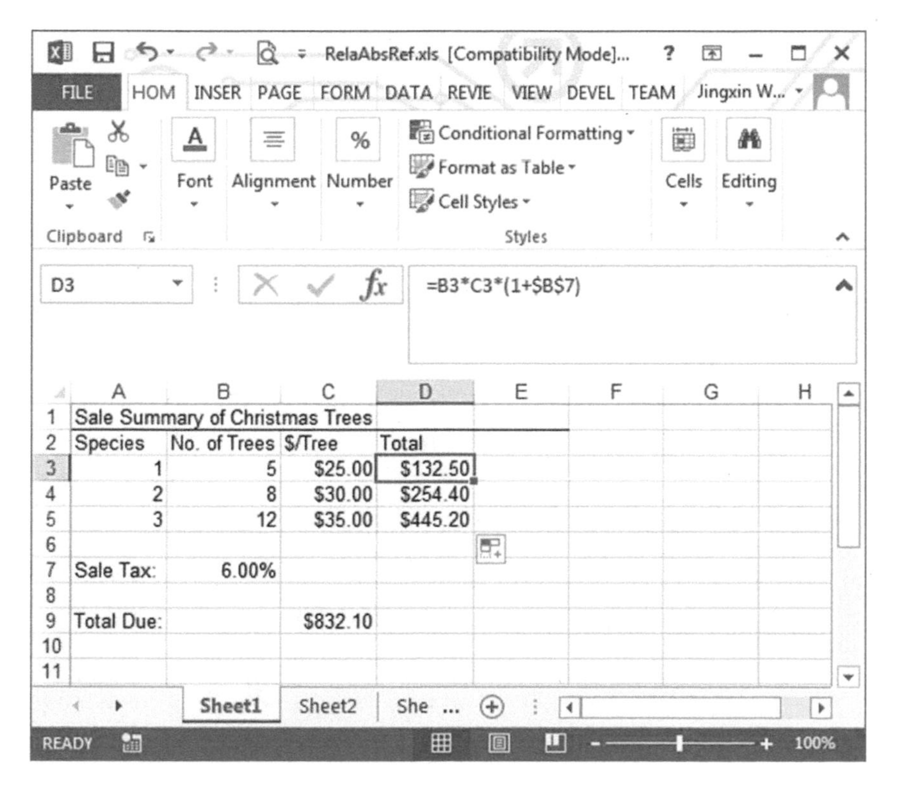

FIGURE 3.2
Formula using absolute cell reference.

In this example, cell B7 contains the sales tax rate. We need to make references to this cell absolute by preceding the column (B) and row (7) with a dollar sign ($). The new formula in cell D3 is as follows:

```
= (B3*C3)*(1+$B$7)
```

These dollar signs indicate to Excel that we want to use an absolute cell reference. Then, when we copy the formula from D3 to D4 and D5, the formula will adjust to the new rows, but the sales tax reference (B7) will remain the same:

- Cell D4: `=(B4*C4)*(1+$B$7)`
- Cell D5: `=(B5*C5)*(1+$B$7)`

So in this example, the relative cell references changed, but the reference to cell B7 did not because it is an absolute reference.

3.2 Excel Functions

Like in other programs, Excel has built-in **functions** that we can use in formulas. The built-in functions can make formulas more powerful and useful. Specifically, Excel functions can help us simplify formulas, allow formulas to perform calculations that are otherwise impossible, and perform conditional execution of formulas (Walkenbach 1999).

Excel built-in functions can simplify a formula significantly. For example, if we calculate the total or sum of the number of Christmas trees sold in 20 different days or the values in 20 cells (A1:A20) without using a function, we need to create a formula:

```
=A1+A2+A3+A4+A5+A6 ... +A20
```

This would be tedious or even impossible if we had more days of sales included in the formula. However, if we use an Excel function, this task is simplified:

```
=SUM(A1:A20)
```

If we need to determine the highest sale for 100 days, a formula cannot work; a function needs to be used. Here's a simple function that returns the highest sale or the highest value in the range A1:A100:

```
=MAX(A1:A100)
```

Suppose that we have a worksheet that calculates sales commissions of Christmas trees. If someone sold more than $500, the commission rate would be 10%; otherwise, the commission rate would be 5%. Without using a function, we would have to create two different formulas and make sure that we use the correct formula for each sale amount. Instead we can use the IF function to ensure the correct commission:

```
=IF(A1>=500, A1*10%, A1*5%)
```

3.2.1 Function Arguments

The parameters used in the parentheses of a function are called **arguments**. A function may use no arguments, one argument, a fixed number of arguments, an indeterminate number of arguments, or optional arguments.

For example, the RAND function, which returns a random number between 0 and 1, does not use an argument. Even so, whether or not a function uses an argument, we still need to provide a set of empty parentheses: `=RAND()`.

A function argument can be a cell reference, a literal text string, or an expression. If a function uses more than one argument, you must separate each argument with a comma. Here are a few examples of functions and their types of arguments:

=SUM(A1:A20)	Cell references
=SUM(TreeSales)	Range name
=SUM(B:B)	Column B
=SUM(6:6)	Row 6
=SQRT(100)	Literal value

A literal argument is a numeric value or a text string that you enter into a function. For example, the SQRT function takes one literal argument. Expressions can be used as an argument in an Excel formula. Excel evaluates the expression first and then uses the result as the argument. For example, an expression of the Pythagorean theorem can be an argument:

```
=SQRT((A1^2)+(A2^2))          Expression
```

3.2.2 Entering Functions

Similar to entering a formula, there are two ways to enter a function into a cell: manually or by inserting it from the *Insert Function* box.

If you know what functions you need to use, entering a function manually may be an efficient way. You simply type the function into a cell. However, if you cannot remember the function name or its format, you can enter it and its arguments using the *Insert Function* dialog box. From the Excel menu,

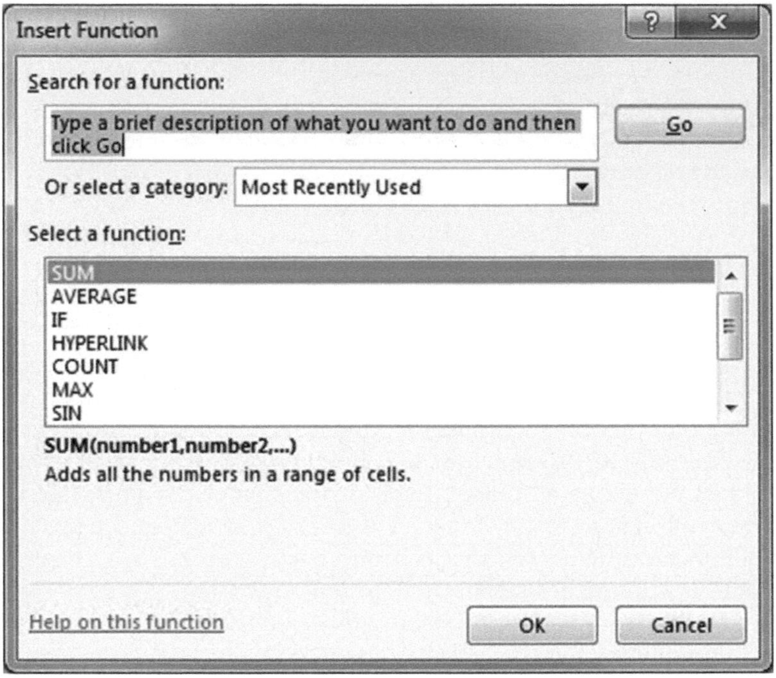

FIGURE 3.3
Insert function dialog box.

click *Formulas* → *Insert Function*, then an *Insert Function* dialog box will be displayed (Figure 3.3).

A drop-down *Or select a category list* allows you choose types of functions to search among. When you select a category, the *Select a function* name list box displays the functions in the selected category. The *Most Recently Used* category lists the functions that you have used most recently.

Once you select a function that you want to use from the list, click *OK*, then a dialog box of Excel *Function Arguments* will appear (Figure 3.4). You can specify the function arguments in that box, or you can easily select a range argument for your spreadsheet by clicking the Collapse Dialog button, then highlighting the range of cells on your sheet, clicking the Expand Dialog button, and then clicking *OK*.

3.3 Major Excel Functions

Excel functions are prewritten formulas. Excel has 11 categories of 328 built-in functions, including database, date and time, engineering, financial, information, logical, lookup and reference, math and trigonometry, statistical,

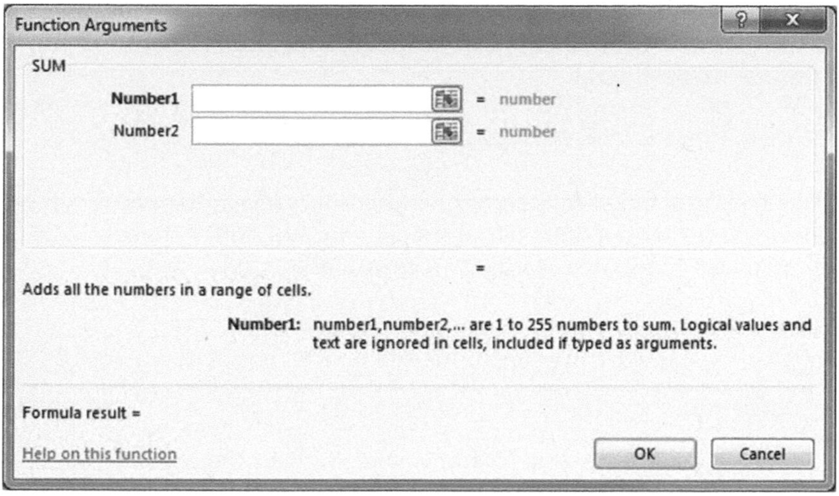

FIGURE 3.4
Function arguments.

and text functions (Microsoft 2013). The following are some commonly used functions from a few of the major function categories.

3.3.1 Mathematical and Trigonometric Functions

Excel provides 59 functions in this category. The category includes common functions, such as SUM, SQRT, SIN, COS, and INT.

3.3.1.1 INT

The INT function takes one argument and returns the integer (nondecimal) portion of a number by truncating all digits after the decimal point. For example,

`=INT(122.88)` *(returns 122).*

3.3.1.2 RAND

This function takes no arguments and returns a uniform random number that is greater than or equal to 0 and less than 1. Being "uniform" means that all numbers have an equal chance of being generated. In the example that follows, the formula returns a random number greater than or equal to 0 and less than 10:

```
=RAND()*10
```

The following formula generates a random integer between two values. The cell named 'Lower' contains the lower bound, and the cell named 'Upper' contains the upper bound:

```
=INT((Upper-Lower+1)*RAND()+Lower)
```

For example, in a forest tree growth simulation project, assuming we want to generate trees with a diameter at breast height (DBH) between 15.24 and 30.48 cm (6 and 12 in.), we can use this formula:

```
=INT(7*RAND()+6)
```

3.3.1.3 ROUND

This function rounds a value to a specified digit to the left or right of the decimal point. It takes two arguments: the first is the value to be rounded while the second is the digit for the number of decimal places. For example,

```
=ROUND(123.456, 0)     (returns 123)
=ROUND(123.456, 1)     (returns 123.5)
=ROUND(123.456, 2)     (returns 123.46)
```

If the second argument is negative, the rounding occurs to the left of the decimal point. For example,

```
=ROUND(123.456, -1)    (returns 120)
=ROUND(123.456, -2)    (returns 100)
=ROUND(123.456, -3)    (returns 0)
```

3.3.1.4 SIN or COS

These Excel functions perform the common trigonometric functions of Sine and Cosine. The SIN function, for example, returns the sine of an angle. SIN takes one argument and the angle expressed in radians. We use the RADIANS function to convert degrees to radians, and we use the DEGREES function to convert radians to degrees. If cell A2 contains an angle in degrees, the formula that follows returns the Sine of that angle:

```
=SIN(RADIANS(A2))
```

3.3.1.5 SQRT

This function returns the square root of its argument. If the argument is negative, this function returns an error.

```
=SQRT(625)             (returns 25)
```

We can use the exponential mark ^ to compute a cube root or raise a value to a certain power. For example,

```
=3^5                    (returns 243)
```

3.3.1.6 SUM

This might be the most commonly used Excel function. It takes from 1 to 30 arguments.

```
=SUM(A1:A10, B1:B4, 20, 50)
=SUM(A1, A2, A3)
=SUM(1, 2, 3)
```

3.3.1.7 SUMIF

This function is for calculating conditional sums. SUMIF takes three arguments. The first argument is the range that you are using in the selection criteria. The second argument is the selection criteria. The third argument is the range of values to sum if the criteria are met. For example, an Excel worksheet contains weekly timber production data by a logger from 1995 to 2010 with a data range from Row 3 to Row 136, while Column C contains the end date of the week and Column L is the weekly production. If we would like to sum this logger's timber production only in 1999, the following function can be used:

```
=SUMIF(C3:C136, "<31-dec-99",L3:L136) -
 SUMIF(C3:C136, "<31-dec-98",L3:L136)
```

3.3.2 Statistical Functions

This category contains 80 functions that perform various statistical calculations. The functions in this category could be very useful for professionals in the fields of forestry and natural resources. From our field data, we always want to calculate the basic statistics of certain variables.

3.3.2.1 AVERAGE, MEDIAN, and MODE

These similar statistical operations have appropriately similar Excel functions. The AVERAGE function returns the average (arithmetic mean) of a range of values. Excel also provides the MEDIAN function (which returns the middle-most value in a range) and the MODE function (which returns the value that appears most frequently in a range):

```
=AVERAGE(A1:A100)
=MEDIAN(A1:A100)
=MODE(A1:A100)
```

If the range argument contains blanks or text, Excel will not include these cells in the calculation. As with the SUM formula, you can supply 1 to 30 arguments in these functions.

3.3.2.2 COUNT

This function counts how many numbers are in a range of values. For example, if all cells of A1:A10 hold numeric values, then

```
=COUNT(A1:A10)          (returns 10)
```

3.3.2.3 COUNTIF

The COUNTIF function counts the number of cells in a range that meet the criteria you specify. This function takes two arguments: the range that contains the values to count and a criterion used to determine what to count. For example, the formula below can count the number of students with grade "A" in column B:

```
=COUNTIF(B:B, "A")
```

Notice that the first argument (B:B) consists of a range reference for the entire Column B instead of a range for certain cells within the column. This enables you to insert new values into cells in Column B without having to change the formula.

3.3.2.4 MAX and MIN

The MAX and MIN functions return the largest value and the smallest value in a range, including numbers, text, and logical values. For example, to find a maximum or minimum value in a range of A1 to A100:

```
=MAX(A1:A100)
=MIN(A1:A100)
```

3.3.2.5 STDEV

This function estimates standard deviation based on a sample (a range of data). It is a measure of how widely values are dispersed from the mean or how spread out values are. STDEV is the square root of the variance. For example, to estimate the standard deviation of a sample of data in the range of A1–A100:

```
=STDEV(A1:A100)
```

3.3.3 Text Functions

For editing values and text on a worksheet, Excel provides 26 built-in text functions that can be used specifically to manipulate text.

3.3.3.1 LEFT and RIGHT

The LEFT or RIGHT function returns the first character or a string of characters of a specified length, beginning at the leftmost or the rightmost position. These functions take one or two arguments. If using two arguments, the first argument is the string of characters and the second argument (optional) is the number of characters (including white spaces) to return. If the second argument is omitted and only one argument is used, the function returns the leftmost or the rightmost character. For example,

```
=LEFT("FOR 240 Intro to Computing", 7)    (returns FOR 240)
=RIGHT("FOR 240 Intro to Computing", 9)   (returns Computing)
=LEFT("FOR 240 Intro to Computing")       (returns F)
=RIGHT("FOR 240 Intro to Computing")      (returns g)
```

3.3.3.2 LEN

The LEN function returns the number of characters in a string of text including white spaces as follows:

```
=LEN("FOR 240 Intro to Computing")     (returns 26)
```

3.3.3.3 MID

The MID function extracts characters from inside a text string. It takes three arguments. The first argument is the text string from which you want to extract the specified number of characters. The second argument is the position at which you want to begin extracting. The third argument is the number of characters that you want to extract. For example,

```
=MID("FOR 240 Intro to Computing", 9, 5)   (returns Intro)
```

3.3.3.4 REPLACE and SUBSTITUTE

The REPLACE function replaces part of a text string with other characters. It takes four arguments. The first argument is the text that contains the characters you want to replace. The second argument is the character position at which you want to start replacing. The third argument is the number of characters to replace. The fourth argument is the new text that will replace the existing text. For example, if we want to replace "Intro" with "Introduction" in "FOR 240 Intro to Computing," we would type:

```
=REPLACE("FOR 240 Intro to Computing", 9, 5, "Introduction")
```

The SUBSTITUTE function is also used to replace part of a text string with other characters. The difference is that SUBSTITUTE replaces specific text with new text while REPLACE replaces any text based on its position and length within a text string. For example, the function below simply replaces "Intro" with "Introduction":

```
=SUBSTITUTE("FOR 240 Intro to Computing", "Intro",
  "Introduction")
```

3.3.3.5 UPPER, LOWER, and PROPER

These functions convert text to upper, lower, or proper case, respectively:

```
=UPPER(A1)
```

3.3.3.6 FIND

This function returns the starting position of a text string from within another text string as follows:

```
=FIND("Intro", "FOR 240 Intro to Computing")    (returns 9)
```

We can nest functions and allow them to work together. This example nests FIND within REPLACE so that "Intro" is found and then replaced with "Introduction":

```
=REPLACE("FOR 240 Intro to Computing", FIND("Intro", "FOR
  240 Intro to Computing"), 5, "Introduction")
```

3.3.4 Logical Functions

This category contains six functions, including AND, FALSE, IF, NOT, OR, and TRUE.

3.3.4.1 IF

This function is one of the most important functions. It can provide us conditional computing with decision-making capability. The IF function takes three arguments. The first argument is a logical test that must return either *TRUE* or *FALSE*. The second argument is the result that you want the function to return if the first argument is true. The third argument is the result that you want the function to return if the first argument is false. In our Christmas tree sale example (Figure 3.2), if we let cell B6 hold the total sale amount, and the seller's goal is $200, the commission rate would be 10% for persons reaching the goal; otherwise, the commission would be 5%:

```
=IF(B6>=200, B6*0.10, B6*0.05)
```

3.4 Build Your Own Functions

Although Excel provides several hundred built-in functions in 11 different categories, we may still need to build our own functions for specific applications. For example, we want to calculate the basal area (BA in ft^2, 1 ft^2 = 0.0929 m^2) of a tree based on its DBH (in inches, 1 in. = 2.54 cm). Here is the formula we can use:

$$BA = 0.005454154*DBH^2$$

This calculation is one we would reuse often, so for efficiency we could build a new BA function on our Excel sheet. We create custom functions using the VBA programming language. To do so, we use Excel's *Developer* tab, which is not present on the menu ribbon by default. To add it, from the Excel menu we click *File* → *Options* → *Customize Ribbon*. Then we check the box for *Developer* under *Main Tabs* and click *OK*.

Using the basal area calculation as an example, here are the steps to create our own function in an Excel Workbook using Developer:

1. From Excel menu ribbon, click *Developer* → *Visual Basic*. A VBA Editor window will open.
2. From the VBA Editor menu, choose *Insert* → *Module* and a VBA Code Editor window will appear.
3. Type the following code in the VBA Code window:

```
Function BA(d As Single) As Single
    BA = 0.005454154 * d * d
End Function
```

4. Close the VBA Editor and go back to your Excel workbook and current worksheet. Now you can use this function as you would for any other Excel built-in functions:

```
=BA(12)          (returns 0.785398)
```

3.5 Charts

Charts have been an integral part of spreadsheets since the early days of Lotus 1-2-3 (Walkenbach 1999). In natural resources, we always expect to present our field data in a graphical format. MS Excel provides us the tools to create a variety of useful charts. Charts are used to display a series of numeric data so that the data and relationships among data sets are easier to understand. Excel enables us to create all the basic chart types: column, line, pie, bar, area, scatter, stock, surface, radar, and combo.

While we typically use the data in a single worksheet for a chart, one chart may contain data from several worksheets within one workbook or even from worksheets in different workbooks. A chart is essentially an object that Excel creates. This object consists of one or more data series, displayed graphically. The appearance of the data series depends on the selected chart type (Walkenbach 1999). You can include a maximum of 255 data series in a chart. The latest version of Excel has no limit on the number of categories (or data points) in a data series for either 2D or 3D charts; only your computer's available memory can limit your chart's categories.

3.5.1 Excel Chart Example

When you create a chart in Excel, you have two options for where to place the chart: (1) Insert the chart directly into a worksheet as an object and (2) create the chart as a new chart sheet in your workbook (right click on the chart → select *Move Chart* → select *New Sheet* in the dialog box that appears).

In the following example, we create a chart that displays the trend of the average harvested tract size by year in West Virginia (Figure 3.5). These data consist of two series: the average tract size harvested annually (1 acre = 0.4 ha) and the year.

Step 1: To create this chart, first select (highlight) the entire range on this worksheet (Figure 3.5). Next click *Insert* on the Excel toolbar, then in the *Charts* section, choose a chart type you want to use. In our case, either a column or line chart should present the data and trend well. Let's click the arrow beside the *Insert Column or Bar Chart* icon, then under *2D Column* select the *Clustered Column*.

Step 2: Now we need to verify the data ranges and specify the orientation of the data (whether it is arranged in rows or columns). The orientation of the data has a drastic effect on the look of your chart. Right click the chart and choose *Select Data* from the drop-down menu. The *Select Data Source* dialog box will pop up (Figure 3.6).

In our example, we need to remove *Series1* data from the *Legend Entries (Series)* box and use it as the *Horizontal (Category) Axis.*

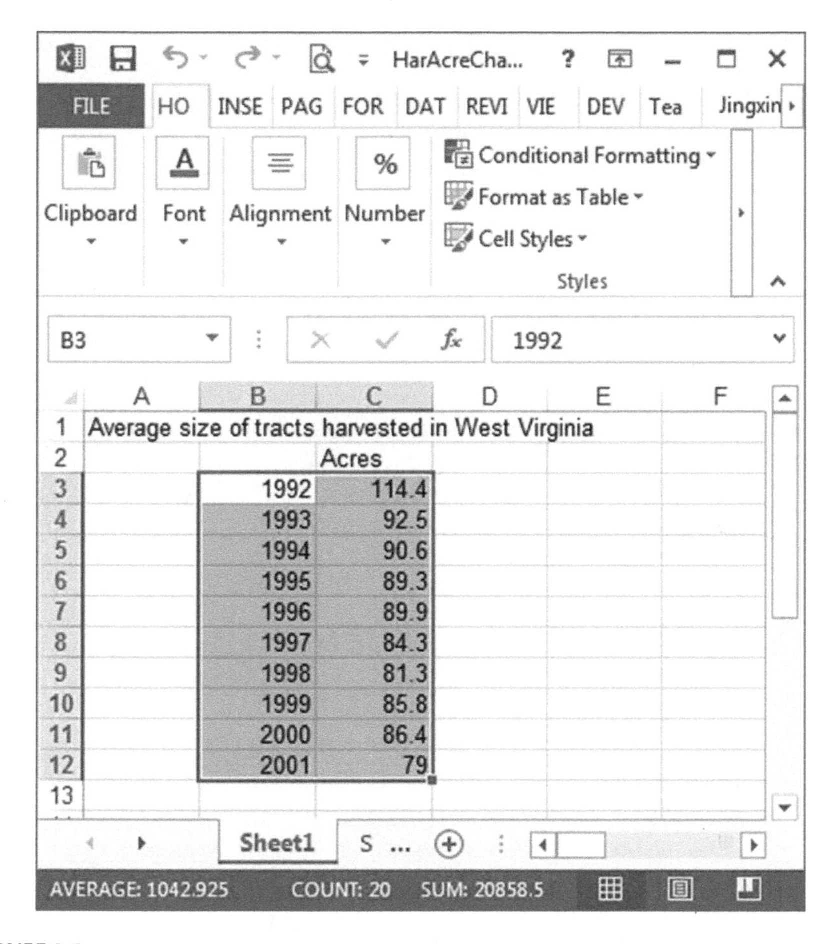

FIGURE 3.5
Data for a chart.

Select *Series1* on the left, then click the *Remove* button. Click the *Edit* button in the right list box, then an *Axis Labels* pop-up dialog box will appear. Select your data series in Column B on your worksheet, and the range will appear in the *Axis Label Range* as shown in Figure 3.7a. Click *OK*. Now the horizontal or category axis labels are set up correctly (Figure 3.7b), so click *OK*.

Step 3: The chart is not yet complete. We need to add a chart title, axes titles, label data, etc. Select the chart, then on the menu click the *Design* tab, then *Add Chart Element*. You will see the choices for labels and axes, including chart title, axis titles, gridlines, legend, data labels, and data table. Keep in mind that you can see these *Chart Tools* only if you select the chart. When you make your choices, click on each element (e.g., x-axis) within the chart to edit the text (Figure 3.8).

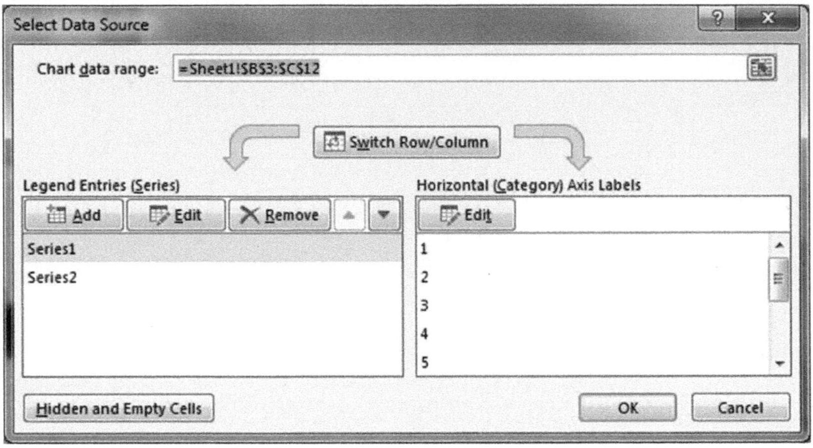

FIGURE 3.6
Verify data series.

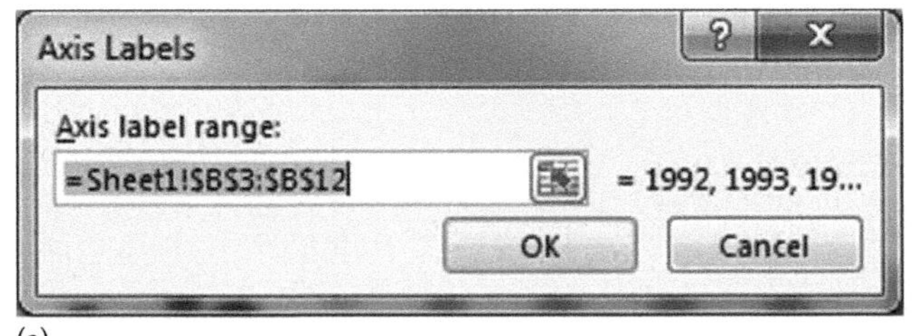

(a)

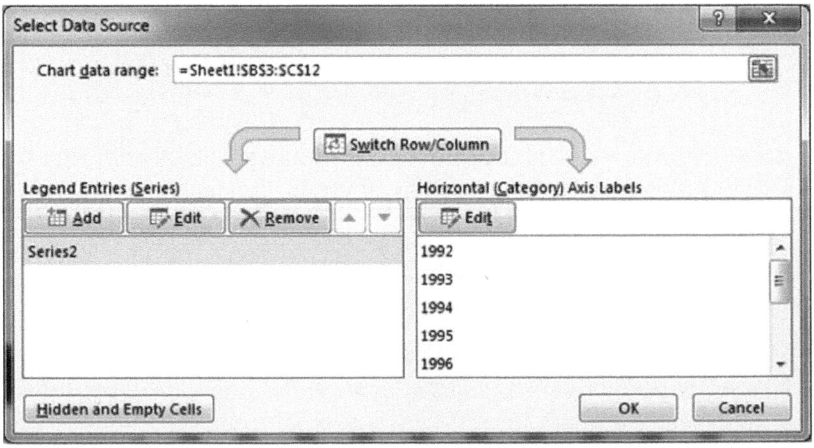

(b)

FIGURE 3.7
Adjustment of data series. (a) Select data range and (b) editing data series and axis labels.

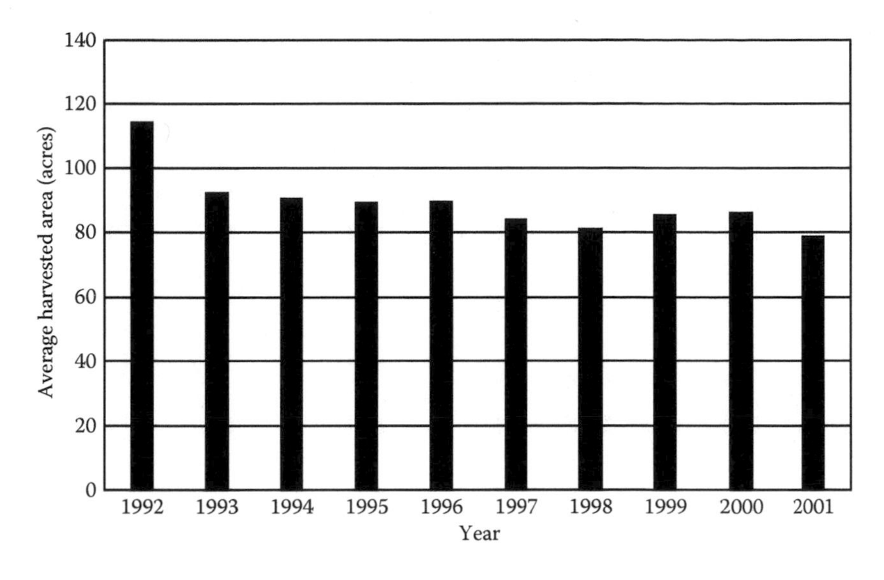

FIGURE 3.8
A chart of average harvested area by year.

Step 4: After we create a chart, we save it and can modify it at any time. Common chart modifications include changing the chart type and chart elements. What we may want to do now is to copy the chart and paste it in either a MS Word file or a PowerPoint slide for a report or presentation.

3.5.2 Combination Chart

A **combination chart** is a chart that combines two or more chart types. For example, you may have a chart that shows both columns and lines. A combination chart also can use a single chart type (e.g., all columns) but include a second value axis.

Creating a combination chart simply involves changing one or more of the data series to a different chart type. Figure 3.9 shows example data for a combination chart for Bell feller-buncher production data—the felling time per tree in minutes vs. tree size (DBH) in inches (1 in. = 2.54 cm). There are two data series: observed time per tree and predicted time per tree based on the model (Time per tree = 0.4051 + 0.01023 * DBH). Suppose we would like to use a Scatter (X–Y) Chart for observed data combined with a Line Chart to represent the prediction of its production by tree size. First, we create the chart as we did in the previous example (by selecting the entire data set on the spreadsheet) (Figure 3.9), clicking *Insert*, then choosing a chart type (in this example *Scatter*). Then within the created chart, click on the data series to which we wish to apply a different chart type (the *Predicted* series). This series should become highlighted. Right click, then select *Change Series Chart Type*, and a new dialog

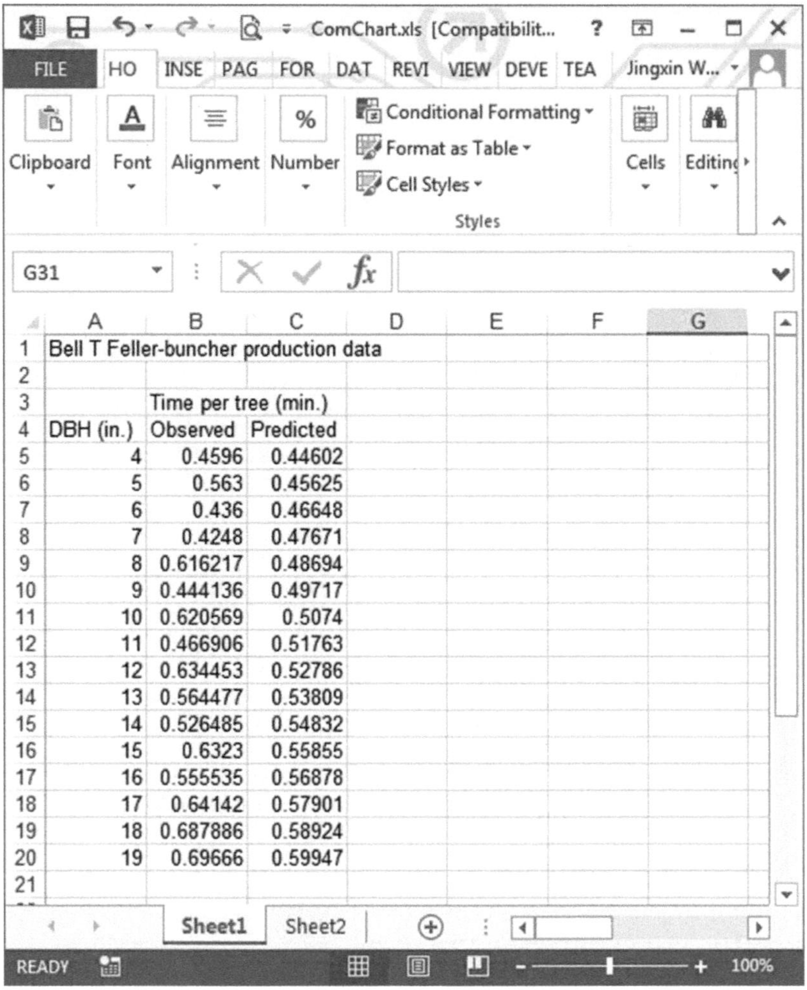

ComChart.xls [Compatibilit...

FILE HO INSE PAG FOR DAT REVI VIEW DEVE TEA Jingxin W...

Clipboard Font Alignment Number Conditional Formatting Format as Table Cell Styles Styles Cells Editing

G31 fx

	A	B	C	D	E	F	G
1	Bell T Feller-buncher production data						
2							
3		Time per tree (min.)					
4	DBH (in.)	Observed	Predicted				
5	4	0.4596	0.44602				
6	5	0.563	0.45625				
7	6	0.436	0.46648				
8	7	0.4248	0.47671				
9	8	0.616217	0.48694				
10	9	0.444136	0.49717				
11	10	0.620569	0.5074				
12	11	0.466906	0.51763				
13	12	0.634453	0.52786				
14	13	0.564477	0.53809				
15	14	0.526485	0.54832				
16	15	0.6323	0.55855				
17	16	0.555535	0.56878				
18	17	0.64142	0.57901				
19	18	0.687886	0.58924				
20	19	0.69666	0.59947				
21							

Sheet1 Sheet2

READY 100%

FIGURE 3.9
Data for a combined chart of a feller-buncher felling vs. tree DBH.

box will open. In the drop-down lists, select the new *Chart Type* for the series you wish to change (in this case, we select *Line*) and then click *OK* (Figure 3.10).

3.5.3 Gantt Chart

Gantt charts display data in a way that is very useful in situations such as developing project timelines. Here they are used to represent the time required to perform each task in a project. As an example, we can use the timeline data for a logger survey project (Figure 3.11) to create a Gantt chart for managing this project (Figure 3.12).

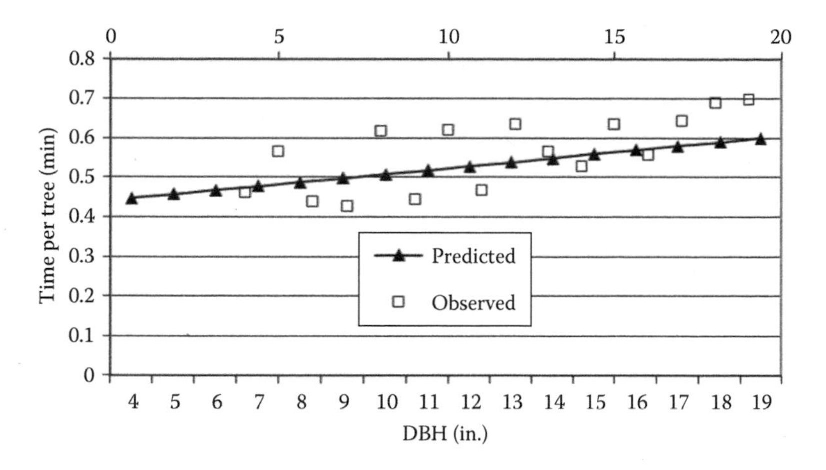

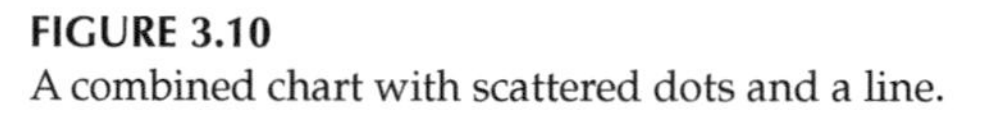

FIGURE 3.10
A combined chart with scattered dots and a line.

GanttChart.xls [Compatibilit...

	A	B	C	D
1	Management Plan for Loggers' Survey			
2	**Task**	**Start Date**	**Duration**	**End Date**
3	Planning Meeting	1/15/2003	1	1/15/2003
4	Develop Survey Form	1/17/2003	30	2/15/2003
5	Print and Mail Forms	2/15/2003	10	2/24/2003
6	Receive Responses	4/3/2003	20	4/22/2003
7	Data Entry	4/20/2003	20	5/9/2003
8	Data Analysis	5/10/2003	20	5/29/2003
9	Write Report	5/25/2003	10	6/3/2003
10	Finalize Report	6/4/2003	5	6/8/2003

FIGURE 3.11
Data for a Gantt chart.

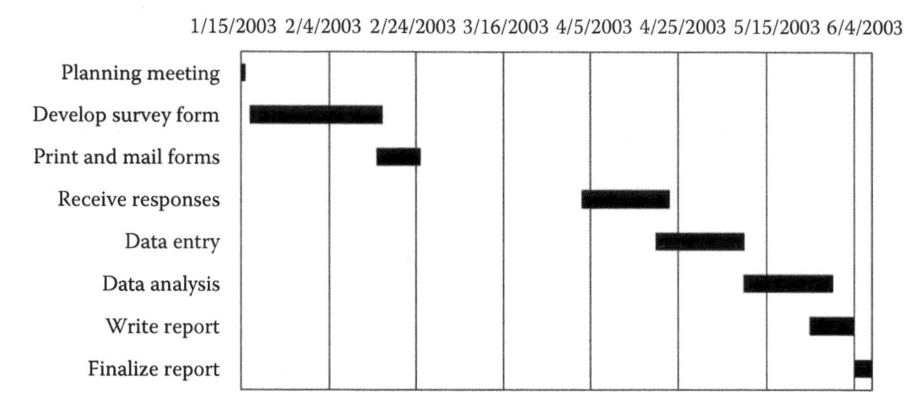

FIGURE 3.12
A timeline chart for a logger's survey project.

To create this Gantt chart, follow these steps:

1. Enter the data as shown in Figure 3.11. The formula in cell D3, which was copied to the rows below it, is =*B3+C3-1*.
2. Select the data range A3:C10, then click the *Insert* tab on the menu ribbon. In the *Chart* group, choose the *Bar Chart* option and then the *2D Stacked Bar* option. Notice that you will get a chart with incorrect category axis labels.
3. To further modify this chart, right click within the chart area and choose *Select Data…* from the drop-down menu. The window *Select Data Source* will pop up. We change the category axis label by clicking the *Edit* button in the right frame box. In the input box, change the axis label range to *A3:A10*, and click *OK*.
4. Continue step (3). To edit the data source of *start date* and *duration*, click *series 1* in the left frame box and click *Edit*. You can define the series name as *Start Date* and series value as *B3:B10*. Now, add a new data series as *Duration* with series value *C3:C10*, and then click *OK*.
5. Adjust the height of the chart so that all the axis labels are visible. You can also accomplish this by using a smaller font size.
6. Right click the horizontal axis to access the *Format Axis* dialog box. Adjust the horizontal axis minimum and maximum scale values to correspond to the earliest and latest dates in the data (note that you can enter a date into the minimum and maximum edit box). For example, enter 1/15/2003 in the minimum edit box and 6/8/2003 in the maximum edit box. You can also change the date space shown in the chart by changing the major unit from the default value to 20 or other values you want. You may also want to change the date format for the axis labels from mm/dd/yyyy to mm/dd under the *Number* section.

7. Access the *Format Axis* dialog box for the vertical axis. In the *Axis Options* tab, check the *categories in reverse order* and set the option for *Horizontal axis crosses* to *At category number: 1* or use the automatic option.
8. The last thing to do is to make the first data series invisible. In the embedded chart, right click the bar corresponding to the first data series, and choose *Format data series* from the drop-down menu. In the *Fill* and *Border color* options under *Fill* tab, set *No fill* and *No line.*
9. Apply other formatting, as desired.

Class Exercises

1. Using the data shown in Figure 3.11, create a simple Gantt chart (Figure 3.12).
2. Using Excel formulas and functions, create a Timber Sale Report that includes:

   A. Data Analysis in Excel: You are given forest inventory and analysis data from three harvest tracts (Tables 3.2 through 3.4). These data have been encoded into an Excel workbook in three different worksheets (1 ft^3 = 0.0283 m^3, 1 acre = 0.40 ha, 1 ft^2 = 0.0929 m^2). Add one more worksheet and name it *Summary* to summarize these tracts for cubic foot volume (CFV), trees per acre (TPA), and basal area per acre (BA/A) with the following statistical measures using built-in functions available in Excel. These functions should be written in the summary sheet and linked to the individual worksheets as appropriate.

      1. Number of plots
      2. Mean
      3. Median
      4. Standard deviation
      5. Coefficient of variation

   B. Statistical Interpretation
   Based on the analyzed data, answer the following questions:

      1. Which tract has the highest variability in TPA (1 acre = 0.4 ha)?
      2. Which tract has the highest mean cubic foot (1 ft^3 = 0.0283 m^3) volume?
      3. Delete Plot #3 from Tract 1 and report the mean CFV of Tract 1 in the summary sheet.
      4. Insert *Plot 7* in Tract 3 and add *CFV = 4500, TPA = 50,* and *BA/A = 100,* then report the mean TPA for this tract.

C. Definitions
   Define the following terms:
   1. Mean
   2. Median
   3. Standard deviation
   4. Coefficient of variation

You should submit your Excel workbook that must include individual worksheets for three tracts and the summary information for Parts A, B, and C. The answers to Parts B and C can be included in the same Excel worksheet.

TABLE 3.2
Tract 1 Inventory Data

Plot	Volume (ft^3)	Trees/Acre	Basal Area (ft^2/Acre)
1	4848.93	213.94	180.00
2	3879.63	249.54	140.00
3	3373.45	255.15	140.00
7	4706.58	435.91	220.00
8	4678.38	350.19	180.00
9	3883.12	204.73	140.00
11	1562.73	102.04	100.00
12	6392.76	368.34	260.00
13	2997.35	151.55	120.00
18	3352.95	189.83	120.00
19	4359.39	148.74	140.00
20	2566.89	242.23	100.00
21	1413.80	118.49	60.00
22	1734.92	99.89	80.00
23	2174.70	47.70	100.00
24	3676.66	159.71	160.00
25	3896.78	203.63	200.00
26	3228.78	84.77	160.00
27	4491.33	110.61	180.00
28	6197.89	388.34	200.00
29	4242.36	160.42	180.00
30	2998.94	71.51	140.00
31	2194.12	32.91	60.00
32	2768.79	42.50	100.00
33	2298.72	45.13	80.00
34	3799.73	131.68	180.00
35	3107.67	104.56	140.00
36	3291.45	148.68	120.00

TABLE 3.3

Tract 2 Inventory Data

Plot	Volume (ft³)	Trees/Acre	Basal Area (ft²/Acre)
1	2158.65	201.03	120.00
2	3478.97	287.03	180.00
3	2511.89	170.63	100.00
4	4578.97	231.54	180.00
5	5237.88	372.67	180.00
6	2273.95	123.13	140.00
7	5326.74	406.87	220.00
8	2804.46	163.15	120.00
10	2729.30	161.56	120.00
11	1605.78	90.22	80.00
12	3087.08	71.53	140.00
13	2227.67	175.59	120.00
14	2286.81	89.55	140.00
15	1062.33	33.61	80.00
19	2065.94	89.34	120.00
20	4613.28	163.35	180.00
21	4061.26	329.91	180.00
22	4848.93	213.94	180.00
23	3879.63	249.54	140.00
24	3373.45	255.15	140.00
25	4706.58	435.91	220.00
26	4678.38	350.19	180.00
27	3883.12	204.73	140.00
28	1562.73	102.04	100.00
29	6392.76	368.34	260.00

TABLE 3.4

Tract 3 Inventory Data

Plot	Volume (ft³)	Trees/Acre	Basal Area (ft²/Acre)
1	4491.33	110.61	180.00
2	6197.89	388.34	200.00
3	4242.36	160.42	180.00
5	2998.94	71.51	140.00
6	2194.12	32.91	60.00
8	2768.79	42.50	100.00
9	2298.72	45.13	80.00
10	3799.73	131.68	180.00
11	3107.67	104.56	140.00
12	3291.45	148.68	120.00
13	5658.78	191.12	220.00
14	3862.88	103.50	160.00
15	4255.19	69.49	160.00
16	3534.86	47.94	140.00
17	3096.07	105.64	120.00
21	4449.36	121.01	180.00
22	3568.32	176.56	160.00
23	2158.65	201.03	120.00
24	3478.97	287.03	180.00

References

Microsoft. 2013. Training courses for Excel 2013. Available online at http://office.microsoft.com/en-us/excel-help/training-courses-for-excel-2013-HA104032083.aspx. Accessed on January 16, 2013.

Walkenbach, J. 1999. *Microsoft Excel 2000 Bible*. IDG Books Worldwide, Inc., Foster City, CA.

4

Statistical Analysis and Mathematical Programming Using Excel

4.1 Data Analysis with Analysis ToolPak

The Analysis ToolPak is an add-in program that provides analytical capabilities or tools for statistical and engineering analyses. Basic tools include correlation, analysis of variance, *t*-test, regression, and others (Table 4.1).

If the *Data Analysis* command is not an option under the *Data* tab on the Excel menu ribbon, we need to install the Analysis ToolPak first in Microsoft Excel. To install it, click the *File* tab, and then click *Options*. An *Excel Options* box will open. In the left-hand box, click *Add-Ins*, and then at the bottom of the box, in the drop-down list beside *Manage*, select *Excel Add-ins* and click *Go*. An *Add-Ins* box will pop up where you can select *Analysis ToolPak*, and then click *OK*.

After you load the Analysis ToolPak, the *Data Analysis* command is available in the *Analysis* group under the *Data* tab on the Excel menu ribbon. You can repeat the above procedures and uncheck the Analysis ToolPak to remove it.

To use the Data Analysis feature, you must arrange the data you want to analyze in columns or rows on your worksheet. Then click the *Data* tab on the Excel menu and click *Data Analysis*. In the *Data Analysis* pop-up box, click the tool you want to use, enter the input range and the output range, and then select the options you want. Let's use examples to illustrate three analysis tools using Excel ToolPak.

4.1.1 Correlation

Correlation is a widely used statistic that measures the degree to which two sets of data vary together or are similar. We use the correlation coefficient to measure the extent to which two measurement variables vary together. That coefficient ranges from −1.0 (a perfect negative correlation) to +1.0 (a perfect positive correlation) while a correlation coefficient of 0 indicates that the

TABLE 4.1

Major Analysis Tools in Excel ToolPak

Analysis of variance	Moving average
Correlation	Random number generation
Covariance	Rank and percentile
Descriptive statistics	Regression
Exponential smoothing	Sampling
F-Test	*t*-Test (three types)
Fourier analysis	*z*-Test
Histogram	

two variables are not correlated. For example, we would like to examine the correlation between tree diameter at breast height (DBH) (in inches, 1 in. = 2.54 cm) and felling time in minutes for a feller-buncher in forest operations (Figure 4.1).

From the Excel menu, click *Data → Data Analysis,* and select *Correlation.* In the *Correlation* dialog box (Figure 4.2), specify the input range *A1:B17,* check *Labels in First Row,* and for *Output options* select *New Worksheet Ply,* then click *OK.*

Figure 4.3 shows the results of a correlation analysis for the two variables *DBH* and *Time Per Tree.* The output consists of a correlation matrix that shows the correlation coefficient for each variable paired with every other variable. The correlation coefficient between *DBH* and *Time Per Tree* is 0.697467 in this example, a positive correlation because it is greater than 0. The correlation between these two variables seems good considering field data quality in forest operations.

4.1.2 Regression

The Regression tool performs linear regression analysis by using the "least squares" method from worksheet data. We can use regression to analyze trends, forecast the future, and build predictive models.

Regression analysis is a typical statistical process for estimating the relationship among variables. It enables us to determine the extent to which one range of data (the dependent variable) varies as a function of the values of one or more other ranges of data (the independent variables). The way to express the relationship among these variables mathematically is called regression.

The Regression tool in Excel can perform simple (one independent variable) and multiple (more independent variables) linear regressions and calculate and standardize residuals automatically. In a simple linear regression for modeling n data points, there is one independent variable x. Mathematically, the relationship between dependent variable y and x can be expressed as:

$$y = \alpha_0 + \alpha_1 x + \varepsilon$$

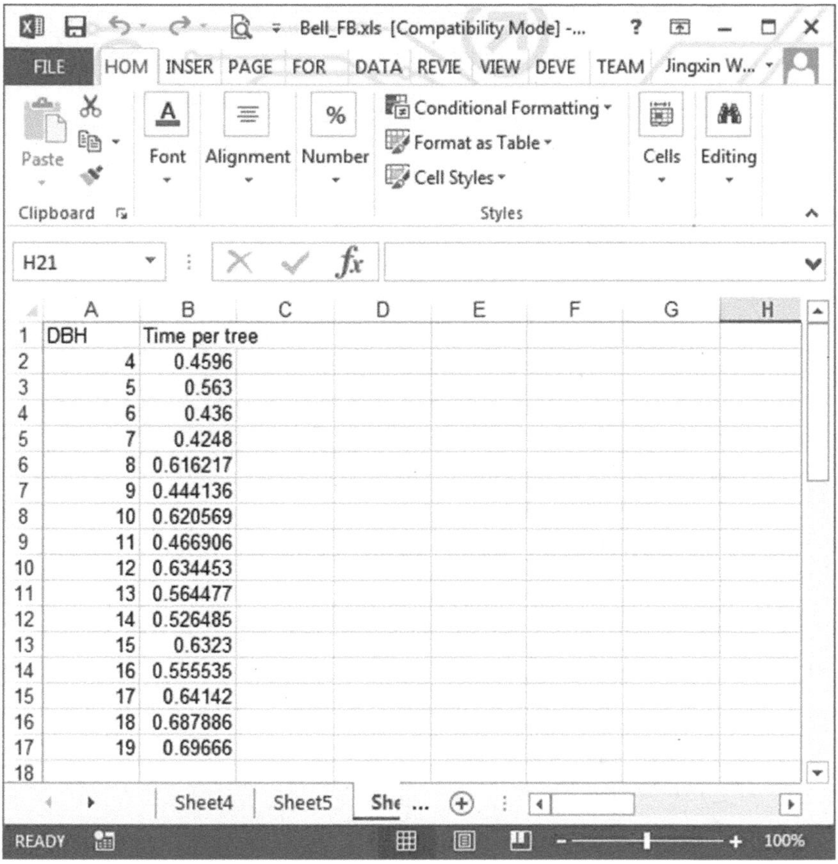

DBH	Time per tree
4	0.4596
5	0.563
6	0.436
7	0.4248
8	0.616217
9	0.444136
10	0.620569
11	0.466906
12	0.634453
13	0.564477
14	0.526485
15	0.6323
16	0.555535
17	0.64142
18	0.687886
19	0.69666

FIGURE 4.1
Data used for the correlation tool.

Similarly, in a multiple linear regression with more independent variables x_1, x_2, …, their generic relationship with dependent variable y is expressed as:

$$y = \alpha_0 + \alpha_1 x_1 + \cdots + \alpha_i x_i + \cdots + \varepsilon \quad i = 1, 2, \ldots, n$$

where
  α_0 is called intercept
  α_i is slope
  ε is an error component

As an example, let's use the same feller-buncher data set (Figure 4.1). Since *DBH* and *Time Per Tree* are correlated well with a correlation coefficient of 0.697467, we might be able to generate a useful regression model between these two variables. In the Excel menu ribbon, click *Data* → *Data*

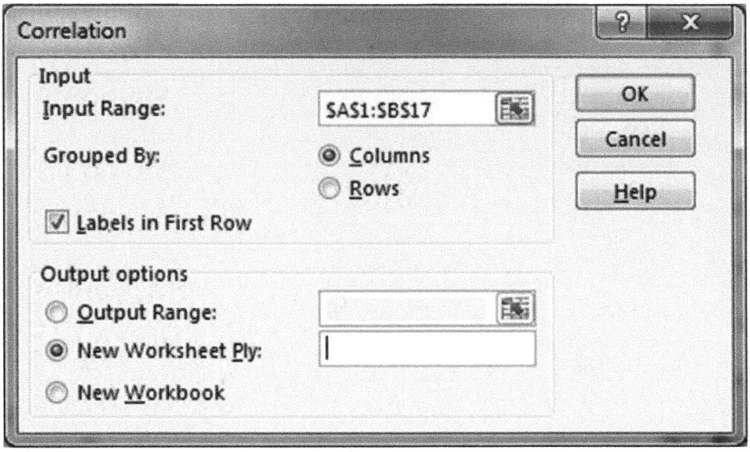

FIGURE 4.2
The correlation dialog box.

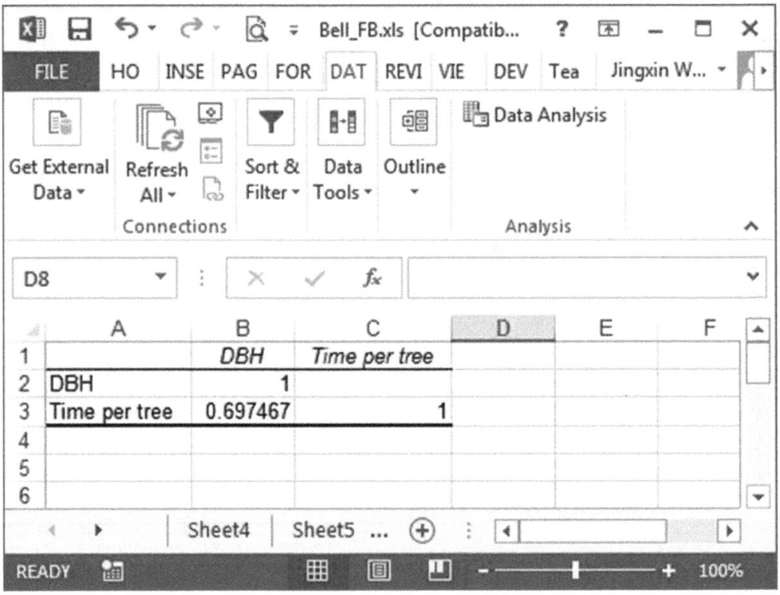

FIGURE 4.3
Results of the correlation analysis.

Analysis, select *Regression* in the *Data Analysis* dialog box, click *OK*, then enter the dependent variable Y range *B1:B17* and independent variable X range *A1:A17*, check the *Labels* and *Confidence Level* boxes, and select the Output Option *New Worksheet Ply*. Additionally, we can specify a few other options. *Constant is Zero* means that the regression line passes through the origin. Clicking *Residuals* will include residuals (the differences between observed and predicted values in which uniform distribution of a band means desirable) in the output. Selecting *Normal Probability Plots* generates a normal probability plot, which is a graphical technique for assessing whether or not a data set is approximately normally distributed (Chambers et al. 1983). A linear pattern of the normal probability plot indicates that the normal distribution is a good model for the data set. Once we select *OK*, we will have a regression summary sheet for this model (Figure 4.4).

Based on the coefficients column in Figure 4.4, we can get the model:

```
Time Per Tree = 0.406 + 0.013*DBH
```

To evaluate whether or not this model is significant or usable, we usually use one or more of the following statistics: *R*-square, MS (mean squared error) or root MS, *F*-value, and *p*-value. In this example, both intercept and DBH are significant at $\alpha = 0.05$ level since their *p*-values are less than 0.05. While our R^2 of 0.486 seems somewhat low, it should be acceptable for

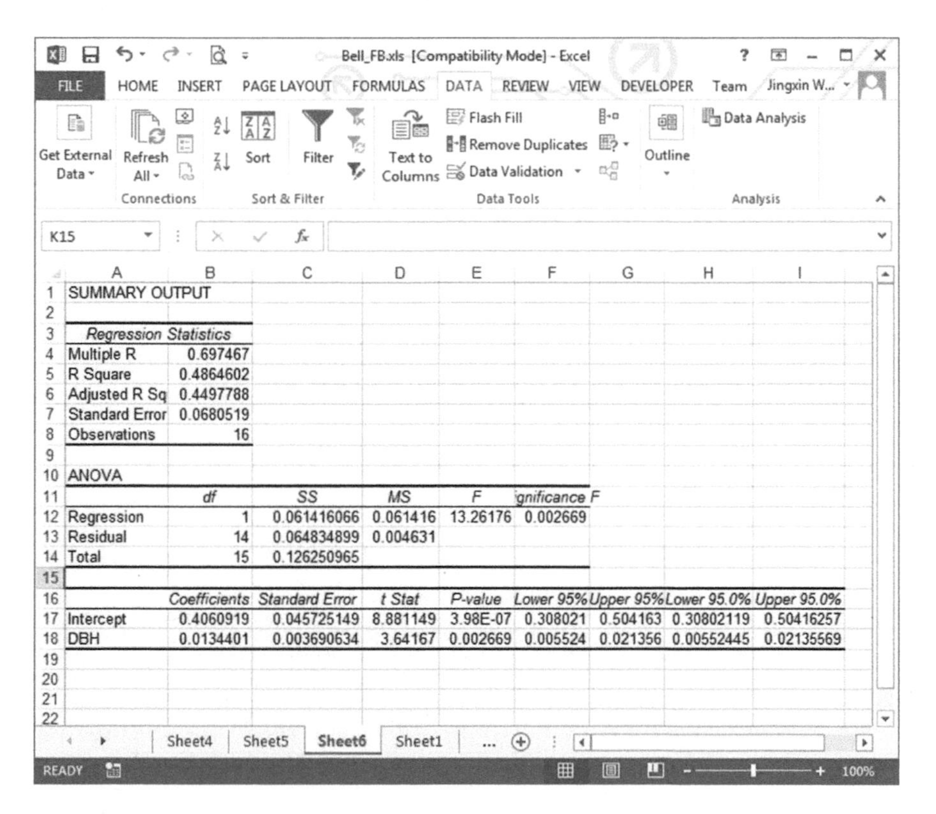

SUMMARY OUTPUT								
Regression Statistics								
Multiple R	0.697467							
R Square	0.4864602							
Adjusted R Sq	0.4497788							
Standard Error	0.0680519							
Observations	16							
ANOVA								
	df	*SS*	*MS*	*F*	*gnificance F*			
Regression	1	0.061416066	0.061416	13.26176	0.002669			
Residual	14	0.064834899	0.004631					
Total	15	0.126250965						
	Coefficients	*Standard Error*	*t Stat*	*P-value*	*Lower 95%*	*Upper 95%*	*Lower 95.0%*	*Upper 95.0%*
Intercept	0.4060919	0.045725149	8.881149	3.98E-07	0.308021	0.504163	0.30802119	0.50416257
DBH	0.0134401	0.003690634	3.64167	0.002669	0.005524	0.021356	0.00552445	0.02135569

FIGURE 4.4
Sample output from the regression tool.

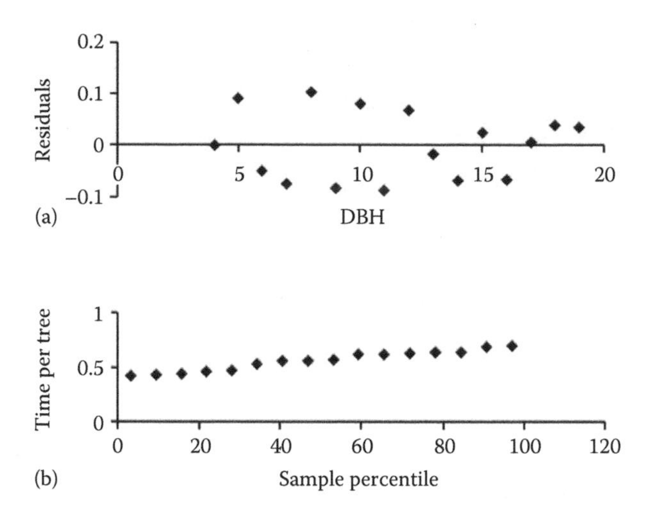

FIGURE 4.5
Plots of (a) DBH residual and (b) normal probability.

forestry applications. Checking our residual plot and normal probability plot (Figure 4.5a and b), this model is acceptable and could be used in applications. It could be improved by including more data and a few more parameters in the model through necessary variable transformations.

4.1.3 *t*-Test

A *t*-test is a statistical hypothesis test in which the test statistic has a Student's *t*-distribution if the null hypothesis is true. It is applied when the population is assumed to be normally distributed but the sample size is small enough that the statistic on which inference is based is not normally distributed. The *t*-test is a statistics test generally used to test whether means of populations are significantly different.

There are three *t*-test statistical functions that you can perform with the Data Analysis option in Excel (Microsoft Office Online 2016):

- Paired Two-Sample for Means
- Two-Sample Assuming Equal Variances
- Two-Sample Assuming Unequal Variances

Two-sample *t*-tests for a difference in means can be either unpaired or paired. So at first, we need to determine if the data are paired or unpaired. For example, suppose we are evaluating the effect of a forest treatment for a tree growth project. We identify 100 trees in the study and then randomize 50 trees to the treatment group and 50 trees to the control group. In this case, we have two independent samples and would use the unpaired form of the *t*-test. If the *t*-test is unpaired, we need to determine whether its variance is equal or unequal by using the statistical function *F-Test Two-Sample for Variance*. This function can be found by clicking *Data → Data Analysis → F-Test Two-Sample for Variance*.

A paired two-sample test is typically applied when there is a natural pairing of observations in the samples, such as when a sample group is tested twice—before and after an experiment. Let's use an example to illustrate this test.

In forest operations, we are concerned with the soil compaction on skid trails after harvesting. So before and after harvest, we collect four soil samples at each of the five measurement points along a skid trail (Figure 4.6).

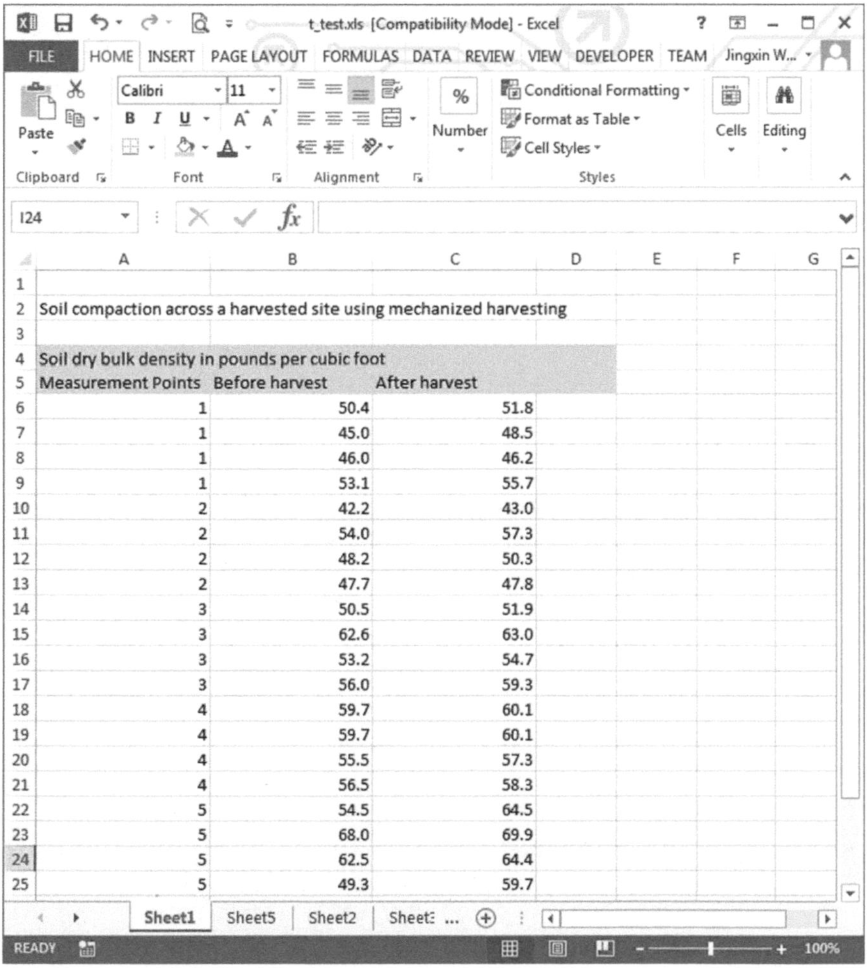

Soil compaction across a harvested site using mechanized harvesting

Soil dry bulk density in pounds per cubic foot

Measurement Points	Before harvest	After harvest
1	50.4	51.8
1	45.0	48.5
1	46.0	46.2
1	53.1	55.7
2	42.2	43.0
2	54.0	57.3
2	48.2	50.3
2	47.7	47.8
3	50.5	51.9
3	62.6	63.0
3	53.2	54.7
3	56.0	59.3
4	59.7	60.1
4	59.7	60.1
4	55.5	57.3
4	56.5	58.3
5	54.5	64.5
5	68.0	69.9
5	62.5	64.4
5	49.3	59.7

FIGURE 4.6
Soil dry bulk densities on a skid trail.

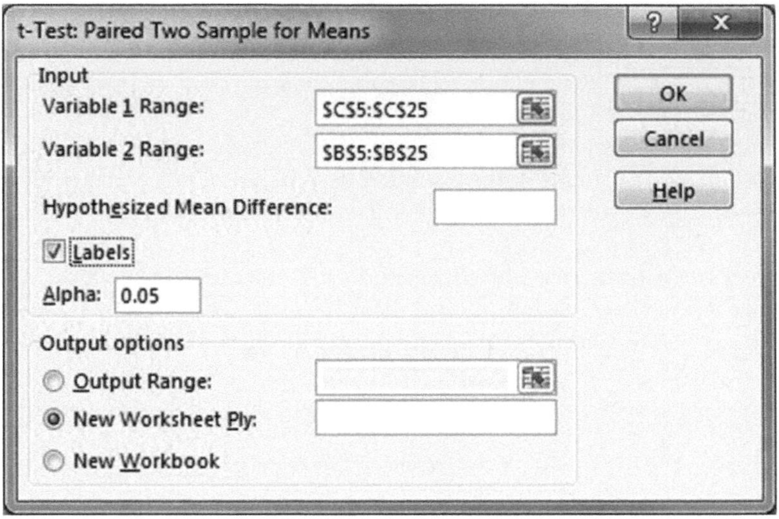

FIGURE 4.7
A *t*-test example of forest soil compaction in Excel.

We want to test if the harvest operations cause a significant compaction. (1 lb = 0.4536 kg, 1 ft^3 = 0.0283 m^3)

Based on the paired *t*-test definition, we know the paired two-sample *t*-test works better for this case. In our example, we specifically determine whether soil dry bulk density has increased after harvest compared to soil dry bulk density before harvest. In this case, the sample size equals 20. For this type of *t*-test, the degrees of freedom is $n - 1 = 19$.

After we enter the data of Figure 4.6, we accessed the *t*-test in Excel as follows: *Data → Data Analysis*, select *t-Test: Paired Two-Sample for Means* and click *OK*. The *t*-Test dialog box will open (Figure 4.7).

The following options need to be specified in the *t*-Test dialog box:

- *Variable 1 Range*: Select everything that is located in the third column, including the label *After harvest*. We want to determine if the soil bulk density after harvest has increased.
- *Variable 2 Range*: Select everything that is located in the second column, including the label *Before harvest*.
- *Hypothesized Mean Difference*: enter *0*
- *Labels*: Check the box because we included the column labels for Variables 1 and 2.
- *Alpha*: This depends on your desired degree of certainty or confidence level: 0.05 is a typical choice if you desire 95% certainty (default); 0.10 if you desire 90% certainty.

- *Output Range*: Select the cell in this worksheet where you want the upper left corner of the output to appear.
- *New Worksheet Ply*: Select this to generate the output on a new worksheet.

Click *OK* to run the analysis, then the following Excel output appears (Table 4.2).

We need to interpret the output table, specifically, to see if soil bulk density before harvest is significantly different from soil bulk density after harvest:

a. The t-value for our example is 3.866971239.
b. One-tailed test: Our t-value (3.866971239) is greater than the critical t-value for a one-tailed test (1.729132812). We can therefore state with 95% certainty that the mean soil dry bulk density increased as a result of the mechanized harvesting. Alternatively, the conclusion could be drawn that the p-value for the one-tailed test is 0.000519123, which is less than alpha (0.05) and is therefore significant.
c. Two-tailed test: The two-tailed test is more stringent because the alpha region of uncertainty is now divided between both outer tails. For the two-tailed test, the t-value needs to be larger (2.093024054) to wind up in the outer 2.5% of either tail and therefore be significant. In this case, our t-value was greater than 2.093024054, therefore large enough to be positioned in the outer 2.5% of either outer tail. This indicates that we may state with 95% certainty that there has been a significant change in the mean soil dry bulk density from before to after harvest. The p-value calculated for the two-tailed test (0.001038245) is also less than alpha (0.05) and is significant. These results support those of our one-tailed t-test, reinforcing our finding of significance.

TABLE 4.2

Statistics of a Paired Two-Sample t-Test for Forest Soil Compaction

	After Harvest	Before Harvest
Mean	56.188	53.73
Variance	47.79509053	43.41063158
Observations	20	20
Pearson Correlation	0.912455727	
Hypothesized Mean Difference	0	
df	19	
t Stat	3.866971239	
$P(T <= t)$ one-tail	0.000519123	
t Critical one-tail	1.729132812	
$P(T <= t)$ two-tail	0.001038245	
t Critical two-tail	2.093024054	

4.2 Mathematical Programming in Forest Management

Mathematical programming involves the use of mathematical models to solve certain types of management science problems. Typical problems may vary from a microanalysis (such as the determination of the best way to cut a tree into logs) to a macro-analysis (such as the evaluation of alternative strategies for managing a forest) (Dykstra 1984). Mathematical programming is an appropriate tool to support optimal decision-making in a multiple-constraint environment and in project management. Mathematical programming is a widely used tool in forest and natural resource management planning. The essence of such programming is to optimize the objective of management by satisfying all the constraining situations. These constraints could arise from factors such as the environment, economics, labor, land use, and policy.

Mathematical programming methods include but are not limited to linear programming (LP), network analysis, multiobjective programming, integer programming, and dynamic programming.

4.2.1 Linear Programming

Linear programming is the most widely used mathematical programming method. It has been broadly applied in natural resource management and related disciplines. Some of the various applications of LP include timber harvest scheduling, biomass harvest and logistics, mill production planning, wildlife management, land use planning, soil loss prevention, and water conservation.

The use of LP requires building a linear model that includes decision variables, objective functions, and constraints. Decision variables are the variables whose values can be controlled and affect the model's performance. The values of the decision variables are not known when you begin a problem. The variables usually represent things that can be adjusted or controlled, for example, the timber or biomass production in a harvest scheduling project. An objective function is a mathematical expression that combines the decision variables to express the goal of your optimization model. The goal could be profit, cost, or revenue. For example, in a forest/biomass harvest scheduling project, your goal could be either to maximize the profits of timber or biomass production or minimize the production cost. The constraints are mathematical expressions that combine the variables to express the resource allocation limits on the possible solutions. For example, the total working hours should be less than the total available hours a harvest crew has, and the various land uses should not exceed the total available land area. Additionally, objective function and variables should be consistent, quantitative, and linear.

Once an LP model is developed, you need to find a way to solve it. The Simplex Algorithm, developed in 1947 by George Dantzig (Winston 2004), has been the most commonly used method to solve this type of optimization problem.

Examples of LP objectives in forest and natural resource management include the following:

- Maximize the net present value of a forest by managing the forest in such a way that it satisfies all the criteria specified in the forest best management practice guidelines.
- Minimize a biorefinery's delivered cost of forest biomass harvesting and processing.
- Minimize the cost of transporting wood products from saw mills to market distribution centers.

The linear relationship in LP implies the following:

1. Proportionality: Decision variables in constraints and objective functions are directly proportional.
2. Additivity: Total contribution of all the variables in the objective function and in the constraints should be the direct sum of the individual contribution of each variable.
3. Certainty: Coefficients in the objectives are deterministic (known for certain).

4.2.2 Network Analysis

Networks arise in numerous settings and in a variety of applications such as in forest product transportation and in piping of oil and shale gas. Network representations are widely used for problems in such diverse areas as production, distribution, planning, facilities' location, and resource management. Tree-stem bucking is a good example of a network model because it presents complex problems. The decision of where to cut depends on various factors, including tree species, size of the tree stem, grades of logs within the stem length, market value for end products, and the number, location, and severity of defects (Bobrowski 1994). These common bucking problems have been solved by employing mathematical programming techniques, including LP, dynamic programming (DP), and, most efficiently, network analysis (Smith and Harrell 1961, Pnevmaticos and Mann 1972, Lawrence 1986, Sessions et al. 1989, Wang et al. 2004). Wang et al. (2009) developed an optimal tree-stem bucking system for Central Appalachian hardwood species using three-dimensional modeling techniques. ActiveX Data Objects were implemented via MS Visual C++/OpenGL to manipulate tree data that were supported by a back end relational data model with five data entity types for tree stems, grades and prices, logs, defects, and stem shapes. A network analysis algorithm was employed to achieve the optimal bucking solution with four alternative stage intervals in the optimization process.

4.2.3 Multi-Objective Programming

Forest and natural resource management problems, especially timber harvest planning, biomass harvest, and logistics modeling, are usually characterized by the need to consider multiple incommensurable objectives over a long time period (Gong 1992). Forests can be managed for multiple uses such as timber, wildlife and habitat, biomass, recreation, and scenic value. Some of these uses might be in conflict, and there is no common measure that can be used to satisfactorily evaluate all of them. A multi-objective or multi-criteria optimization model can be constructed to solve the multiple-use problem. Many multi-objective optimization techniques have been introduced and applied in forest resource management (Dykstra 1984).

Goal programming is the most widely used technique for general multi-objective programming and has been most extensively applied to natural resource management problems (Charnes and Cooper 1961). Goal programming minimizes deviations from multiple goals, or objectives, subject to goal constraints and other physical constraints (Dykstra 1984). In contrast to physical constraints, the goal constraints are satisfied as closely as possible but need not all be met completely. Typical applications of goal programming in natural resource management include timber production, the management of small woodlands, land use planning, and the evaluation of trade-offs between timber management, outdoor recreation, grazing, and production of game animals for hunting (Field 1973, Bell 1976, Rustagi 1976, Schuler et al. 1977).

4.2.4 Integer Programming

An integer programming model is the same as an LP model except that some or all of the variables take on integer values (Winston 2004). For example, in a biomass harvest and logistics scheduling project, we use integers for the number of biomass supply locations and the number of biomass refinery facilities. If only some of the variables are constrained to take on integer values, then the model is called mixed integer LP. Integer programming problems are significantly more difficult computationally than the equivalent LP problems. A few studies that have utilized integer programming in natural resource management include timber harvest scheduling, forest road engineering, design of forest cutting units, and locations of forest products' manufacturing plants (Bare and Norman 1969, Kirby 1973, Bonita 1977, Dykstra and Riggs 1977). Wu et al. (2011) developed a mixed integer programming model to estimate the delivery cost of woody biomass. The model was designed to optimize a woody biomass-based biofuel facility's location with the objective of minimizing the total annual delivery cost of woody biomass under resource and operational constraints.

4.2.5 Dynamic Programming

As you can imagine, quite a few decision problems in forest and natural resource management involve making a sequence of interrelated decisions in such a way that overall effectiveness is maximized (Dykstra 1984). DP is a technique for solving such problems and achieving solutions by working backward from the end of a problem to the beginning, thus breaking up a large problem into a series of smaller problems (Winston 2004). It provides a systematic procedure for determining the optimal combination of decisions. A dynamic problem can be divided into stages with a decision required at each stage (Figure 4.8). Associated with each stage is a number of states. The effect of the decision at each stage is to transform the state at that stage into a state associated with the subsequent stage.

For example, in Figure 4.8, there are four stages (numbered across the x-axis), state nodes are numbered within the blue squares, and alternative states are denoted by arcs (arrows) and include numeric weights. Therefore, at stage 0, we have state node 0 and three possible states (arcs) for piping alternatives: from node 0 to either node 1, node 2, or node 3 depending on the weight of each arc. The weight of an arc (in this case, from nodes 0 to 1 the weight would be 4) could represent the length of pipe between these two nodes. The states and weights of these arcs are deterministic, and this DP is called deterministic DP. If the states and weights of arcs are random in the DP

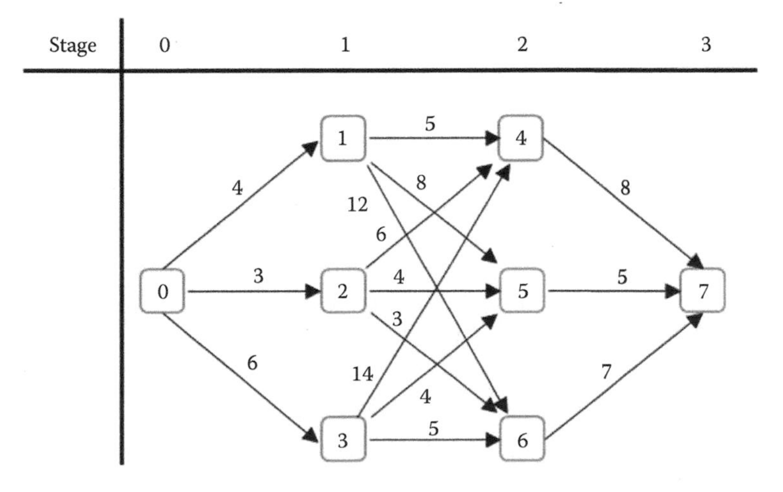

FIGURE 4.8
A graphical illustration of DP: a piping problem's stages, state nodes (blue squares), and weights of arcs (arrows).

process, it is called probabilistic dynamic programming (Winston 2004). In this section, we will primarily focus on deterministic dynamic programming.

Many forest and natural resource management problems are time-oriented, such as forest stand growth and dynamics, water quantity and quality, insect or wildlife populations, and temporal scheduling of forest and biomass harvest. Decisions, therefore, must be implemented sequentially over time (Dykstra 1984). DP has been applied in natural resource management and related fields, including timber management, tree-stem bucking, log or lumber sawing optimization, pest management, biomass transportation scheduling, shale gas pipe route layout, and forest fire detection. DP is one of the most widely used stand-level optimization techniques. From the 1970s through the 1990s, significant advances occurred in the application of DP in forest and natural resource management.

The basis of DP is a divide-and-conquer process (Figure 4.9). It is an exhaustive search method, including three basic steps: (1) partition the problem into smaller (independent) subproblems, (2) solve the subproblems recursively (repeats Step 1), and (3) combine the subproblems' solutions to solve the original problem. We use DP when the subproblems are not unique (subproblems share subproblems), if we use the straight divide-and-conquer approach, you would repeatedly solve the same problems. Through using recursive functions, you solve a subproblem once and store the solution in a table (data structure) to reference when needed.

Here is an example of DP formulation. Suppose the potential cutting points along a tree stem from butt to top are denoted by X_i ($i = 1, 2, \ldots, n$), the interval (or stage interval) between two potential cutting points is identified as Y_k ($k = 1, 2, \ldots, K$), and the weight on the arc (W_{ij}) is denoted by the value of log, then the directed graph (S) for tree-stem bucking can be

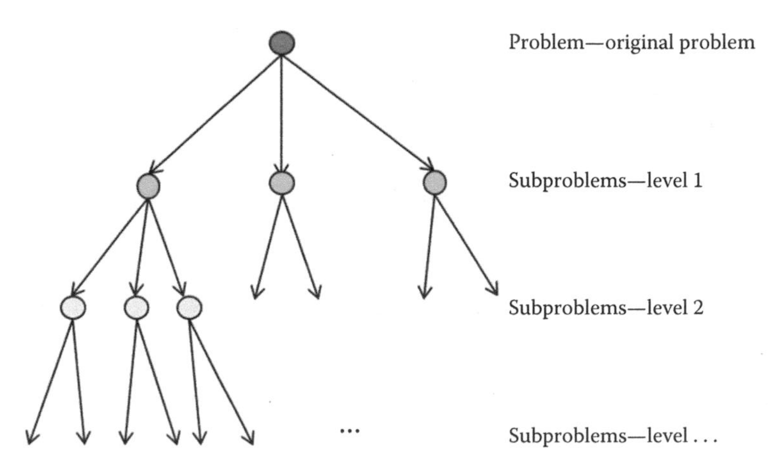

FIGURE 4.9
Recursive tree and bottom-up solution.

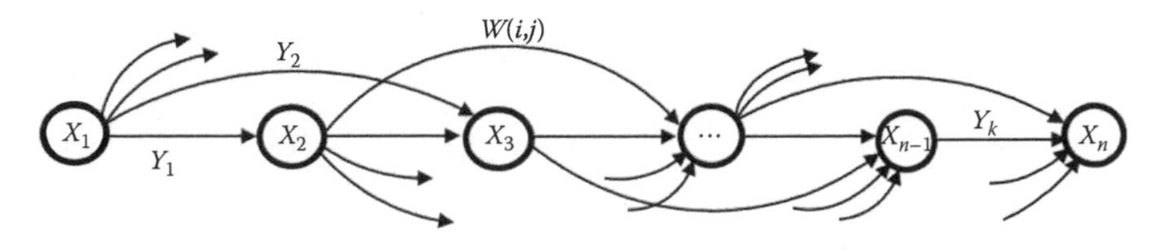

FIGURE 4.10
Directed graph of tree-stem bucking model.

expressed by the point set $(X_1, X_2, \ldots, X_n) \in X$, the arc set $(Y_1, Y_2, \ldots, Y_K) \in Y$, and the weight set $W_{ij} \in W$, i.e. $S = \{X, Y, W\}$ (Figure 4.10). Here, n refers to the maximum number of potential cutting points, and K to the total number of potential arcs.

Paths from X_1 to X_n depended on the interval between two cutting points and log length. The interval between the two closest potential cutting points is called "stage interval." The stage interval is a very important factor in the model. If the stage interval is longer, the solution time will be less and it may not meet the requirements of accuracy, and vice versa. If the increments for scaling length are 0.2, 0.5, and 1.0 m in the log product standard, the stage interval is determined based on these values.

The Dijkstra algorithm, known as the labeling algorithm, has been shown to be among the fastest algorithms available for solving minimum path problems and is particularly well suited to being programmed on a computer (Dykstra 1984). The principle of the Dijkstra algorithm for the shortest path problem was adopted to find the longest path in the directed graph of tree-stem bucking. The objective of this model is to find the longest path from butt to top of the tree.

In the node set X of $S = \{X, Y, W\}$, suppose Z is a subset of X, and $Z_0, Z_1, Z_u, Z_v \in Z$. If $\{Z_0, Z_v\}$ is in the longest path of Z_0–Z_v, then $\{Z_0, Z_u\}$ is also the longest path of Z_0–Z_u. Since all the arcs in the directed graph of tree-stem bucking ($S = \{X, Y, W\}$) are forward and $W_{ij} \geq 0$, the values on the arcs from butt (X_1) to top (X_n) can be expressed as follows:

$$L(X_1,X_n) = \max\left[L(X_1,X_{n-1}) + W(n-1,n)\right]$$
$$L(X_1,X_{n-1}) = \max\left[L(X_1,X_{n-2}) + W(n-1,n-2)\right]$$
$$\ldots$$
$$L(X_1,X_2) = \max\left[L(X_1,X_1) + W(1,2)\right]$$

Based on the above principle, in order to search for the longest path from the origin X_1 to the destination X_n, the farthest point around the X_1 should be found first. Then the searching radius around X_1 is increased gradually until the longest path from X_1 to X_n is found.

4.3 Linear Programming Formulation

It is important to define a real problem in detail before formulating the linear model. First, observe the situation carefully and pinpoint the optimization problem that you need to solve. Remember to try to find the most important parts of the problem, remain open-minded about the problem, and be specific when thinking about the problem (Dykstra 1984). Once you understand the whole problem and what you are trying to solve, you may follow the steps below to formulate your model (Dykstra 1984):

1. Identify the activity set and define a set of decision variables
2. Define the objective function using the decision variables
3. Allocate the resources that will be needed in conducting the activity and the restrictions associated with any potential solutions

After you've followed the procedures of LP formulation, you may begin LP. Let's start with a simple example. A forest landowner owns 36.42 ha (90 acres) of lands: 16.19 ha (40 acres) of red pine and 20.23 ha (50 acres) of northern hardwoods. The estimates of the yearly revenues generated by each type of forest are $200/ha/year ($80/acre/year) for red pine and $275/ha/year ($110/acre/year) for northern hardwoods. The landowner doesn't want to spend more than 180 days per year managing the forestlands. Red pine takes 5 days/ha/year (2 days/acre/year) of labor, while northern hardwoods require 7.5 days/ha/year (3 days/acre/year). Provide the landowner an optimal plan of how many acres (1 acre = 0.4 ha) of each type of forest should be managed per year in order to maximize the total revenues.

To solve this problem, you have to define decision variables, find your objective, and construct model constraints. First, let `X1` be the number of acres (1 acre = 0.4 ha) of red pine to manage, `X2` be the number of acres (1 acre = 0.4 ha) of northern hardwoods to manage, and `Z` be the total revenue.

The objective function expresses the relationship between `Z` (the total revenue) and the decision variables `X1` and `X2`. To formulate the objective function, you need the revenue information for each type of forest: $200/ha/year ($80/acre/year) for red pine and $275/ha/year ($110/acre/year) for northern hardwoods. Also, you may note that this is a maximization problem. Now, we can construct the objective function as:

```
Max Z = 80 X1 + 110 X2
```

To achieve the solution, we also need to determine what constraints limit the decisions. In this example, we have three types of constraints: land constraints, working hours or time constraints, and nonnegative constraints.

1. Land constraints: The area managed in each type of forest cannot exceed the area available. Therefore

```
X1 ≤ 40 acres (1 acre = 0.4 ha) of red pine
X2 ≤ 50 acres (1 acre = 0.4 ha) of northern hardwoods
```

2. Working hour constraints: The landowner only has 180 days available to land management. The forests require 5 days/ha/year (2 days/acre/year) of labor (red pine) and 7.5 days/ha/year (3 days/acre/year) (northern hardwoods), respectively. So, the time constraints can be expressed as:

```
2 X1 + 3 X2 ≤ 180
```

3. Nonnegative constraints: The last constraints that need to be considered are nonnegative constraints. Since the decision variables X1 and X2 refer to areas, none of them may be negative. Therefore

```
X1 ≥ 0
X2 ≥ 0
```

In summary, we can formulate this forest management problem as the following LP model:

```
Max Z = 80 X1 + 110 X2
```

Subject to:

```
X1 ≤ 40
X2 ≤ 50
2 X1 + 3 X2 ≤ 180
X1 ≥ 0
X2 ≥ 0
```

4.4 Solve Mathematical Models in Forest Management Using Excel Solver

4.4.1 Example of Optimization Problem

In the previous section, we learned how to formulate a linear model. In this section, we will use Microsoft Excel to solve LP problems. Let's try another example. WVUTIMBER produces both sawlogs and pulpwood from its forest. It takes 242.81 ha (600 acres) of forest to produce 100 truckloads of sawlogs and 161.87 ha (400 acres) to produce 100 trucks of pulpwood. Let's assume 100 truckloads is one standard delivery. The company has 971.25 ha (2400 acres) of forest lands. Also, harvesting the products for sawlogs takes 100 labor hours, while harvesting for pulpwood requires 200 labor hours per standard delivery. The total amount of labor hours with the company is 600 for a given harvesting season. Assume that the demand for pulpwood at the processing plant is not 100 truckloads more than that of sawlogs. Also, the maximum demand for pulpwood is 200 truckloads per season. Let's say that profit from one standard delivery (100 truckloads) of sawlogs is $50,000 and that from the same of pulpwood is $40,000. WVUTIMBER wants to determine the highest profit combination of sawlogs and pulpwood to harvest from its forest. This problem can be formulated as LP. Let's define the decision variables first.

```
X1 = # of 100 Truckloads of sawlogs per season
X2 = # of 100 Truckloads of pulpwood per season
```

The objective is to maximize the profit (`Z`) of this timber sale including pulpwood and sawlogs.

```
Maximize Z = 50,000X1 + 40,000X2
```

Next, we need to define our constraints.

Acres:

Total available acres (1 acre = 0.4 ha) = 2400

100 truckloads of sawlogs require 242.81 ha (600 acres).

100 truckloads of pulpwood require 161.87 ha (400 acres).

Hence, `600X1 + 400X2 ≤ 2400` (we can cut all the forest lands or leave portions uncut if harvesting is not profitable at the present time). We cannot, however, cut more than what we have.

Labor hours:

Total available labor hours: 600

Labor requirement per 100 sawlog truckloads = 100

Labor requirement per 100 pulpwood truckloads = 200

Therefore

```
100X1 + 200X2 ≤ 600
```

Market restriction:

Truckloads of pulpwood shouldn't exceed truckloads of sawlogs by more than 1 standard delivery (100 truckloads). Therefore

```
X2 - X1 ≤ 1
```

Demand restriction:

Maximum demand for pulpwood is 200 truckloads (2 standard deliveries).

$X2 \leq 2$

As in any LP, the variables in this example cannot be negative, so $X1,\ X2 \geq 0$.

This problem can be written in the following format:

$\text{Max: } Z = 50000X1 + 40000X2$ .... *(OBJECTIVE FUNCTION)*
$600X1 + 400X2 \leq 2400$ .... *(CONSTRAINT 1)*
$100X1 + 200X2 \leq 600$ .... *(CONSTRAINT 2)*
$X2 - X1 \leq 1$ .... *(CONSTRAINT 3)*
$X2 \leq 2$ .... *(CONSTRAINT 4)*
$X1 \geq 0$ .... *(NONNEGATIVE CONSTRAINT)*
$X2 \geq 0$ .... *(NONNEGATIVE CONSTRAINT)*

This model can be solved in Excel using Solver. The layout and data input of the model are detailed in Figure 4.11.

In MS Excel, nonnegative constraints may be specified either in the format of other constraints in Figure 4.11 or by using the Solver's nonnegative option. Note that we entered numbers or values for our problem in the cells

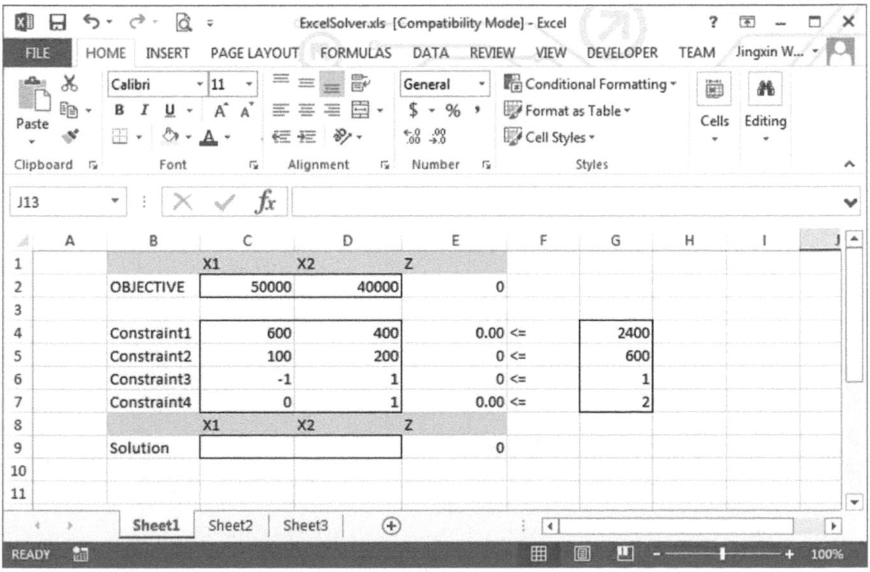

	A	B	C	D	E	F	G
1			X1	X2	Z		
2		OBJECTIVE	50000	40000	0		
3							
4		Constraint1	600	400	0.00	<=	2400
5		Constraint2	100	200	0	<=	600
6		Constraint3	-1	1	0	<=	1
7		Constraint4	0	1	0.00	<=	2
8			X1	X2	Z		
9		Solution			0		

FIGURE 4.11
Input of timber harvesting LP optimization model in MS Excel.

TABLE 4.3

Formulas for Cells for Problem Shown in Figure 4.11

Description	CELL Name	FORUMLA
Objective function	E2	=SUMPRODUCT($C$9:$D$9,C2:D2)
Constraint 1	E4	=SUMPRODUCT($C$9:$D$9,C4:D4)
Constraint 2	E5	=SUMPRODUCT($C$9:$D$9,C5:D5)
Constraint 3	E6	=SUMPRODUCT($C$9:$D$9,C6:D6)
Constraint 4	E7	=SUMPRODUCT($C$9:$D$9,C7:D7)
Solution	E9	=E2

surrounded by solid line boxes. Cells that are not bordered contain formulas. The logical comparison signs (<=) in column F are included only to improve the figure's readability. Often, objective functions can be denoted by the symbol "Z." Once the problem is solved, the cells will be populated with values by Solver. The Excel formulas for Column E are listed in Table 4.3.

In Excel, the *SUMPRODUCT* function multiplies components in the given arrays and returns the sum of these products. Its syntax is:

```
=SUMPRODUCT(array1, array2, array3, ...)
```

Array1, array2, array3, ... are 2–255 arrays whose components you want to multiply and then add. The array arguments must have the same dimensions. If they do not, SUMPRODUCT returns the #VALUE! error. The SUMPRODUCT function treats array entries that are not numeric as if they were zeros. For example,

Array1			Array2		
Row	A	B	Row	D	E
2	2	5	2	3	8
3	3	6	3	5	9
4	4	7	4	7	10

the function

```
=SUMPRODUCT(A2:B4, D2:E4)
```

will return the products of 2 * 3 + 5 * 8 + 3 * 5 + 6 * 9 + 4 * 7 + 7 * 10 = 213.

In our example, `=SUMPRODUCT($C$9:$D$9,C2:D2)` (from the *Objective Function* row in Table 4.3) returns the products of `50000*X1+40000*X2`, which is the objective function of our optimization problem.

4.4.2 Activate Excel Solver

To use Excel Solver, we first need to activate it. Here are the procedures:

- Click *File* on the Excel menu and then click *Options*.
- In the *Excel Options* dialogue box that opens, click the *Add-Ins* category on the left.
- Select *Excel Add-ins* in the *Manage* dropdown box at the bottom, then click *Go*.
- A new *Add-Ins* window will open at this time. Check *Solver Add-in* and click *OK*.

Now, Solver is ready to use in Excel. To view this add-in, click the *Data* tab on Excel's menu ribbon. The Solver add-in should be at the right of the tool bar.

4.4.3 Use Excel Solver

Once the data are entered into an Excel worksheet (as in Figure 4.11) with the formulas described in Table 4.3 also included, we can click *Solver* on the *Data* tab in the Excel main menu ribbon. This will open a *Solver Parameters* window where we will need to specify the parameters (Figure 4.12).

First we *Set Objective* in the cell where the objective function resides. We specify that as cell *$E$2*. Notice that we use the absolute reference of cell *E2* with dollar signs preceding both column E and row 2. Our objective function is to maximize the profit, thus we will click the *Max* radio button.

The problem should be solved by changing the values of *X1* and *X2* in four of our constraints. Therefore, we specify *$C$9:$D$9* as the coefficients of constraints in *By Changing Variable Cells*. Next we need to specify our constraints. These constraints can be added by clicking the *Add* button. You can also edit your existing constraints by clicking *Change* or *Delete*.

Cell references are the cells where we used our formula for each of the constraints, and the default equality is <= which can be changed. Constraint cells are the right-hand side values in constraint equations (Figure 4.13). In this figure, we added four constraints all together in one statement. You may add them one by one.

While the Solver is typically for linear models, it can also handle nonlinear models. Regarding a model's linearity and non-negativity, we check *Make Unconstrained Variables Non-Negative* to force the solver to implement non-negativity constraints (Figure 4.12). If a model is nonlinear, this should be specified by clicking the *Options* button in the Solver window. We will not further discuss nonlinear optimization since no examples were formulated.

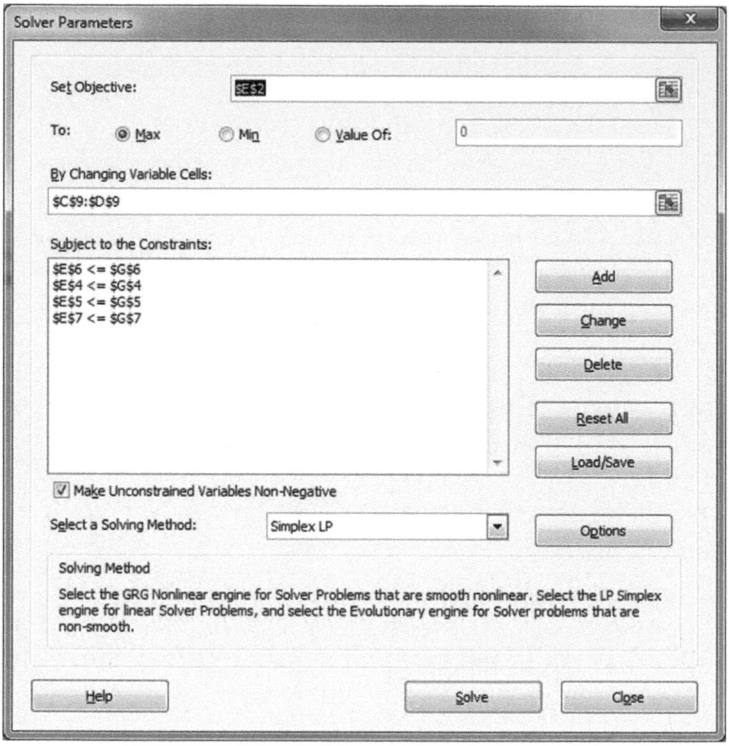

FIGURE 4.12
Solver Parameter specification window.

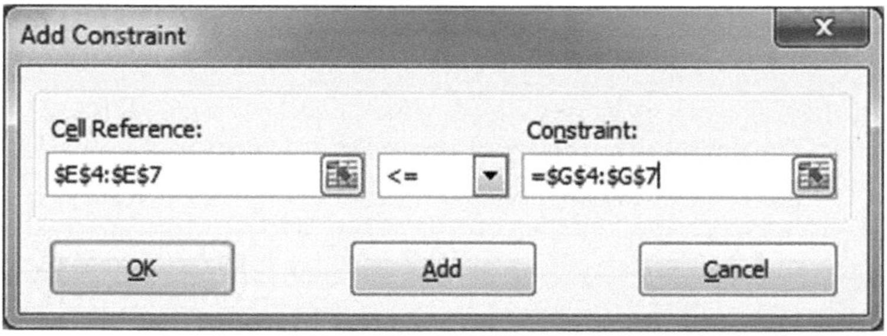

FIGURE 4.13
Adding constraints in Solver.

To run the Solver, click the *Solve* button. If the Solver finds a solution, a *Solver Results* window with the appropriate message for results will appear (Figure 4.14).

We can select *Answer*, *Sensitivity*, and/or *Limits* to further analyze the problem. A simple solution can be viewed by checking *Keep Solver Solution* and clicking *OK*. Values of *X1* and *X2* will populate cells *C9* and *D9* of the problem's spreadsheet as shown in Figure 4.15.

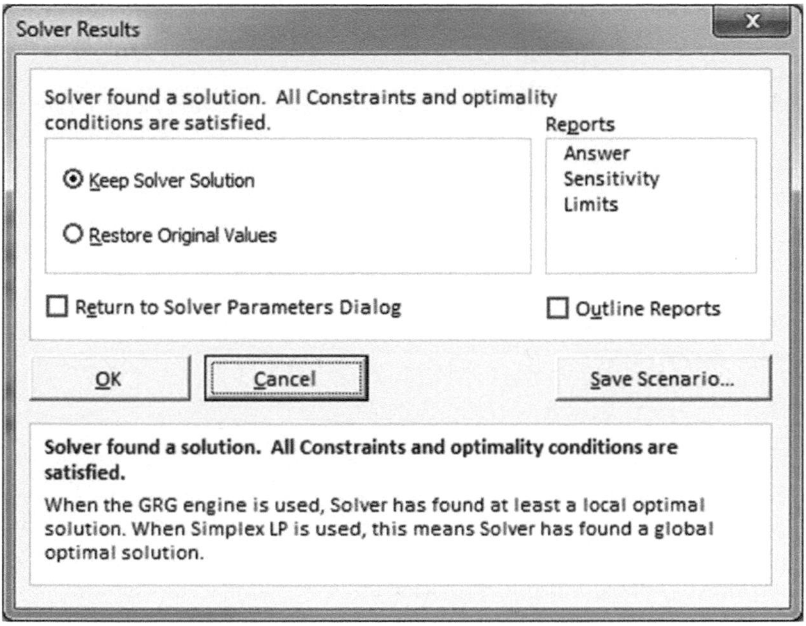

FIGURE 4.14
Solver Results window.

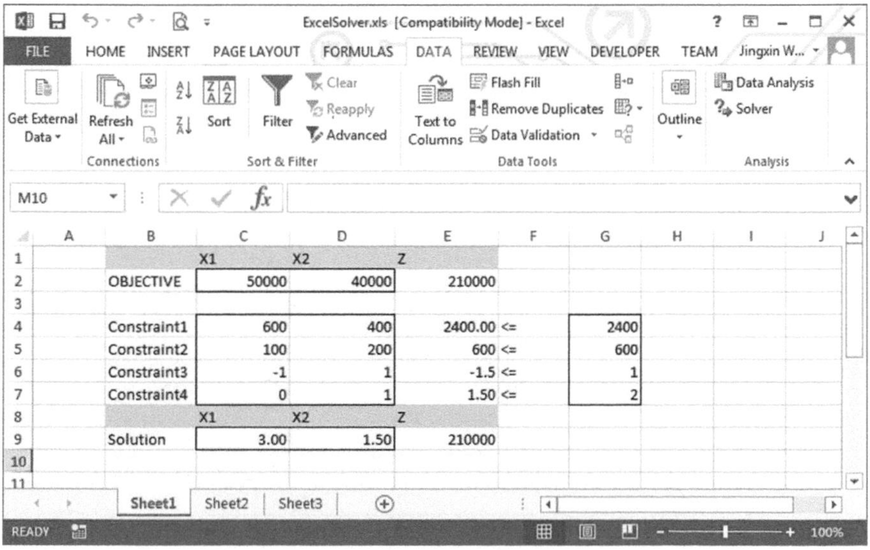

	B	C	D	E	F	G
1		X1	X2	Z		
2	OBJECTIVE	50000	40000	210000		
3						
4	Constraint1	600	400	2400.00	<=	2400
5	Constraint2	100	200	600	<=	600
6	Constraint3	-1	1	-1.5	<=	1
7	Constraint4	0	1	1.50	<=	2
8		X1	X2	Z		
9	Solution	3.00	1.50	210000		

FIGURE 4.15
Solver solution of the timber harvesting optimization problem.

The Solver provides the solution in Solution (row 9) on the Excel worksheet (Figure 4.14) and it assigned *X1* as *3.00*, *X2* as *1.50*, and *Z* value as *210,000*. This means that under the given constraints, if we make 3 standard deliveries (300 truckloads) of sawlogs and 1.5 standard deliveries (150 truckloads) of pulpwood, we are able to make $210,000 in profit, which is the maximum that can be achieved. If you desire, you can perform sensitivity analysis for further insights into the problem.

Class Exercises

1. Analysis of DBH and Merchantable Height of Trees with Excel Regression and Chart Tools: Table 4.4 is a small portion of a typical forest cruising data set and presents the relationship between DBH in inches (1 in. = 2.54 cm) and merchantable height (MHT) (in the number of 4.88 m (16 ft) logs) for 31 trees.

   Create an Excel workbook (save it as DBH_MHT.xls) that includes a worksheet that contains the data in Table 4.4.

   Once you have created this workbook, you need to accomplish the following things:

   a. Draw an *X–Y* Scatter Chart to check the relationship between these two variables visually.
   b. Use the Correlation Analysis tool to verify if your visual assessment of the relationship between the DBH and the MHT is correct.
   c. Use the Regression Tool to obtain a mathematical model of MHT = $a + b$ * DBH. List the parameters in the model. Evaluate the fitness of the model based on: *R*-square, MS or root MSE, *F*-value, and *p*-value, and explain how well the model fits.
   d. Create a Line Chart of the predicted MHT vs. DBH.

2. Forest Harvest and Management Planning Optimization: Consider the forest in Figure 4.16. Individual stands are identified (1–8) in the circles, and the area (in acres, 1 acre = 0.4 ha) of each stand is also indicated. For example, 33.26 ha (82.18 acres) is the size of stand number 1. Our objective is to harvest the maximum possible area without cutting adjoining stands. This is called a *common adjacency constraint* and is often required to reduce a harvest's impact on soil, water, and forest aesthetics. Formulate a linear program for this problem and solve it using the Excel Solver. Use the approach we followed in our earlier example that had 3 stands instead of 8.

TABLE 4.4
MHT vs. DBH

DBH (in.)	Observed MHT (16 ft Logs)
7	0
8	0
10	0.5
11	0.5
12	0.5
12	1
12	1.5
12	1
12	1.5
13	2
14	1.5
14	2
15	2
15	1
16	1.5
17	2
17	2
18	1.5
18	2
19	2
19	2.5
22	2.5
22	2.5
22	2.5
23	2.5
23	2
23	2
26	3
27	3
28	3.5
30	4

Hint: If I only had Stands 1, 2, and 3 instead of 8 stands, my objective function would be:

```
Max Z = 82.18X1 + 81.75X2 + 58.80X3
```

This is an integer programming problem. The decision variables (`X1`, `X2`, and `X3`) are binary variables (with values of 0 or 1). Therefore, `X1` takes a value of 1 if Stand 1 is harvested and 0 otherwise. There are several ways we can formulate constraints to solve this 3-stand problem.

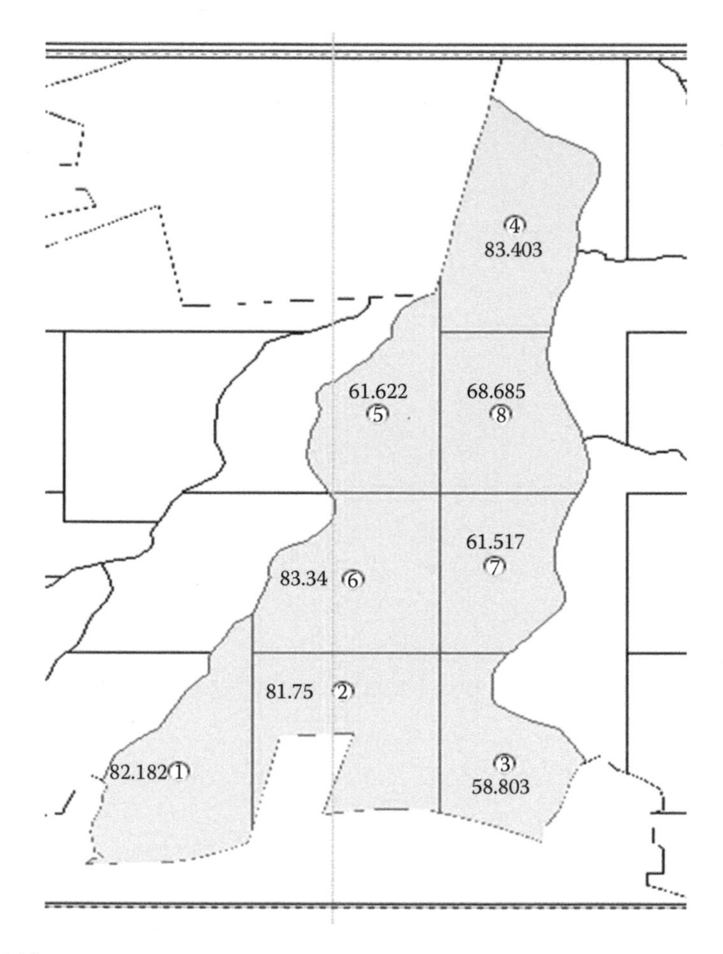

FIGURE 4.16
Forest stand with area in acres.

One simple way to formulate these spatial or adjacency constraints is with the following approach:

Harvest stand 1 if stand 2 is not harvested and vice versa.

Harvest stand 2 if stand 3 is not harvested and vice versa.

Subject to:

```
X1 + X2 <= 1;
X2 + X3 <= 1,
```

where `X1` and `X2` are binary variables (they can take a value of 0 or 1).

These binary variables can be specified by bounding the values within 0 and 1 and specifying the variable as integer type or simply by specifying binary type while assigning constraints.

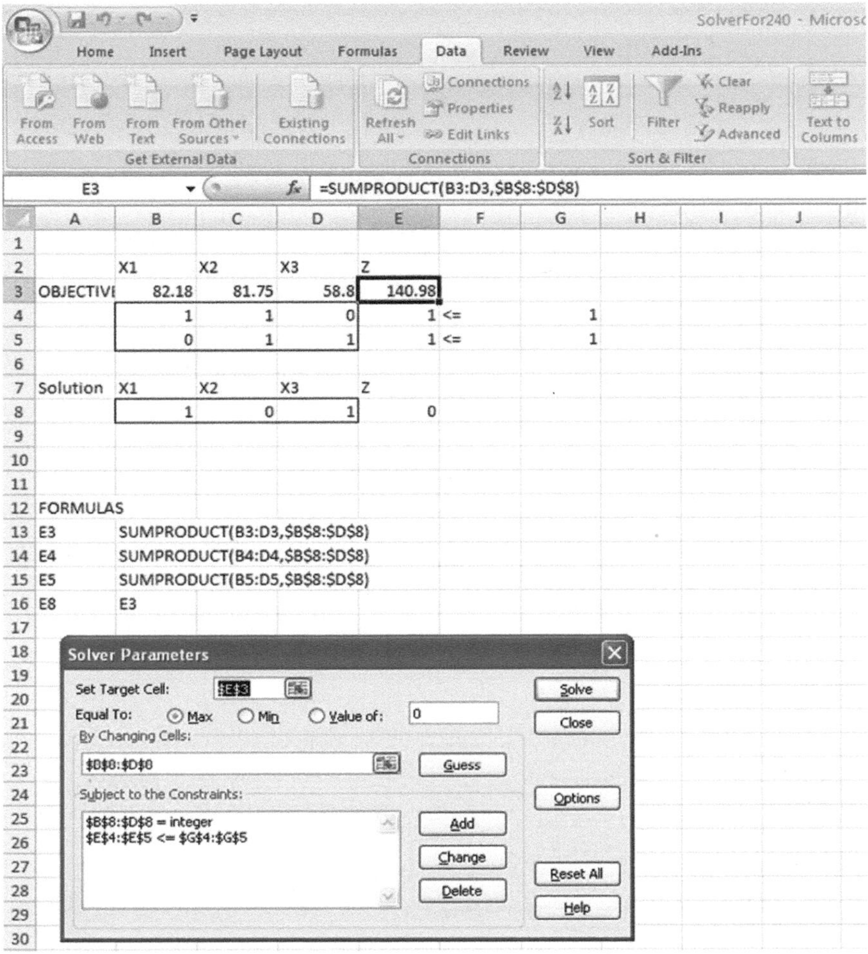

FIGURE 4.17
Problem, formula, and solution for harvesting 3 stands (a problem related to this exercise).

This problem can be specified in Excel with constraints as shown in Figure 4.17. The same figure shows that the model, once solved, will tell us that harvesting stands 1 and 3 (depicted by the value 1 for `X1` and `X3`, and 0 for `X2`) will yield approximately 56.66 ha (140 acres) of harvested stands.

3. Using Excel Solver to Solve a Linear Model of Forest Management
   The XYZ Sawmill Company's CEO asks to see next month's log-hauling schedule for his four sawmills. He wants to make sure he keeps a steady, adequate flow of logs to his sawmills to capitalize on the good lumber market. Minimizing the cost of transportation is something that is also important to him.

TABLE 4.5

Round Trip Cost of Hauling One Truck Load[a]

	Round Trip Cost (in $)				
Logging Site	**Mill A**	**Mill B**	**Mill C**	**Mill D**	**Max. Truck Loads/Day**
1	19	7	3	21	10
2	15	21	18	6	30
3	11	14	15	22	20
Mill demand	15	10	20	15	

[a] Includes the number of trucks each mill demands per day as well as the maximum number of truck loads available from each site per day.

The harvesting group plans to haul from three logging sites. The average haul cost is $2 per mile (1 mile = 1.61 km) for both loaded and empty trucks. The logging supervisor estimated the maximum number of truckloads of logs coming off each harvest site daily (max. truck loads/day). The number of truckloads varies because terrain and cutting patterns are unique for each site. Finally, the sawmill managers have estimated the truckloads of logs their mills need each day (mill demand). All these estimates are listed in Table 4.5. Define an LP model and solve it using the Excel Solver.

Hint: *Decision variables*: Let Xij be the haul cost from logging site i to mill j, where Logging Site i = 1, 2, 3 and Mill j = Mill A, Mill B, Mill C, Mill D. Therefore, you need to define 3 × 4 decision variables.

Objective: Minimize the haul costs from each site to each mill.

Constraints:

1. Loads to each mill (4 constraints)
2. Total truckloads from each logging site (3 constraints)
3. Nonnegative constraints for all the 3 * 4 variables

Additional data manipulations:

1. Suppose Logging Site 3 hired an extra driver and is now able to produce a maximum of 27 truckloads per day. How would the optimal hauling cost change? How would the logging schedule change? (Except for the change in maximum truckloads, assume all other factors remain the same.)
2. Suppose a bridge closed between Logging Site 2 and Mill A, so the truck has to take a detour. The overall cost of hauling from Logging Site 2 to Mill A therefore increased from $15 to $20. How would the optimal hauling cost change? How would the logging schedule change?

References

Bare, B.B. and E.L. Norman. 1969. An evaluation of integer programming in forest production scheduling problems. Agricultural Experiment Station, Purdue University, Lafayette, IN. Research Bulletin No. 847.

Bell, E.F. 1976. Goal programming for land use planning. Pacific Northwest Forest and Range Experiment Station, USDA Forest Service, Portland, OR. General Technical Report PNW-53.

Bobrowski, P.M. 1994. The effects of modeling on log bucking solution techniques. *Journal of the Operational Research Society* 45(6): 624–634.

Bonita, M.L. 1977. Location of forest industries in the Philippines. In *FAO/Norway Seminar on Storage, Transport and Shipping of Wood*. Food and Agriculture Organization of the United Nations, Rome, Italy. Publication No. FOI:TF-RAS38(NOR), pp. 45–61.

Chambers, J., W. Cleveland, B. Kleiner, and P. Tukey. 1983. *Graphical Methods for Data Analysis*. Wadsworth, Boston, MA.

Charnes, A. and W. Cooper. 1961. *A Survey of Goal Programming*. Wiley, New York.

Dykstra, D.P. 1984. *Mathematical Programming for Natural Resource Management*. McGraw-Hill, New York, 318pp.

Dykstra, D.P. and J.L. Riggs. 1977. An application of facilities location theory to the design of forest harvesting areas. *AIIE Transactions* 9: 271–277.

Field, D.B. 1973. Goal programming for forest management. *Forest Science* 19: 125–135.

Gong, P. 1992. Multiobjective dynamic programming for forest resource management. *Forest Ecology and Management* 48: 43–54.

Kirby, M. 1973. An example of optimal planning for forest roads and projects. In *Planning and Decision-Making as Applied to Forest Harvesting*. J.E. O'Leary (ed.). Forest Research Laboratory, Oregon State University, Corvallis, OR, pp. 75–83.

Lawrence, M. 1986. Optimal bucking: A review of the literature. Forest Research Institute, Rotoura, New Zealand. IEA/Bioenergy Project CPC-9, Report No. 1.

Microsoft Office Online. 2016. Use the Analysis ToolPak to perform complete data analysis. http://office.microsoft.com/en-us/excel-help/use-the-analysis-toolpak-to-perform-complex-data-analysis-HA102748996.aspx#_Toc340479281. Accessed on January 25, 2016.

Pnevmaticos, S.M. and S.H. Mann. 1972. Dynamic programming in tree bucking. *Forest Products Journal* 22(2): 26–30.

Rustagi, K.P. 1976. Forest management planning for timber production: A goal programming approach. School of Forestry and Environmental Studies, Yale University, New Haven, CT. Bulletin 89.

Schuler, A.T., H.H. Webster, and J.C. Meadows. 1977. Goal programming in forest management. *Journal of Forestry* 75: 320–324.

Sessions, J., J. Garland, and E. Olsen. 1989. Testing computer-aided bucking at the stump. *Journal of Forestry*, 87(4): 43–45.

Smith, G.W. and G. Harrell. 1961. Linear programming in log production. *Forest Products Journal* 11(1): 8–11.

Wang, J., C. LeDoux, and J. McNeel. 2004. Optimal tree-stem bucking of northeastern species of China. *Forest Products Journal* 54(2): 45–52.

Wang, J., J. Liu, and C. LeDoux. 2009. A 3D bucking system for optimal bucking of central Appalachian Hardwoods. *International Journal of Forest Engineering* 20(2): 26–35.

Winston, W.L. 2004. *Operations Research—Applications and Algorithm* (4th Edition). Brooks/Cole Cengage Learning, Belmont, CA, 1418pp.

Wu, J., J. Wang, and J. McNeel. 2011. Economic modeling of woody biomass utilization for bioenergy and its application in central Appalachia, USA. *Canadian Journal of Forest Research* 41: 1–15.

5

Visual Basic for Applications in Microsoft Excel

5.1 Introduction to VBA

Visual Basic for Applications (VBA) is the programming language for Microsoft Office and its associated applications, such as Excel, Word, and Access (Microsoft 2016). There are many advantages to using VBA, including building user-defined functions and automating repetitive processes. VBA uses the Visual Basic Runtime Library, but it normally runs code within a host application such as Excel rather than as a stand-alone program (Microsoft 2016). To implement functions beyond a regular Excel spreadsheet, a Visual Basic programming environment is installed with Microsoft Excel. In this chapter, we will learn how to use Visual Basic Programming in Microsoft Excel.

5.1.1 Visual Basic Editor in Excel

For the latest version of Excel, we open an Excel workbook and then click *Developer* on the main menu ribbon. If the *Developer* tab is not visible on the ribbon, you can add it through the following steps: click the *File* menu, choose *Option*, and then in the *Excel Options* dialog box, choose the *Customize Ribbon* button. Under the main tabs list, check *Developer*. This should place the tab onto your Excel main menu ribbon. Click *Developer*, then click on the *Macro Security* button, and check the second-level *Disable all Macros with Notification* and click *OK*. Now from the main menu, click *Developer* → *Visual Basic* and a Visual Basic Editor (VBE) will appear (Figure 5.1).

Within the VBE, there is a *Project Explorer* window and a *Properties* window (Figure 5.1). The *Project* window shows all the workbooks that are open (*Book1*) and their objects (*Sheet1, … ,* and *ThisWorkbook*) (Figure 5.1). The *Properties* window shows the properties of the selected object in the Project window. For example, Figure 5.1 shows the properties of *Sheet1* in the *Properties* window.

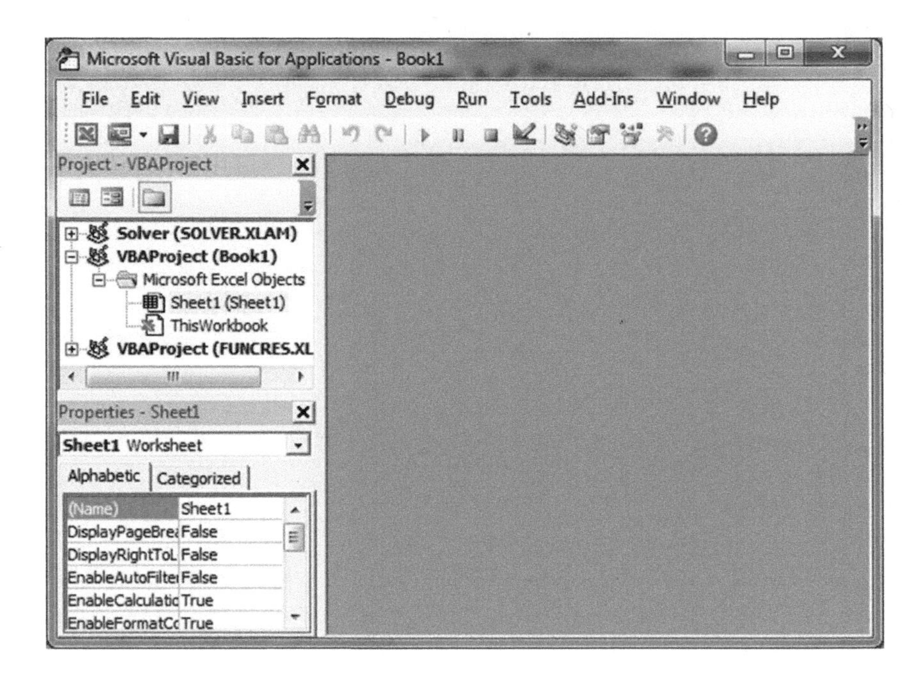

FIGURE 5.1
Visual Basic Editor in Excel.

You can go to the VBE menu bar and click *View* and then *Project Explorer* or *Properties window* to show or hide these two windows. To show the code window, you just have to double-click on the name of an Excel object in the *Project* window (*Sheet1* or *ThisWorkbook* in this case) and its code window will appear.

5.1.2 Writing Code in VB Editor

Let's use a simple example to illustrate coding with VBA. Suppose that we want to get the sum of timber sales automatically. Instead of just using the *SUM* function we learned in Excel, we can automate this calculation or an even more complex computing problem by using VBA functions or procedures. We can begin with a blank Excel spreadsheet.

a. From the Excel menu, click *Developer* → *Visual Basic*, and a VBE will open.
b. In the VBE, double-click on *Sheet1* in the *Project* window. A code window of *Sheet1* appears.
c. Type the following code in the code window:

```
Sub TimberSaleSum()
        Range("C3").Value = 5000
```

```
        Range("C4").Value = 12000
        Range(“C5”).Value = 7500
        Range("C8").Formula = "=C3+C4+C5"
        Range("A1").Select
End Sub
```

d. Go to the menu bar at the top of the VBE, click the *Run* tab, and then select *RunSub/UserForm*. A Macros window will open, in which you should select *Sheet1.TimberSaleSum*, and then click *Run* to run the program. Now, go back to your Excel worksheet to view the results (Figure 5.2).

e. Values have been filled into cells *C3*, *C4*, *C5*, and *C8*. Now *Sheet1* is a worksheet with automating functionality. You can change the values in cells *C3*, *C4*, and *C5*, and the sum of them will be computed automatically.

5.1.3 Running Events within Excel

There are several ways to run events within Excel, including using menu, keyboard, and command buttons/text boxes (Excel Macros & Programming 2010). Approximately 90% of macros are run by clicking on a button. The button can be on the worksheet or on a user form that is developed. Events can also be opening a workbook, selecting a sheet, changing the value of a cell, etc. To create a button using a text box as an

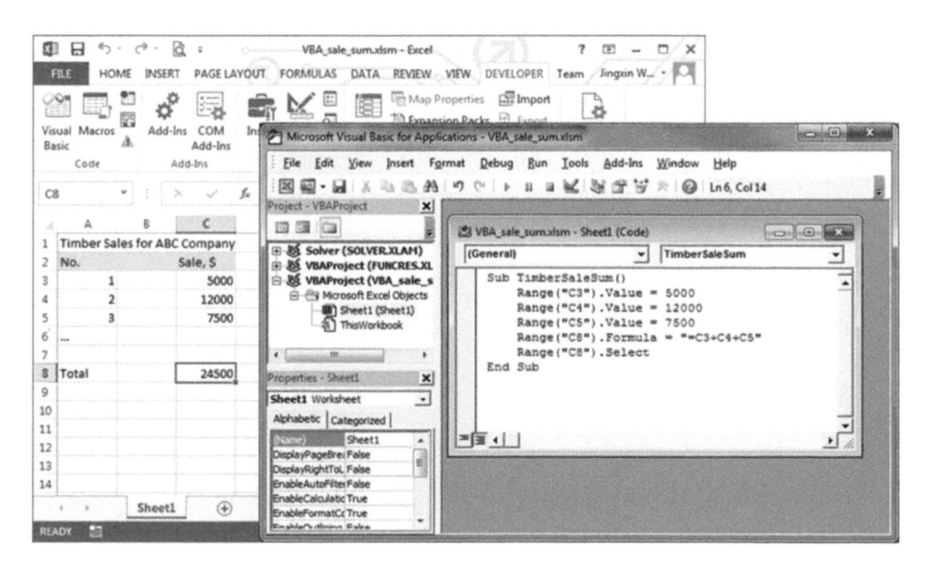

FIGURE 5.2
Running results of the VBA example.

event trigger for our example (Figures 5.1 and 5.2), we return to Sheet1 and then follow the steps below:

a. From the Excel menu, click *Insert*, and then in the *Text* section click *Text Box*.
b. Move the cursor to a desired cell location on *Sheet1*, then left click and hold to drag the cursor, and stretch the text box to the desired size. Type "*Refresh*" into the text box to name the button and fill in a color of your preference (Figure 5.3).
c. Right click the text box, select *Assign Macro* from the drop-down menu, then select the macro *Sheet1.TimberSaleSum* from the Assign Macro list box, and click *OK*.
d. Now, you can click this *Refresh* button (text box) to reset the cell values to default.

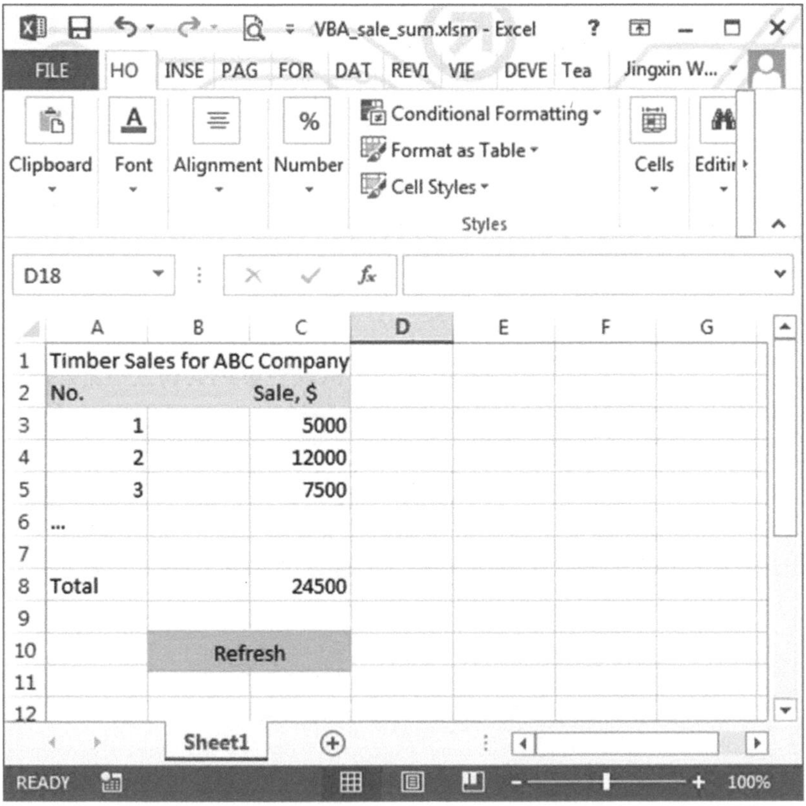

FIGURE 5.3
Create a Refresh button using a text box to trigger a VBA event.

5.2 VBA Fundamentals

5.2.1 Variables and Data Types

Like Visual Basic or any other programming language, VBA uses variables to store temporary information that is used for execution within a procedure, module, or workbook. We typically follow the general conventions to name variables, such as start with a letter, don't use Excel keywords, and stay below 250 characters. When you declare a variable, you tell the computer to reserve space in memory for later use. As we will explain in Section 14.2, a declared variable can be any one of the following major data types (Table 14.2):

- Boolean: A data type with only two possible values, True or False.
- Integer: A data type holding integer values stored as 4-byte whole numbers.
- Single: A data type storing single-precision floating-point variables as 4-byte floating-point numbers.
- Double: A data type that stores double-precision floating-point variables as 8-byte floating-point numbers.
- String: A data type consisting of a sequence of contiguous characters that represent the characters themselves rather than their numeric values. A string can be any combination of letters, numbers, spaces, punctuation, or other symbols/characters.
- Object: A universal data type that holds addresses that refer to any object. Object variables are stored as 4-byte.

To declare a variable, use a Dim (Dimension) statement followed by the variable name and then "*As*" followed by the variable type. For example, here is a declaration for a string variable:

```
Dim strSpecies As String
```

5.2.2 Modules

Modules refer to a related set or block of declarations and procedures. A module is a separate object in VBA associated with a workbook. A standard module can be added to a project via the *Insert* menu of the VBE. Standard modules are contained within a separate folder in the *Project Explorer* and contain variable declarations and procedures or functions. For more details, please refer to Section 14.4.

5.2.3 Variable Scope

The variable scope is the life span when a variable is available to a program. When a variable is in its scope, it can be accessed and/or manipulated. Like in VB.NET, variables in VBA can also be declared at either the procedural level, the module level, or the class level. As will be addressed in Section 14.1, private and public keywords can be used to declare variables or procedures for their scopes (Table 14.1). *Public* can be used to make module- or class-level variables available outside of the module and class.

- Procedural-level variables are only private to procedures in which they were declared.
- Module-level variables are private to the modules in which they appear.
- Class-level variables are private to the class in which they appear.

5.3 Harvesting Machine Rate Spreadsheet Program

5.3.1 Machine Rate

A **machine rate** is a calculated hourly charge for owning and operating a piece of capital equipment (USDA Forest Service 2006). The machine rate method is commonly used in calculating hourly costs of machines in construction, agriculture, and forestry. This classical approach in forestry was first defined by Matthews (1942) and was then revised by Miyata (1980). Costs are averaged over the ownership life of the asset to estimate a constant hourly charge. The machine rate calculations are simple, easy to understand, do not require detailed cost history, and are constant over the life of the machine. They are particularly useful for generic comparisons of equipment and operations. The major components of machine costs are (a) ownership cost (fixed), (b) operating cost (variable), and (c) labor cost.

a. Ownership Cost = $D + IIT$
   i. Depreciation (D)
      Depreciation (D) is defined as a "decline in value of a machine due to wear, obsolescence, and weathering" (Warren 1977) or "loss in value associated with the production of a unit of output" (Stuart 1977). Three methods used to compute depreciation are (1) straight line, (2) declining balance, and (3) sum of year's digits. We'll use the straight line method

(i.e., value of equipment depreciates at a constant rate) to calculate depreciation using the formula:

$$D(\$/SMH) = \frac{P-S}{N \times SMH}$$

where
D—Depreciation
P—Initial investment cost of equipment
S—Salvage value (amount equipment can be sold for at disposal time)
N—Economic life in years
SMH—Scheduled machine hours per year

ii. Interest, Insurance, and Taxes (*IIT*)

$$IIT(\$/SMH) = \frac{\left[\frac{(P-S)\times(N+1)}{2N} + S\right]\times \%IIT}{SMH}$$

b. Variable cost ($\$/PMH$) = *M&R* + *F&L*
   i. Maintenance and Repair (M&R)

$$M\,\&\,R(\$/PMH)\frac{(\%M\,\&\,R)\times D}{UT}$$

where
PMH—Productive Machine Hours
%M&R—maintenance and repair, assumed value or from records
UT—utilization rate, $UT = \frac{PMH}{SMH} \times 100\%$

   ii. Fuel and Lubricant (*F&L*)

$$F\,\&\,L(\$/PMH) = \text{Consumption rate} \times \text{price}$$

c. Labor cost ($\$/SMH$) = Wage × (1.0 + fringe benefit rate)
d. Total hourly cost ($\$/PMH$) = Fixed cost + variable cost + labor cost

5.3.2 Example of Machine Rate Program with VBA

Let's develop an Excel spreadsheet program to automate hourly cost calculations of harvesting machines using the machine rate method and VBA (Figure 5.4). There are a few major steps we need to follow.

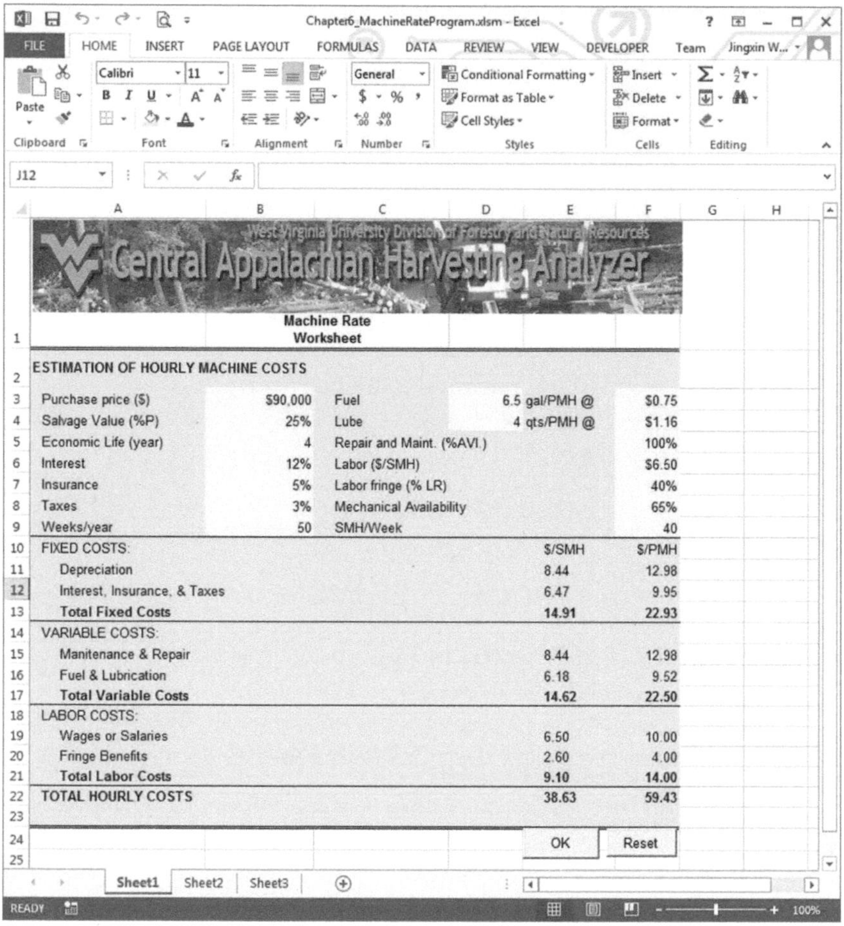

Central Appalachian Harvesting Analyzer — Machine Rate Worksheet					
ESTIMATION OF HOURLY MACHINE COSTS					
Purchase price ($)	$90,000	Fuel		6.5 gal/PMH @	$0.75
Salvage Value (%P)	25%	Lube		4 qts/PMH @	$1.16
Economic Life (year)	4	Repair and Maint. (%AVI.)			100%
Interest	12%	Labor ($/SMH)			$6.50
Insurance	5%	Labor fringe (% LR)			40%
Taxes	3%	Mechanical Availability			65%
Weeks/year	50	SMH/Week			40
FIXED COSTS:				$/SMH	$/PMH
Depreciation				8.44	12.98
Interest, Insurance, & Taxes				6.47	9.95
Total Fixed Costs				**14.91**	**22.93**
VARIABLE COSTS:					
Manitenance & Repair				8.44	12.98
Fuel & Lubrication				6.18	9.52
Total Variable Costs				**14.62**	**22.50**
LABOR COSTS:					
Wages or Salaries				6.50	10.00
Fringe Benefits				2.60	4.00
Total Labor Costs				**9.10**	**14.00**
TOTAL HOURLY COSTS				**38.63**	**59.43**
				OK	Reset

FIGURE 5.4
A spreadsheet program for harvesting cost calculation.

5.3.2.1 Design Interface

Open a new Excel workbook and a worksheet (e.g., *Sheet1*) where we'll create an interface as in Figure 5.4. To add the two command buttons (*OK* and *Reset*), you need to click *Developer* in the Excel menu ribbon, then in the *Controls* group click *Insert*, and from *ActiveX Controls* click the *command button icon*. Move the cursor to a desired cell location on *Sheet1*, then click and hold the left button of the mouse, and stretch the command button to the desired size.

Right click the command button, and select *Properties* from the drop-down menu. You can change the name of the button to "*cmdok*" and the caption to "*OK*." Similarly, you can create another command button and then rename it "*cmdReset*" and caption it "*Reset*."

5.3.2.2 Write Code

Once you've finished the interface design, you can add code behind the command buttons. Right click the *OK* command button and select *View Code* from the drop-down menu, or just double-click the button. The code window appears with the following empty procedure:

```
Private Sub cmdok_Click()
  ......

End Sub
```

Similarly, when coding the Reset button, you will find this empty procedure:

```
Private Sub cmdReset_Click()
  ......

End Sub
```

Now we can use the calculation procedures for the machine rate discussed in the previous section to complete the coding for these two procedures:

```
Private Sub cmdok_Click()
     Range("E11").Value = (Range("B3").Value - Range("B3").
       Value * Range("B4").Value) / (Range("B5").Value *
       Range("B9").Value * Range("F9").Value)

     Range("E12").Value = (((Range("B3").Value - Range("B3").
       Value * Range("B4").Value) * (Range("B5").Value + 1) /
       (2 * Range("B5").Value) + Range("B3").Value *
       Range("B4").Value) * (Range("B6").Value + Range("B7").
       Value + Range("B8").Value) / (Range("B9").Value *
       Range("F9").Value))

     Range("E13").Value = Range("E11").Value + Range("E12").
       Value
     Range("F11").Value = Range("E11").Value / Range("F8").
       Value
     Range("F12").Value = Range("E12").Value / Range("F8").
       Value
     Range("F13").Value = Range("F11").Value + Range("F12").
       Value
     Range("F15").Value = Range("E11").Value * Range("F5").
       Value / Range("F8").Value
     Range("F16").Value = Range("D3").Value * Range("F3").
       Value + Range("D4").Value * Range("F4").Value
     Range("F17").Value = Range("F15").Value + Range("F16").
       Value
```

```
    Range("E15").Value = Range("F15").Value * Range("F8").
      Value
    Range("E16").Value = Range("F16").Value * Range("F8").
      Value
    Range("E17").Value = Range("E15").Value + Range("E16").
      Value

    Range("E19").Value = Range("F6").Value
    Range("E20").Value = Range("F6").Value * Range("F7").
      Value
    Range("E21").Value = Range("E19").Value + Range("E20").
      Value
    Range("E22").Value = Range("E13").Value + Range("E17").
      Value + Range("E21").Value

    Range("F19").Value = Range("E19").Value / Range("F8").
      Value
    Range("F20").Value = Range("E20").Value / Range("F8").
      Value
    Range("F21").Value = Range("F19").Value + Range("F20").
      Value
    Range("F22").Value = Range("F13").Value + Range("F17").
      Value + Range("F21").Value

End Sub

Private Sub cmdReset_Click()
          Range("B3").Value = 90000
          Range("B4").Value = 0.25
          Range("B5").Value = 4
          Range("B6").Value = 0.12
          Range("B7").Value = 0.05
          Range("B8").Value = 0.03
          Range("B9").Value = 50

          Range("D3").Value = 6.5
          Range("D4").Value = 4
          Range("F3").Value = 0.75
          Range("F4").Value = 1.16
          Range("F5").Value = 1
          Range("F6").Value = 6.5
          Range("F7").Value = 0.4
          Range("F8").Value = 0.65
          Range("F9").Value = 40

          Range("E11", "F13").Value = 0
          Range("E15", "F17").Value = 0
          Range("E19", "F22").Value = 0
End Sub
```

After coding for the two buttons, we turn off the *Design Mode* under *Developer* and save the program. Then we go back to the Excel worksheet, input or change any parameters, and then click *OK* or *Reset* to run the program.

5.4 VBA User Forms and Controls

Like forms in VB.NET, Excel UserForm is a programmable container for ActiveX controls. It enables us to build customized windows to serve as a user interface in any VBA applications (Birnbaum and Vine 2007). The ActiveX controls on the form have properties, methods, and events to control the appearance and behavior of the interface window. In this section, we will use another example to learn how to design a *UserForm* using ActiveX controls in VBA to calculate timber sale statistics.

In this example, we start with VBE, then click *Insert* → *UserForm*. On this form, we use controls from the VBE ToolBox, including *RefEdit*, *List Box*, *Label*, and *Text Box* (Figure 5.5). When we run it (by clicking the *Calculate*

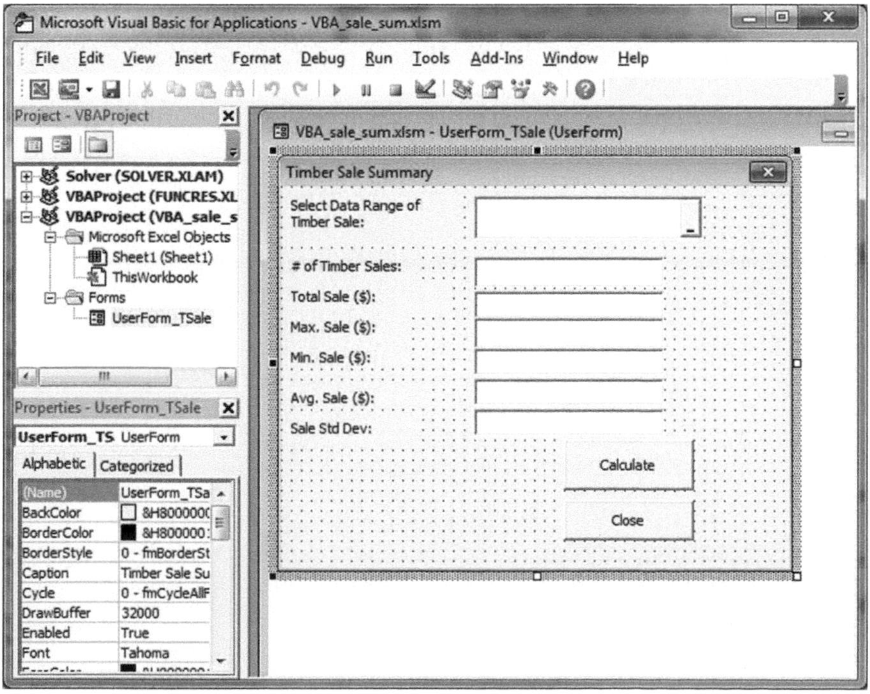

FIGURE 5.5
Design interface in VBA for timber sale summary.

TABLE 5.1

Property Settings of VBA Controls

Control	Property	Setting
TextBox	Name	txtNumSale, txtTotSale, txtMaxSale, txtMinSale, txtAvgSale, txtStdSale
Label	BorderStyle	fmBorderStyleSingle
Command Button	Name	cmdCalSaleStats, cmdClose
Command Button	Caption	"Calculate", "Close"
Command Button	Enabled	True for all the Command Buttons
RefEdit	Name	refStats

button), the program will compute and display the basic statistics of timber sales described in Figure 5.3, including the number, total, maximum, minimum, average, and standard deviation of these timber sales.

Table 5.1 summarizes the properties of the ActiveX controls used in this VBA project. You should be able to set up different fonts, colors, and borders to suit your personal preference.

The code for this project is contained entirely within its form module. All program code is entered into several event procedures of the ActiveX controls on the form. These procedures are as follows:

```
Private Sub cmdCalSaleStats_Click()

    Const NUMFORMAT = "#.00"

    On Error GoTo ErrorHandler
    txtNumSale.Text = Application.WorksheetFunction.Count _
                        (Range(refStats.Text))
    txtTotSale.Text = Application.WorksheetFunction.Sum _
                        (Range(refStats.Text))
    txtMaxSale.Text = Application.WorksheetFunction.Max _
                        (Range(refStats.Text))
    txtMinSale.Text = Application.WorksheetFunction.Min _
                        (Range(refStats.Text))
    txtAvgSale.Text = Format(Application.
                       WorksheetFunction.Average _
                        (Range(refStats.Text)), NUMFORMAT)
    txtStdSale.Text = Format(Application.
                       WorksheetFunction.StDevP _
                        (Range(refStats.Text)), NUMFORMAT)

    Exit Sub

    ErrorHandler:
            MsgBox "An error was encountered while
              calculating the statistics. " _
```

```
                          & vbCrLf & Err.Description & vbCrLf & _
                         "Check for a valid range selection and
                           try again." & vbCrLf, _
                         vbCritical, "Error " & Err.Number

End Sub

Private Sub cmdClose_Click()

    Unload UserForm_TSale
    End

End Sub
```

Several event procedures are associated with any VBA controls. For example, "Click()" is used for the command button *cmdCalSaleStats*, and the command button *cmdClose* in this example. Some other event procedures that could be used for *UserForm*, *RefEdit*, and *ComboBox* worksheets include the "Initialize()" event for the UserForm object, the "Change()" event of the *ComboBox* control and the *List Box* control, and the "DropButtonClick()" and "Enter()" event procedures in this program to clear text from the *RefEdit* control.

To run the project from the VBE menu, select the data range of timber sales from Excel *Sheet1* and then click the *Calculate* button. The program in running mode will look like Figure 5.6. Click the *Close* button to end the program.

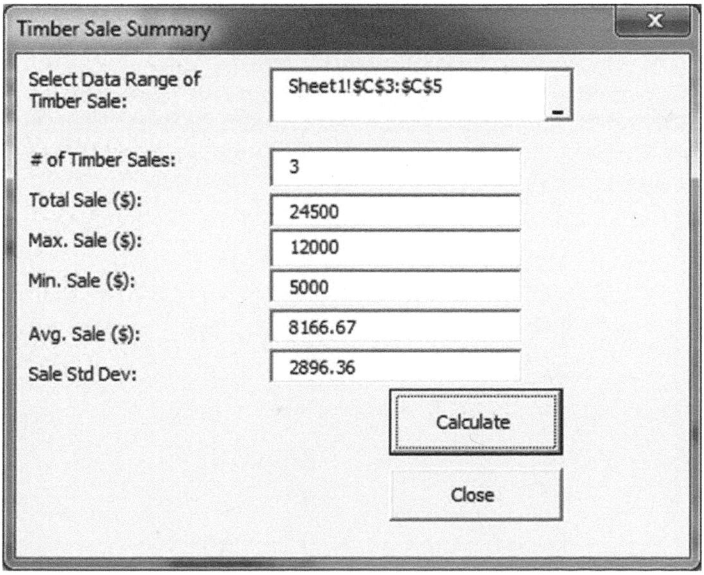

FIGURE 5.6
Interface at running mode of the VBA project for timber sale.

Class Exercises

1. What is VBA? Why is it useful for field data manipulations in forest and natural resource management?
2. Expand the timber sales in Figure 5.3 from 3 to 30, and implement a VBA project for these sales following the procedures and forms used in Section 5.4.

References

Birnbaum, D. and M. Vine. 2007. *Microsoft VBA Programming* (3rd Edition). Thomson Course Technology, Boston, MA, 214pp.

Excel Macros & Programming. 2010. Excel Macros (VBA). http://www.excel-vba.com/. Accessed on April 4, 2010.

Matthews, D.M. 1942. *Cost Control in the Logging Industry*. McGraw-Hill, New York, 374pp.

Microsoft. 2016. Introducing Visual Basic for applications. http://msdn.microsoft.com/en-us/library/office/aa188202(v=office.10).aspx. Accessed on August 29, 2016.

Miyata, S. 1980. Determining fixed and operating costs of logging equipment. USDA Forest Service General Technical Report NC-55, St. Paul, MN.

Stuart, B. 1977. An unpublished paper presented at the *Virginia Polytechnical Institute's Cost Analysis and Evaluation Seminar*, November 15–17, Blacksburg, VA, 1977.

USDA Forest Service. 2006. Machine costs—The cost of forest operations. http://www.srs.fs.usda.gov/forestops/mach_costs.htm. Accessed on August 29, 2016.

Warren, J. 1977. Analyzing logging equipment costs, in logging cost and production analysis. Timber Harvesting Report No. 4. Compiled by LSU/MSU Logging and Forestry Operations Center, Bay St. Louis, MI, 108pp.

Section III

Database Management

6

Database Concepts and the Entity-Relationship Model

6.1 Fundamental Database Concepts

6.1.1 Database Management System

A **Database Management System (DBMS)** is a software program that provides efficient access to persistent data for concurrent uses, such as MS Access, Oracle, Sybase, DB2, and others (Atkins 2001). The first DBMS appeared in the late 1960s having evolved from file systems (Ullman and Widow 1997). Basic functions of a modern DBMS typically include the following: create and remove a database (DB), create and remove tables, populate tables in a DB, modify and delete data in a DB, and query a DB.

Additional functions of a commercial DBMS could include the implementation of referential integrity constraints, semantic integrity constraints, business rules, and indexing. Like MS Excel, a DBMS typically provides several categories of built-in functions, such as statistical, mathematical, and financial functions.

The most commonly used DB model for DBMS is the relational model (O'Neil and O'Neil 2001). A DBMS that utilizes the relational model is called a relational database management system (**RDBMS**). A RDBMS contains not only data, but also **metadata**, which is information about the structure of the data (Ullman and Widow 1997). It also maintains indexes for the data, which are the data structures that enhance quick and efficient searching.

6.1.2 Database

A **database** is a collection of related data concerning a certain topic or business application. For example, if we develop a DB containing timber cruising data for a forest product company, two of the tables in the DB might have the following schemes:

```
Plot(PlotNo, PlotType, Crew, Date, Forest)
Tree(TreeNo, PlotNo, Species, DBH, Height, Product)
```

Here, *Plot* and *Tree* are the table names. *PlotNo, PlotType, Crew, Date,* and *Forest* are called attributes of table *Plot*. *PlotNo* is the primary key in *Plot* table while it is a foreign key in *Tree* table (primary and foreign keys are defined in Section 6.2.2). Typical records in the above tables might be:

```
Plot(1, DBH and Height, 1, 01/20/14, Univ Forest)
Tree(1, 1, Yellow poplar, 12, 40, Sawlog)
```

We must note that there is a relationship between these two tables, which is implemented by using *PlotNo* in the *Tree* table. Therefore, the *PlotNo* value in *Tree* table must match the *PlotNo* value in *Plot* table or be NULL.

Typical queries or questions that we could ask in these tables include the following: (1) list trees that are red oak, (2) list trees with diameter at breast height (DBH) greater than or equal to 20.32 cm (8 in.), and (3) summarize the volume of trees by *Species* or *PlotNo*.

6.1.3 Table, Record, and Field

A **table** stores raw data in logical groupings (the Students table, e.g., contains data about students) and organizes the data into rows and columns.

The table or datasheet is divided into rows called **records** and columns called **fields**. For example, in the Students DB, the data shown in the student table has columns of similar information, such as *StudentID, SName,* and *GPA*; these columns of data items are fields. Each field has a name and is identified as a certain type of data (text or number) and has a specified length. The rows of data within a table are its records. Each row of data is considered a separate **entity** that can be accessed (in our case, an individual student). All the fields of information concerning a certain student are contained within a specific record for a specific student entity.

At the intersection of a row (record) and a column (field) is a **value**, the actual data element. For example, Smith, the Student Name of a record is a data value.

6.1.4 MS Access Database

Microsoft Access follows traditional DB terminology. The terms database, table, record, field, and value indicate a hierarchy from largest to smallest (Prague et al. 1999). An Access DB includes these objects or components: tables, queries, forms, reports, macros, modules, and Visual Basic for Applications.

MS Access can work with only one DB at a time. Within a single Access DB, however, you can have hundreds of tables, forms, queries, reports, macros, and modules. They all are stored in a single file with the file extension .MDB (multiple database) for Access 2003 or earlier, .accdb for Access 2007 or later version, or .ADP if you are using SQL Server.

6.2 Relational Databases and the Entity-Relationship Model

The early DBMS used several data models for describing the structure of the information in a DB, such as the hierarchical or tree-based model and the graph-based network model (Ullman and Widow 1997). The relational DB model was introduced by E.F. Codd with IBM in the early 1970s. The relational DB model is predefined as the mathematical notion of a relation. The fundamental object in the relation model is the table or relation.

The following example demonstrates the simplicity of the relational model, and the subsequent Structured Query Language (SQL) sample suggests how the relational model promotes queries written at a very high level. (We will discuss SQL in Chapter 8.) The table or relation (called *Tree*) records *Species, DBH, Height,* and *Product type* (Table 6.1).

If we would like to know the potential DBH and product types of cruised trees of *Red oak,* we could perform a query using SQL in a DBMS as follows:

```
Select DBH, Product
From Tree
Where Species ='Red oak';
```

Basically, the above query performed three actions. It (1) examined all the records of the table *Tree* mentioned in the *From* clause, (2) filtered out those records by some criteria indicated in the *Where* clause, and (3) presented an answer to certain attributes of those records, as indicated in the *Select* clause.

As a programmer or DB administrator, we always use a tool to design a DB. The entity-relationship (ER) model is one such tool that has been popularly used for DB design. The ER model, one of the schematic data models, was proposed by P. P. Chen in the mid-1970s (Chen 1976). A few major components in the ER model are discussed in the following sections.

TABLE 6.1

A Relational Table of Trees

TreeNo	PlotNo	Species	DBH	Height	Product
1	1	Red oak	12	30	Pulp
2	1	Yellow poplar	26	45	Sawlog
...		...	...		
1	2	Sugar maple	16	34	Sawlog
2	2	Red oak	18	38	Veneer
...		...	...		

6.2.1 Entity

An **entity** is an object and has properties (or attributes) that are descriptive of the entity. For example, a student entity has *StudentID, name, major, address, GPA*, etc. A collection of entities, all of which share the same set of attributes, is called an **entity type**. For example, a student table in a university DB is an entity type.

An **attribute** of an entity is a descriptive property of the entity. Each attribute has a value (or null) that should come from some domain. An attribute is composite if it can be decomposed into two or more components; otherwise, it is simple. For example, *address* for student entity may be composite with components of street, city, state, and zip (Atkins 2001). An attribute is derivable if its value can be computed from other attributes. For example, age may be a derived attribute if there is another attribute called birthdate for a student.

6.2.2 Keys

A key for an entity type consists of one or more attributes that can distinguish the entities. There are two types of keys: primary and foreign keys. A **primary key** of a relational table uniquely identifies each record in the table, and it may consist of one or more fields or attributes. A **foreign key** is a field in a relational table that matches the primary key column of another table. The foreign key can be used to cross-reference tables.

For example, considering the following two entity types:

```
Trees(TreeNo, PlotNo, DBH, Height, Species)
Plot(PlotNo, PlotType, Crew, Date)
```

PlotNo is a primary key in the *Plot* table. One primary key in the *Trees* table is *TreeNo* and the other could be *TreeNo* and *PlotNo* together. *PlotNo* in the *Trees* table is a foreign key that is used to cross-reference between table *Plot* and table *Trees* for relationships.

6.2.3 Relationships

A **relationship** is an association between two entity types or tables. For example, let's still use the above two entity types of *Plot* and *Trees*:

```
Plot(PlotNo, PlotType, Crew, Date)
Trees(TreeNo, PlotNo, Species, DBH, Height)
```

The underlined *PlotNo* and *TreeNo* refer to the primary key in each table. We can define a relationship, *Contains*, between *Plot* and *Trees*. It simply means entity type *Plot* contains entity type *Trees*. Like in the real world, a field plot typically contains several trees.

6.2.4 Types of Relationships

There are three major types of relationships between tables: one-to-one, one-to-many, and many-to-many. Let E$\{e_1, e_2, \ldots, e_n\}$ and F$\{f_1, f_2, \ldots, f_n\}$ represent entity types, and R be the relationship between them (Atkins 2001). Graphical representations of these two entities can be illustrated as follows:

One-to-One (*1–1*): A relationship (R) is one-to-one (1–1) if whenever the pair (*e*, *f*) is in R, then neither *e* nor *f* participates in any other pair (Figure 6.1).

For a real-world example, let's consider two entity types: Managers and Sawmills for the forest products industry in a state. We want to associate a mill manager with the sawmill she or he manages. We know that a manager only manages one mill, and a single mill is only being managed by one manager. In this case, the relationship between entity types of Managers and Sawmills is 1–1.

One-to-Many (*Many-to-One*) (*n–1*): A relationship (R) is many-to-one (n–1) (or one-to-many) from E to F, if whenever the pair (*e*, *f*) appears in R, then *e* may not participate in any other pairs of *f*, but *f* may participate in other pairs of *e* (Figure 6.2).

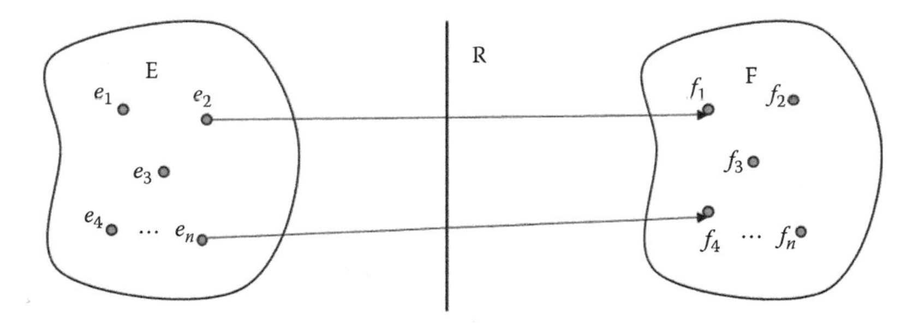

FIGURE 6.1
One-to-one relationship between entity types E and F.

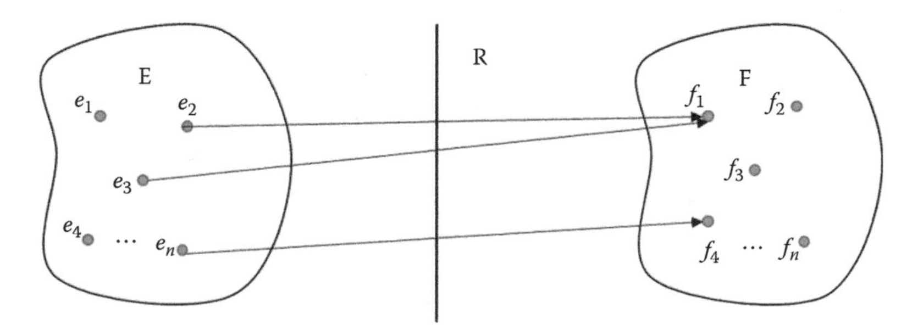

FIGURE 6.2
Many-to-one relationship.

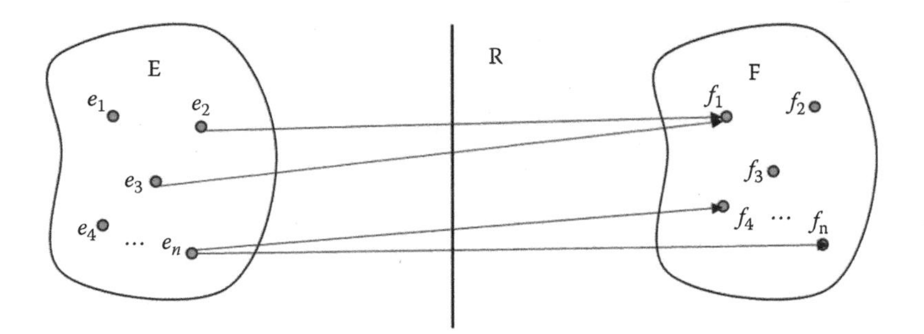

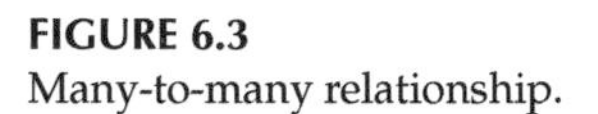
FIGURE 6.3
Many-to-many relationship.

For example, consider the *Plot* and *Trees* tables. Trees to Plot is $n - 1$ since several trees can be in one particular plot but a particular tree can be located in only one plot.

Many-to-Many (n–n): A relationship (R) is many-to-many if whenever (e, f) appears in R, then e may participate in other pairs of f and vice versa (Figure 6.3).

A many-to-many relationship refers to a relationship between tables in a DB when a parent row in one table contains several child rows in the second table, and vice versa. For example, in a student DB, a *Student* table to *Course* table is n–n, since one student can take many courses and one course can be taken by many students.

Relationship implementations: Once we understand the relationships between tables, we always ask ourselves how we can implement these three major relationships of tables in a DBMS. Some procedures we typically follow are: (1) a nonweak entity type with single value and simple attributes (such as student, tree, or plot) is simply translated into a table; (2) a weak entity type with incomplete attributes is typically translated into a table with the borrowed key plus partial key; (3) a one-to-one relationship is translated by making the foreign key unique; (4) a one-to-many relationship from E to F is represented with a foreign key (representing E) in the table F; and (5) a many-to-many relationship is represented by a table whose key is the set of key attributes from participating tables, plus any attributes associated with the relationship.

Here is an example of the implementation of a one-to-many relationship between *Plot* and *Trees*. Key *PlotNo* in *Trees* table is the borrowed key to implement this 1–n relationship between Plot and *Trees*.

```
Plot(PlotNo, PlotType, Crew, Date)
Trees(TreeNo, PlotNo, DBH, Height, Species)
```

To show an example of a many-to-many implementation, a student DB contains a table for students (*tblStudent*) and a table for courses (*tblCourses*). There is a many-to-many relationship between these two tables. To build this many-to-many relationship, we need to create a third table (*tblStudentCourse*) to represent it.

```
tblStudent(StudentID, Sname, GPA)
tblCourse(CourseID, CourseTitle, CreditHour, Description)
tblStudentCourse(StudentID, CourseID, DateTaken, Status)
```

6.3 ER Model Examples in Forest Operations

6.3.1 ER Model Notations

The ER model diagram is a visual representation of data or entity types that describes how these data or entity types are related to each other. There are a few ways to represent the ER model, including Chen's model, Crow's Foot notation, IDEF1X, and others. The notations used in this chapter are based on P. Chen's model (Chen 1976). Entity, attribute, and relationship are represented by specific notations (Figure 6.4).

6.3.2 ER Model for Timber Cruising

The ER model for timber cruising developed by Wang et al. (2004) has four data entity types—*Plot, Tree, Transect,* and *Species* in the model (Figure 6.5). Each entity type has its own attributes. For example, the *Plot* entity type has

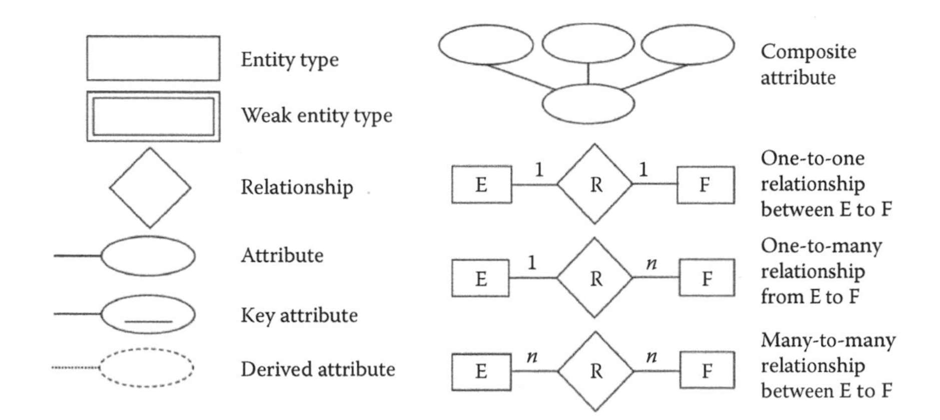

FIGURE 6.4
Some notations for an ER model diagram.

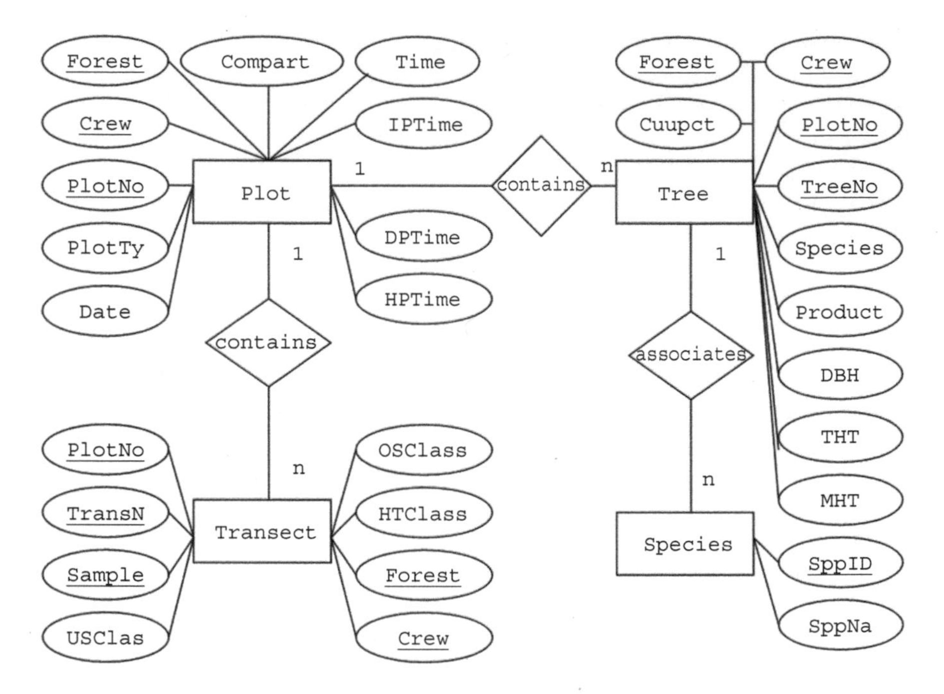

FIGURE 6.5
ER model of an integrated timber cruising system.

forest, crew, plotNo, plotTy, and other attributes. Entities are related to each other using relationships such as *contains* and *associates* in the model. The relationship between *Plot* and *Tree* or *Transect* is one-to-many and simply means that the plot *contains* many trees or transects. The many-to-one relationship is applied between *Tree* and *Species* entity types, which implies each species *associates* with many trees in the Tree table. Attributes belonging to a key are underlined for an entity type. For example, *forest, crew,* and *plotNo* together is a primary key of entity type *Plot,* while *forest, crew, plotNo,* and *treeNo* combined is a primary key of entity type *Tree.* The foreign key of *forest, crew,* and *plotNo* for entity *Tree* and *Transect* represents the one-to-many relationships of *Plot* to *Tree* and *Plot* to *Transect*, respectively. The foreign key of *SppID* for entity type *Tree* represents many-to-one relationship between *Tree* and *Species.*

6.3.3 ER Model for Time Study of Timber Harvesting

In forest operations, time and motion studies of harvesting machines are typical field practices. A computer-based time study system was developed for timber harvesting operations (Wang et al. 2003). For field data collection and data storage, a relational DB model was developed and used for holding harvesting functions, variables, and time study data in the handheld system (Wang et al. 2003), which was implemented based on

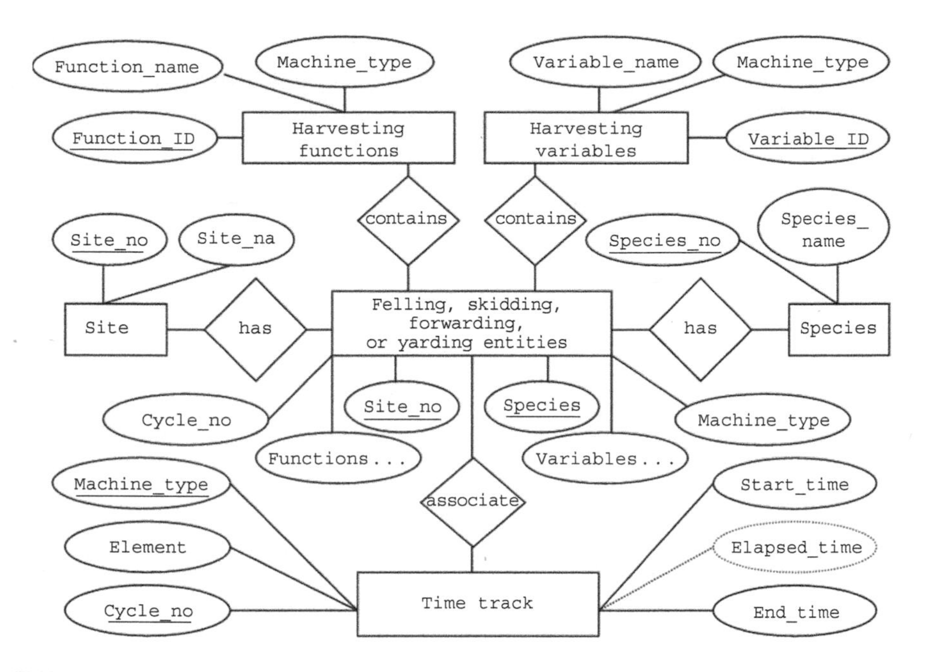

FIGURE 6.6
ER model of a time study system for timber harvesting operations.

the ER model (Figure 6.6). Basically, there are six data entity types in the model: *Variables, Site, Species, Time Track, Felling/Skidding/Forwarding/Yarding,* and *Harvesting Functions*. Each entity type has its own attributes. For example, the *Harvesting Functions* entity type has *function_ID, function_name,* and *machine_type* attributes. Entities are related using relationships such as *has, contains,* and *associates* in the model. *Elapsed_time,* an attribute derived from *start_time* and *end_time,* was used in the *Time Track* entity. Similarly, attributes belonging to a key are underlined for an entity type. *Cycle number, machine type, function start time, end time, elapsed time,* and associated harvesting variables are automatically recorded.

Harvesting functions, variable data, and *species* are stored in separate data tables in the design module, which are identified by their primary keys and harvesting machine types. In the *Collect* module, harvesting functions and variables can be queried and retrieved for a specific machine type on which another data table is created for storing functions, variables, and elemental times. Species information is also retrieved for each data entry. The site data table contains general information such as site number, name, location, slope, and weather conditions about the logging site. Site number is used as a foreign key to associate site information with other data tables created in the collect module.

The *Time Track* entity type is used to track the start and end times of each element in a work cycle. It can also be used as a backup data table for felling, skidding/forwarding, or yarding data entities. The *Felling/Skidding/Forwarding/Yarding* entity type is designed to store time study data of felling, skidding, forwarding, or yarding depending on the type of logging operations being studied. Data schemas of the main data storage on the desktop PC are the same as those used in the handheld system in order to facilitate the data transfer.

Class Exercises

Upon completion of this chapter, you need to understand the following key concepts:

1. What are DBMSs, databases, tables, and queries?
2. What are the primary and foreign keys?
3. What is an entity-relationship?
4. How many types of relationships between tables are there? How can you implement them in a forest and natural resource management project?

References

Atkins, J. 2001. *Database Management* (Lecture Notes). Department of Computer Science, West Virginia University, Morgantown, WV.

Chen, P.P. March 1976. The entity-relationship model: Toward to a unified view of data. *ACM Transactions on Database Systems* 1(1): 9–36.

O'Neil, P. and E. O'Neil. 2001. *Database Principles and Programming Performance.* Morgan Kaufmann Publishers, New York, 870pp.

Prague, C., D. Roche, and M. Irwin. 1999. *Microsoft Access 2000 Bible*. John Wiley & Sons, Incorporated, 1272pp.

Ullman, J. and J. Widow. 1997. *A First Course in Database Systems*. Prentice Hall, Upper Saddle River, NJ, 470pp.

Wang, J., S. Grushecky, and J. Brooks. 2004. An integrated computer-based cruising system for central Appalachian hardwoods. *Computers and Electronics in Agriculture* 45(2004): 133–138.

Wang, J., J. McNeel, and J. Baumgras. 2003. A computer-based time study system for timber harvesting operations. *Forest Products Journal* 53(3): 47–53.

7

Introduction to MS Access

MS Access is a relational database management system (RDBMS). Like other DBMSs, Access can store and retrieve data, present information, and automate repetitive tasks. Additionally, we can use Access to create user-friendly Windows forms and generate powerful reports for potential customers. MS Access can work seamlessly with other Microsoft products such as MS Office (Word, Excel, PowerPoint) and MS Visual Studio (Visual Basic and Visual C++, and .NET) by using ActiveX (formerly called Object Linking and Embedding) objects.

MS Access has evolved over the years from earlier versions (Access 2, Access 95, Access 97, Access 2000, Access 2003, and Access 2007) to Access 2010, Access 2013, Access 2016, and later versions. Earlier versions of Access databases must be converted to the format of later versions before they are usable.

7.1 MS Access Usability and Functionality

7.1.1 MS Access Usability

A triangular hierarchy indicates MS Access's usability from the bottom of basic tables and queries applications to functions and expressions, to macros, to Visual Basic for Applications (VBA), and to the top of Windows Application Programming Interface (API) (Prague et al. 1999). There are five major components or usability levels in MS Access:

1. *Objects*: They include tables, queries, forms, and reports. We can create, modify, and delete these objects.
2. *Functions/Expressions*: We can use functions/expressions to simplify data processing and manipulations, such as data validation and implementations of business rules.
3. *Macros*: A macro is a stored series of commands that can be used to automate repetitive work without programming.
4. *VBA*: VBA allows us to program complex processes.

5. *Windows API*: Through Windows API, we can call functions, procedures, and Dynamic Link Libraries written in other languages such as C, Java, or Visual Basic. Similarly, we can write the interfaces from other applications to connect to the Access database. In fact, many of the tools in Access (such as Wizards and built-in functions) are written in VBA.

7.1.2 Major MS Access Functionality

Relational database management: Access is a RDBMS. It can create tables, define keys, create queries, and generate reports. Access has full referential integrity at the database engine level, which can be defined to prevent inconsistent data updates or deletions.

Wizards: Access features quite a few Wizards to design tables, queries, forms, and reports. The Wizards can walk us step-by-step through the process of database object design.

Importing and exporting data: Access allows us to import from or export to many common formats of data, including dBase, Excel, SQL Server, Oracle, ASCII text, and HTML. Importing typically creates an Access table while exporting an Access table creates a file in the format of the destination program.

Multiple-table queries and relationships: Tables are typically related in Access. Therefore, we can define table relationships and use multiple tables to create queries, forms, and reports.

Dynamic data exchange: By using the Dynamic Data Exchange, new objects can be added to Access forms and reports, such as images, text, or documents.

Built-in functions: Like MS Excel, MS Access also provides more than 200 built-in functions in 17 categories, such as mathematical, statistical, financial, and database management.

7.1.3 Why Use More than One Table?

As we have learned, a relational database typically utilizes several related tables to present the information efficiently. Three main advantages to an application using multiple tables are that it can (a) manipulate data more efficiently, (b) reduce data redundancy, and (c) simplify data entries.

For example, if we develop a student database containing student and course data, it is better to have two related tables, student and course, so that we would not need to store course information in the student table. Then if a college professor tries to retrieve some specified student records with the name, course, and date (since 9/1/2001) when the course was taken, she/he would need to use a query because she/he cannot obtain all the information she/he needs from one table. Instead of asking the question in actual English, she/he would use a method known as Query by Example (QBE) in MS Access.

When you enter instructions into the QBE window, the query translates the instructions and retrieves the desired data. In this example, the query first combines data from the *Student*, the *Course*, and the *StudentCourse* tables. Then it retrieves the required fields. Access then filters the records, selecting only those in which the value of *DateTaken* is later than 9/1/2001. It sorts the resulting records first by student ID and then by student name within the student IDs that are alike. Finally, the records appear on-screen in a datasheet.

The selected records from these three tables form a new data set and are called a **dynaset**—a dynamic set of data that can be changed according to the raw data in the original tables. After you run a query, the resulting data set can be used in a form that can be displayed on-screen in a specified format, or printed in a report.

7.2 Access Tables and Queries

7.2.1 Access Tables

In this section, we will provide instructions for performing several tasks in MS Access. Please note that for these instructions we have used the 2013 or 2016 version of MS Access, and that the steps may vary slightly if you are using a different Access version.

Open *MS Access* and then click *Blank desktop database*. This will open a dialog box where you can enter a database name (let's use "dbStudent") and select a folder to store it, then click *Create*. An MS Access window will be displayed (Figure 7.1). Click the *Save* icon at the top of the menu bar.

In Access, we can create a table, create a query, design a form, and write a report. Suppose we would like to create a database (*dbStudent*) of students and the courses they complete. *Student* and *Course* would be the two entity types (or two tables), and the relationship between them would be many-to-many. In the ER Model section, we mentioned that a new, third table needs to be created to represent this many-to-many relationship, and this table's primary key should be the keys from both tables. Therefore, we need to create the third table called *tblStudentCourse* for the relationship. The three tables in *dbStudent* will then be:

```
tblStudent(StudentID, Sname, GPA)
tblCourse(CourseID, CourseTitle, CreditHour, Description)
tblStudentCourse(StudentID, CourseID, DateTaken, Status)
```

Figure 7.1 shows the design windows for these three tables. To create one of these tables (e.g., *tblStudent*) in the *dbStudent* database, click the *Create* tab, and

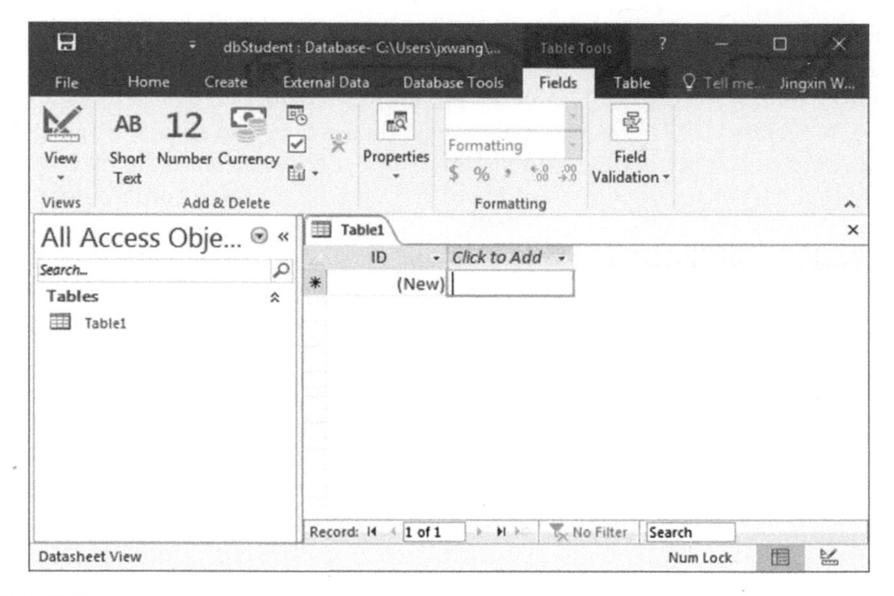

FIGURE 7.1
MS Access table design view.

then in the *Tables* group, click *Table Design*. A new table is inserted in the database, and the table opens in the *Design View* so that you may proceed easily to adding fields and setting design keys, etc. Type in the *Field Names* as shown in Figure 7.2 for the respective tables, and set the *Data Types* from the drop-down options. To set the primary key(s), highlight the row(s) of the intended primary key(s) and then click the *Primary Key* button (it is under the *Design* tab in the *Tools* section). One of the several ways to save your new tables is to simply right click on the new table's tab, then click *Save*. A *Save As* window will open where you can type in a name for your table, then click *OK*. Repeat this process to create the remaining tables (*tblCourse* and *tblStudentCourse*).

When you have created the three tables, enter the data shown in Figure 7.3. To enter data, click on the tab for a particular table (or double-click the table name in the list of tables in the left-hand box), then click the *Home* tab on the main menu, and in the *View* section, select *Datasheet View* from the drop-down options. You can then enter data into the appropriate cells of the datasheet. Be sure to click *Save* after completing the data entry for each datasheet.

7.2.2 Relationships between Tables

Recall that there are three types of relationships between tables: one-to-one, one-to-many, and many-to-many. To demonstrate how relationships are built in Access, we can use the tables we just created in the *dbStudent* database. Click the *Database Tools* menu tab and then click *Relationships*. A *Show Table* dialog box (Figure 7.4) will appear in which you can add tables, queries, or both, and then build the relationships for them.

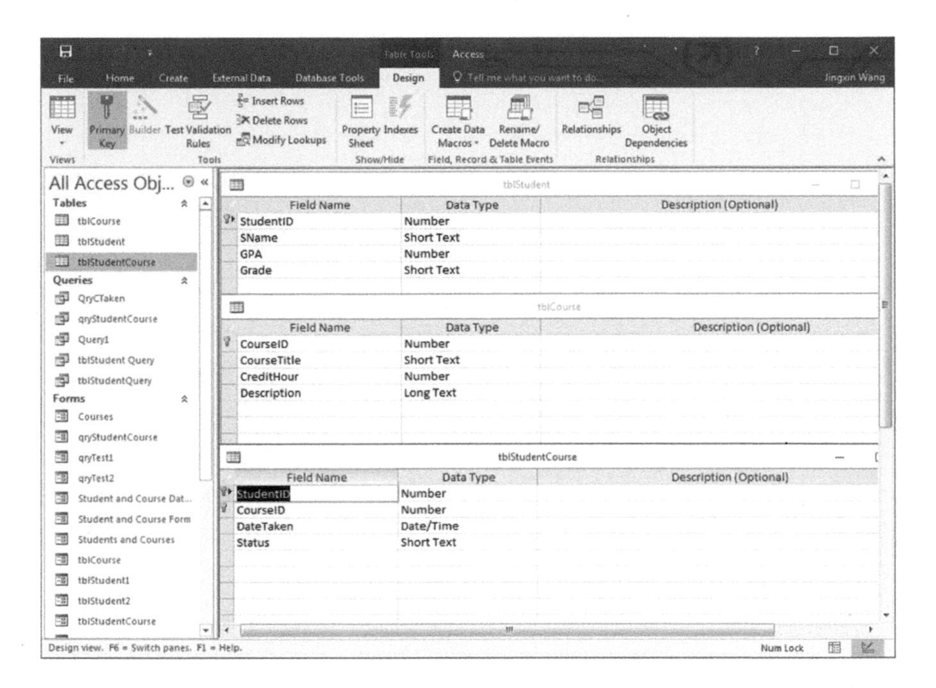

FIGURE 7.2
Design windows of tables: `tblStudent`, `tblCourse`, and `tblStudentCourse` in `dbStudent` database.

Add these three tables consecutively by selecting a table and then clicking *Add.* Close the dialog box, then use the mouse to point to *StudentID* in *tblStudent* and simply drag it to the *tblStudentCourse* table. An *Edit Relationships* box will open with *Student ID* in both columns. When you click *Create,* the box will close and a line will appear between *StudentID* in the two tables. Repeat this process, dragging *CourseID* from *tblCourse* to *tblStudentCourse,* to complete building the many-to-many relationship (Figure 7.5). Depending on the order in which you selected the tables, your Relationship boxes may appear in a different order. Note that for a connection or relationship to be established, the field names (*StudentID, CourseID,* etc.) do not have to match, but they must contain the same data type (number, short text, date/time, etc.). Once relationships are built, you can edit or delete them via the *Edit Relationships* button.

7.2.3 Access Queries

A **query** is used to extract information from tables, queries, or both in a database. It can select and define a group of records that fulfill a certain condition. Using an Access query, we can answer a specific question to perform data manipulation, to combine data from different tables, or even to modify data in different tables or queries. The assembled data can then be used for a form or a report.

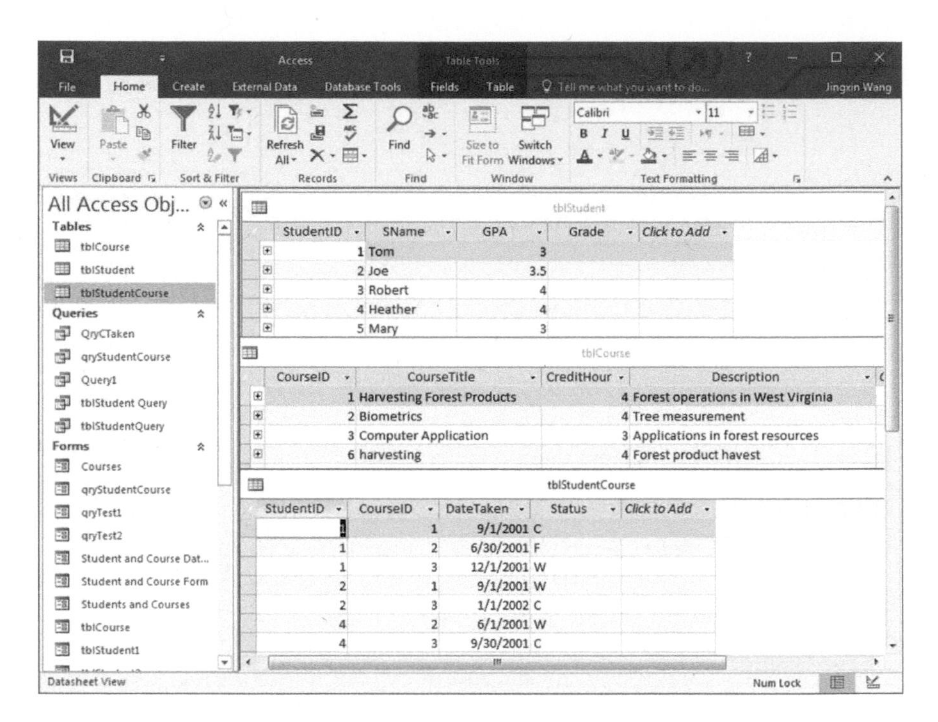

FIGURE 7.3
Data entries for the three tables in `dbStudent` database.

Show Table

Tables | Queries | Both

tblCourse
tblStudent
tblStudentCourse

Add | Close

FIGURE 7.4
A `Show Table` dialog box for building a relationship.

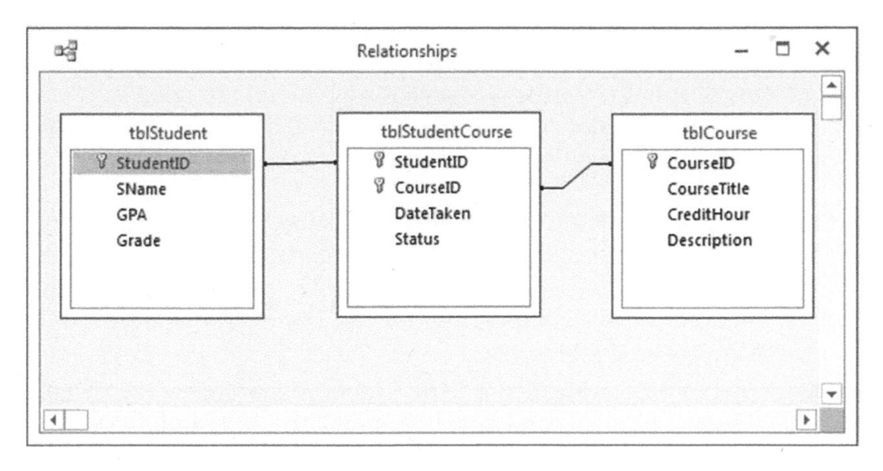

FIGURE 7.5
Relationships being built for tables in `dbStudent` database.

To perform a query, click *Create* from the Access menu ribbon. In the *Queries* section, you will see options to create a query either using the *Query Wizard* or using *Query Design*. For a typical query, we just select *Query Design*, then when the *Show Table* box pops up (Figure 7.4), we can add the tables (or queries or both) on which we will build a query.

After adding our three *dbStudent* database tables consecutively (by selecting a table and then clicking *Add*), we close the *Show Table* box and the added tables will display in a QBE window (Figure 7.6).

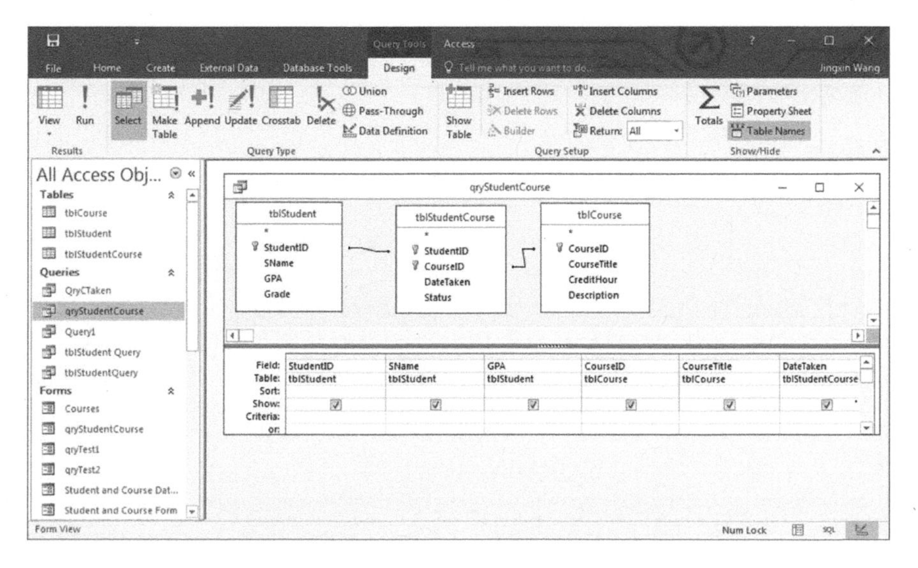

FIGURE 7.6
QBE window for designing a query.

Suppose we would like to list the following information for a specific student: Student ID, Name, GPA, Course ID, Course Title, and Date Taken. We cannot gather all this information from a single table; we need to create a query and join three tables together to display the required information. There are two ways to create a query:

1. Query manually—Drag the data fields you need (*StudentID, SName,* and *GPA* from *tblStudent; CourseID* and *CourseTitle* from *tblCourse; DateTaken* from *tblStudentCourse*) to the display area in the QBE window. Then click the *View* menu under the *Design* tab, and you will have a datasheet of this query.
2. Query using SQL—From the *Home* tab on the menu bar, Click *View* → *SQL View*, then type the following SQL commands:

```
SELECT tblStudent.StudentID, tblStudent.SName,
  tblStudent.GPA, tblCourse.CourseID, tblCourse.
  CourseTitle, tblStudentCourse.DateTaken
FROM tblStudent INNER JOIN (tblCourse INNER JOIN
  tblStudentCourse ON tblCourse.CourseID =
  tblStudentCourse.CourseID) ON tblStudent.StudentID =
  tblStudentCourse.StudentID;
```

Save this query and name it "qryStudentCourse" in the database.

7.3 Access Forms and Reports

7.3.1 Access Forms

A form is one of the basic objects in Access. It can serve as an interface to help users add, edit, or display data from a database in a quick, easy, and accurate manner. Access forms provide a more structured view of the data than does a datasheet such as an Excel worksheet (Prague et al. 1999). From this structured view, database records can be viewed, added, modified, or deleted. Entering data through forms might be the most efficient way to enter data into database tables, especially if the database has multiple users. Well-designed Access forms are essential for efficiency and accuracy (Microsoft Corporation 2016). They can also be implemented with business rules to restrict access to certain fields within the table and check the validity of data before they are accepted into the database table.

To create an Access form, click the *Create* tab on the menu ribbon, then select *Form Wizard*, which will display a *Form Wizard* dialog box (Figure 7.7).

Select the query we just created (*qryStudentCourse*) and then select data fields from the list box of *Available Fields* and add them to the list box of

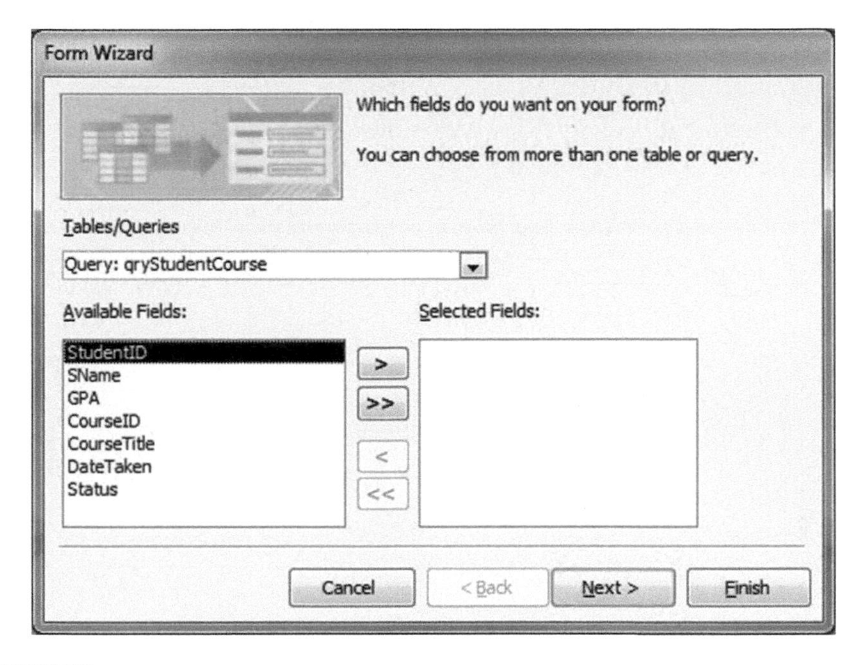

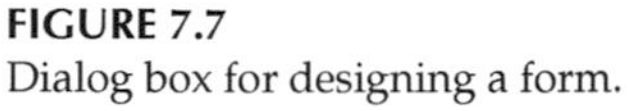
FIGURE 7.7
Dialog box for designing a form.

Selected Fields by clicking the arrow (or add them all by clicking the double arrow) and then click *Next* (Figure 7.7). From there, the Wizard will walk you through a set of questions that will help you determine the design of your form. For our purposes, when asked "How do you want to view your data?" click *by tblStudent* and select *Form with subform(s)*, then *Next* in the first pop-up box. In the next pop-up Wizard window, select *Datasheet* as the layout you would like for your subform, then click *Next*. In the final Wizard window, name your form and subform *tblStudent* and *tblStudentCourse Subform*, respectively, and select *Open the form to view or enter information*, then click *Finish*. You will get the following form, which will allow you to edit, add, and delete the data in the tables (Figure 7.8).

7.3.2 Access Reports

Access reports format and present data from a database in a printed format. There are several ways to generate a report in Access. For example, a report can list all the records in a given table, such as a student table. We can also create a report that summarizes only the students who meet a given criterion, such as all of those who have a GPA of 3.5 and above. We do this by incorporating a query into our report design. A report can combine multiple tables to present complex relationships among different sets of data. For example,

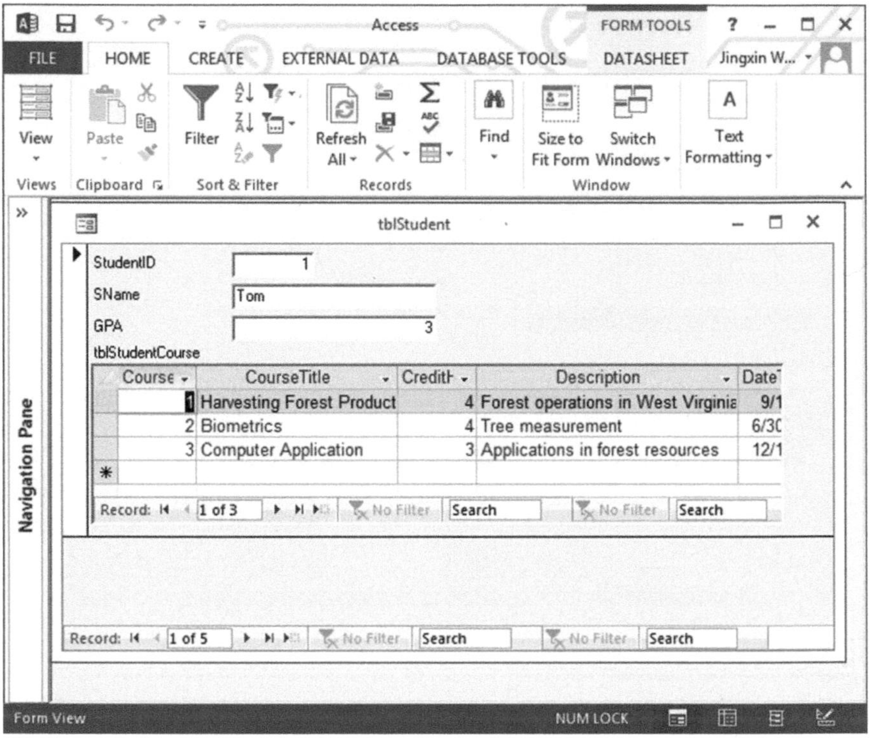

FIGURE 7.8
An Access data entry form.

a report can be generated based on the query we just created for three tables in the *dbStudent* database.

To generate an Access report, click *Create* on the menu ribbon, and then click *Report Wizard* in the *Reports* group. A new *Report Wizard* dialog box will pop up (Figure 7.9).

Select the query we just created (*qryStudentCourse*) and add the desired data fields by selecting them and clicking the arrows. For our example, we selected all the fields by clicking the double arrow button, then clicked *Next.*

As with forms, the Wizard will pose a set of questions that will help you determine the design of your report. For our purposes, when asked "How do you want to view your data?" select *by tblStudent* and then click *Next*. In the next pop-up Report Wizard window, select *StudentID* as the grouping level, then click *Next*. In the next Report Wizard for sorting order, you may simply click *Next*. You may then choose among options of how to format the report and then click *Next*. In the final Wizard window, name your report "Student" and "Course Report," select *Preview the report*, and then click *Finish*. You will get the following report, which will allow you to edit, modify, and delete data fields on the report (Figure 7.10).

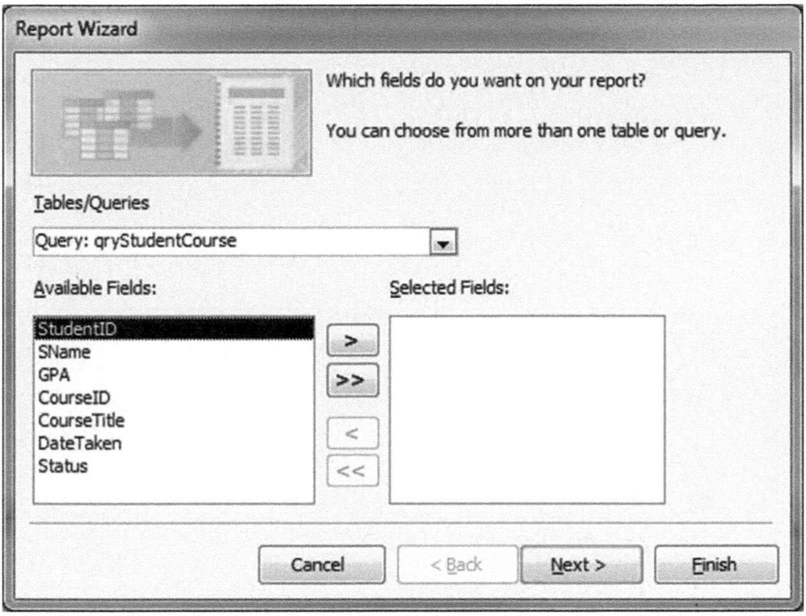

Report Wizard

Which fields do you want on your report?

You can choose from more than one table or query.

Tables/Queries

Query: qryStudentCourse

Available Fields:

StudentID
SName
GPA
CourseID
CourseTitle
DateTaken
Status

>
>>
<
<<

Selected Fields:

Cancel | < Back | Next > | Finish

FIGURE 7.9
A dialog box for report.

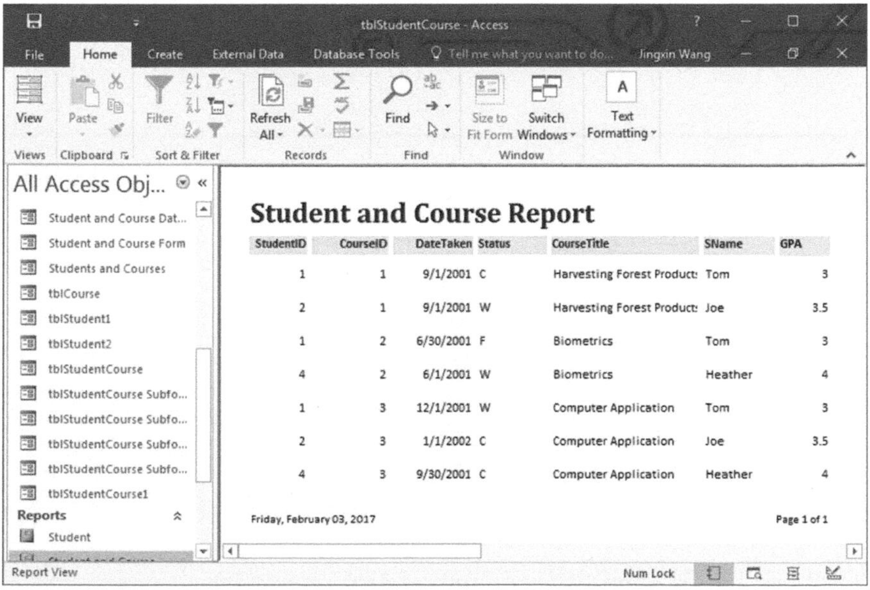

tblStudentCourse - Access

File | Home | Create | External Data | Database Tools | Tell me what you want to do... | Jingxin Wang

View | Paste | Filter | Refresh All | Find | Size to Fit Form | Switch Windows | Text Formatting

Views | Clipboard | Sort & Filter | Records | Find | Window

All Access Obj...

Student and Course Dat...
Student and Course Form
Students and Courses
tblCourse
tblStudent1
tblStudent2
tblStudentCourse
tblStudentCourse Subfo...
tblStudentCourse Subfo...
tblStudentCourse Subfo...
tblStudentCourse Subfo...
tblStudentCourse1

Reports

Student

Student and Course Report

StudentID	CourseID	DateTaken	Status	CourseTitle	SName	GPA
1	1	9/1/2001	C	Harvesting Forest Product:	Tom	3
2	1	9/1/2001	W	Harvesting Forest Product:	Joe	3.5
1	2	6/30/2001	F	Biometrics	Tom	3
4	2	6/1/2001	W	Biometrics	Heather	4
1	3	12/1/2001	W	Computer Application	Tom	3
2	3	1/1/2002	C	Computer Application	Joe	3.5
4	3	9/30/2001	C	Computer Application	Heather	4

Friday, February 03, 2017 | Page 1 of 1

Report View | Num Lock

FIGURE 7.10
A sample Access report on student and course data.

Class Exercises

Timber Cruising Database Applications

In this lab exercise, you are required to use Microsoft Access and the database design knowledge you have learned to design a database for timber cruising. You will need to create a database, name it *TimberCruising.accdb*, and then create the following database objects:

A. Tables—Build the following tables (data for these tables can be found in Tables 7.1 through 7.3):

1. Table Plot (`tblPlot`) and its fields

`PlotID`	Number → Integer, primary key
`PlotType`	Text
`Forest`	Text
`Crew`	Text
`Date`	Date/Time

2. Table Trees (`tblTrees`) and its fields

`TreeID`	Number → Integer
`PlotID`	Number → Integer, Required: YES (refers to `PlotID` in `tblPlot`)
`SpeciesID`	Number → Integer
`DBH`	Number → Single
`MHT`	Number → Single
`THT`	Number → Single

3. Table Species (`tblSpecies`) and its fields

`SpeciesID`	Number → Integer, primary key
`Species`	Text

B. Queries—Design the following queries for generating reports:

1. Basal Area and Volume summary by Plot ID
2. Basal Area and Volume summary by Species ID

You can use the following equations to compute the basal area and volume for each individual tree:

```
BA = SUM(0.005454154*DBH²)
VOL = SUM((0.04 - 0.02* (MHT)2 + 0.59 * (MHT)) * DBH)
```

where:

BA = basal area, ft^2 (1 ft^2 = 0.0929 m^2)
VOL = volume, ft^3 (1 ft^3 = 0.0283 m^3)
DBH = diameter at breast height, inches (1 in. = 2.54 cm)
MHT = merchantable height, the number of 4.88 m (16 ft) logs

TABLE 7.1

Data for Plot Table

PlotID	PlotType	Forest	Crew	Date
1	VRP	Univ.For	1	10/5/2001
2	VRP	Univ.For	1	10/5/2001
3	VRP	Univ.For	1	10/5/2001
4	VRP	Univ.For	1	11/10/2001
5	VRP	Univ.For	1	11/10/2001

TABLE 7.2

Data for Species Table

SpeciesID	Species
1	Birch, yellow
2	Cherry, black
3	Maple, red
4	Maple, sugar
5	Oak, chestnut
6	Oak, northern red
7	Yellow poplar
8	Other hardwood

C. Forms—Forms are used as the interfaces, while reports are the final output results for the database applications.

1. Design forms to manipulate the data in the above three tables, including edit, add, and delete.
2. Enter the data of Tables 7.1 through 7.3 forms.

TABLE 7.3
Data for Tree Table

TreeID	PlotID	SppID	DBH	MHT	THT
1	1	3	34	1.5	117
2	1	1	21	0	0
3	1	1	13	0	0
4	1	1	18	1.5	97
5	1	4	12	0.5	83
6	1	8	2	0	0
7	1	4	19	3	102
8	1	1	18	0	0
9	2	3	14	1.5	0
10	2	6	23	2.5	0
11	2	6	15	2.5	0
12	2	3	8	0	0
13	2	5	14	2	0
14	2	6	19	2.5	0
15	3	5	13	2	75
16	3	6	10	0	70
17	3	6	10	0	68
18	3	5	6	0	0
19	3	6	9	0	58
20	3	5	8	0	64
21	3	6	8	0	65
22	3	3	3	0	0
23	3	6	15	1	77
24	3	6	10	0	70
25	3	3	9	0	37
26	4	3	8	0	0
27	4	2	17	2	0
28	4	2	22	2.5	0
29	4	6	12	1	0
30	4	6	12	1.5	0
31	4	2	23	2	0
32	5	7	16	0	0
33	5	7	16	2	93
34	5	7	18	3	95
35	5	4	27	1.5	100
36	5	7	20	0	0
37	5	7	18	4	106

D. Reports—Generate the following reports:
   1. Basal Area and Volume summary by Plot ID
   2. Basal Area and Volume summary by Species
E. Questions—Please provide answers to the following questions:
   1. What is a database management system (DBMS)?
   2. Explain what basic features and functions a DBMS can provide.
   3. Give an example of a DBMS.
   4. Why would we want to use MS Access?

References

Microsoft Corporation. 2016. Create a form in Access. https://support.office.com/en-us/article/Create-a-form-in-Access-5d550a3d-92e1-4f38-9772-7e7e21e80c6b. Accessed on September 4, 2016.

Prague, C., D. Roche, and M. Irwin. 1999. *Microsoft Access 2000 Bible*. John Wiley & Sons, Incorporated, 1272pp.

8

Structured Query Language and Access Query Examples

8.1 Structured Query Language

Structured Query Language (SQL) allows us to access and execute queries on a database, including table creation or deletion, and data retrieval, addition, update, or deletion. SQL is an American National Standards Institute standard for accessing database systems (SQLCourse.com 2017). It works with any Database Management System (DBMS) such as MS Access, DB2, MS SQL Server, Oracle, and Sybase, although it varies slightly from system to system.

8.1.1 SQL Statements and Clauses

SQL provides many statements in the following categories: Data Definition Language (DDL), Data Manipulation Language (DML), Transaction Control, Session Control, System Control, and Embedded SQL. In this chapter, we only focus on SQL related to DML and DDL (Table 8.1).

There are many SQL clauses that can be used together with SQL statements. Here are some commonly used SQL clauses (Table 8.2).

8.1.2 SQL Syntax

The basic syntax of the SQL commands is:

```
SELECT attributes [[AS] alias], ...
FROM table(s)
[WHERE Boolean conditions]
[GROUP BY attributes]
[HAVING ...]
[ORDER BY ...];
```

The above syntax consists of a `SELECT` statement, a `FROM` clause, and a `WHERE` clause for conditional selection. A query typically ends with a semicolon.

TABLE 8.1

SQL Statements in DML and DDL Categories

DML		DDL	
Statement	**Description**	**Statement**	**Description**
`SELECT`	Extracts data from a database table	`CREATE`	Creates a new database table or index
`UPDATE`	Updates data in a database table	`DROP`	Drops a table or index
`DELETE`	Deletes data from a database table	`ALTER`	Alters table's design
`INSERT`	Inserts new data into a database table		

TABLE 8.2

Commonly Used SQL Clauses

Clause	Description
`FROM`	Specify the tables on which the query performs
`WHERE`	Specify the selection condition(s)
`GROUP BY`	Group a result into subsets
`HAVING`	Provides Boolean condition for `GROUP BY` clause
`ORDER BY`	Specify the order in which results display

The result from a SQL query is stored in a result set. Most DBMSs allow navigation of the result set with database functions, such as `MoveFirst`, `MoveNext`, `MoveLast`, and others. We will discuss these functions in the Visual Basic .NET programming chapter (Chapter 14).

While a semicolon is the standard way to end a query or separate each SQL statement in DBMSs, it is optional in MS Access and SQL Server. We do not have to use a semicolon after each SQL statement in MS Access.

To conditionally select data from a table, a `WHERE` clause can be used or included in the `SELECT` statement. The following operators can be used in the `WHERE` clause for logical comparisons (Table 8.3).

SQL also allows us to use subqueries or nested queries. A SQL query returns a set of records; thus, we should be able to apply a set of operations to test, for example, if a membership is in a record set (Atkins 2001). The test should determine if an attribute value is related to a set of values. The syntax for a query with a subquery is:

```
SELECT *
FROM table(s)
WHERE {attribute values}  =, <>, >,< {any or all (set
  of values from a subquery)};
```

TABLE 8.3

Operators Used in the WHERE Clause

Operator	Description
=	Equal
<>	Not equal
>, <	Greater or less than
>=, <=	Greater than or equal; less than or equal
LIKE	Search for a pattern
AND, OR	Join two or more conditions
BETWEEN ... AND	Select a range of data between two values
NOT	Select data outside a defined data range

8.2 Basic SQL Examples

Let's use an example of a table called *Trees* (Table 8.4) to illustrate how to use SQL. This table has three records (one for each tree) and four fields (TreeID, Species, DBH, and TotalHeight).

8.2.1 Statements and Clauses

We use SELECT to select columns of data. For example, we can use this SQL command to select TreeID and Species in Table 8.4:

```
SELECT TreeID, Species FROM Trees;
```

which will give us this result set:

TreeID	Species
1	Oak
2	Poplar
3	Maple

TABLE 8.4

A Database Table of Trees

TreeID	Species	DBH	TotalHeight
1	Oak	10	25
2	Poplar	23	50
3	Maple	20	40

To select all columns from the *Trees* table, we list all the fields and separate them using a comma, or use a * symbol instead of column names. The result set will be the same as Table 8.4.

```
SELECT * FROM Trees;
```

To select only the trees with a species name `Oak`, we add a `WHERE` clause to the `SELECT` statement:

```
SELECT * FROM Trees WHERE Species='Oak';
```

The result set is:

TreeID	Species	DBH	TotalHeight
1	Oak	10	25

We used single quotes around the conditional values in this example for `'Oak.'` Typically, SQL uses single quotes around text values. Some DBMS such as MS Access also accept double quotes.

We can use this SQL command to select trees with a diameter at breast height (DBH) of greater than 25.4 cm (10 in.):

```
SELECT * FROM Trees WHERE DBH>10;
```

We can have multiple conditions in the `WHERE` clause using `AND`. For example, we can use `AND` to display trees with species of `Oak`, and a `DBH` of 20. We will generate an empty result set.

```
SELECT * FROM Trees
WHERE Species='Oak' AND DBH=20;
```

Sometimes we would like to select our result in a range of data between two values that could be numeric, text, or date values. In that case, we can use either the `BETWEEN ... AND` operator or other logical comparison operators. For example, in order to display trees with the total height between 25 and 35, we use the following SQL:

```
SELECT * FROM Trees
WHERE TotalHeight BETWEEN 25 AND 35;
```

Or we use:

```
SELECT * FROM Trees
WHERE TotalHeight >= 25 AND TotalHeight <= 35;
```

Both of these SQL commands would generate the same result set:

TreeID	Species	DBH	TotalHeight
1	Oak	10	25
4	Oak	15	30
5	Maple	15	32

In Access, the BETWEEN ... AND operator selects fields that are between and including the test values. Therefore, trees with a total height of 7.62 or 10.67 m (25 or 35 ft) will be listed. In some other DBMSs, the BETWEEN ... AND operator may only select fields that are between the test values, excluding trees with a total height of 25 or 35.

To list the trees outside the range used in the above example, we can use the NOT operator.

```
SELECT * FROM Trees
WHERE TotalHeight NOT BETWEEN 25 AND 35;
```

GROUP BY and ORDER BY are two other clauses commonly used together with the SELECT statement. GROUP BY categorizes the result into groups while ORDER BY is used to sort and order the results.

The DISTINCT keyword can be used in the SELECT statement to select only distinct (different) values for a selected field(s) or attribute(s). For example, suppose we added a few more trees to Table 8.4 and have a new tree list (Table 8.5).

If we select species from Table 8.5, this SQL statement may be used:

```
SELECT Species FROM Trees;
```

TABLE 8.5

A New Database Table of Trees

TreeID	Species	DBH	TotalHeight
1	Oak	10	25
2	Poplar	23	50
3	Maple	20	40
4	Oak	15	30
5	Maple	15	32

It will return this result set:

Species
Oak
Poplar
Maple
Oak
Maple

Notice that the species `Oak` and `Maple` are each listed twice in the result set. If instead we want to have a list of unique species, we can use a SQL statement that will list only three unique species:

```
SELECT DISTINCT Species FROM Trees;
```

8.2.2 SQL Functions

SQL can use many built-in functions provided by a DBMS for data manipulation. Some functions in the Aggregate category include `Sum`, `Count`, `Max`, `Min`, `Avg`, and `StDev`. Mathematical (`SIN`, `COS`, `LOG`, and others), Logical (`IIF`, `Choose`), and String (`Char`, `Varchar`) functions are included in the Scalar category.

The `AVG` function calculates the average of a field or the average of combined fields with the `GROUP BY` clause. For example, if we want to calculate the average DBH of trees in Table 8.5 by species, we use this SQL command:

```
SELECT Species, avg(DBH) as Average_DBH FROM Trees
GROUP BY Species
OODER BY Species;
```

And get the result:

Species	Average _ DBH
Maple	17.5
Oak	12.5
Poplar	23

Similarly, the `SUM` function will total the values in a field or a group of fields. For example, we calculate the basal area (BA) of each tree using this expression `BA = 0.005454154*DBH*DBH` (DBH in inches, 1 in. = 2.54 cm) and summarize the BA by species. Here is the query:

```
SELECT Species, sum(0.005454154*DBH*DBH) as BA FROM Trees
GROUP BY Species
OODER BY Species;
```

We will get this result set:

Species	BA
Maple	3.4088
Oak	1.7726
Poplar	2.8852

We can sort and order the results by more than one field and need to separate these groupings using commas. The sorting and ordering are in ascending (ASC) by default. If we would like to list the result in a descending order by a field, we use the DESC keyword right after this field in the ORDER BY clause. Here are two examples of how to use the ORDER BY clause:

```
SELECT Species, DBH FROM Trees
ORDER BY Species, DBH;

SELECT Species, DBH FROM Trees
ORDER BY Species DESC;
```

Like in MS Excel, we can also use the COUNT function to calculate the number of records. We use either the COUNT(*) function for a whole table or COUNT(field) for a specific field. The COUNT function does not count records that have null fields unless the argument in the COUNT function is the asterisk (*). Let's use Table 8.5 as an example.

```
SELECT COUNT(*) FROM Trees;                (returns 5)
```

The example below returns the number of trees with a DBH larger than 20:

```
SELECT COUNT(*) FROM Trees WHERE DBH>20;      (returns 1)
```

The COUNT(field) function returns the number of rows without a null value in the specified column. If Table 8.5 were modified to include a few null values (Table 8.6), the following query would return a "3":

```
SELECT COUNT(DBH) FROM Trees;              (returns 3)
```

TABLE 8.6

Modified Database Table of Trees

TreeID	Species	DBH	TotalHeight
1	Oak	10	25
2	Poplar		50
3	Maple		
4	Oak	15	30
5	Maple	15	32

8.3 MS Access Queries

Let's consider other examples of MS Access queries. First we build a database and name it "dbForProdSale". It will hold the data in two tables for forest products and forest products companies that supply these products. The schemas for these two entity types are as follows:

a. tblFPCompany(FPC_ID, FPCName, Status, City)
b. tblFProduct(FP_ID, FPName, FPUse, Quantity, UnitPrice)
   In the tblFPCompany table, FPC_ID represents the forest products company's ID, FPCName is the company's name, Status indicates the status of its production, and City is the company's location. Similarly, for tblFProduct table, we use five fields or attributes to represent the ID, name, use, quantity, and unit price of a forest product. Since the relationship between Company and Product is many-to-many, we need to create another table to represent this relationship.
c. tblFPComProd(FPC_ID, FP_ID, Quantity)
   Figure 8.1 shows a design view of these tables, while Figure 8.2 shows how relationship tools may be used to designate relationships among them.
   Once this many-to-many relationship is built, enter some data into the tables as shown in Figure 8.3.

The following queries can be performed using the tables we just created for forest products sales (they were modified based on some query examples by Dr. J. Atkins (2001)). Again, we can use either QBE or SQL with MS Access. But be aware that sometimes it is difficult or impossible to create a complex query using QBE.

1. List the forest products company ID, products ID, quantity, and quantity supplied if increased by 20%.

```
SELECT FPC_ID, FP_ID, Quantity, Quantity*1.2 AS NewQuantity
FROM tblFPComProd;
```

2. List all the records in the Forest Products table.

```
Select * from tblFProduct;
```

3. List the IDs of the forest products companies that can supply at least one forest product in a quantity between 20 and 50.

```
SELECT FPC_ID, Quantity
FROM tblFPComProd
WHERE Quantity between 20 and 50;
```

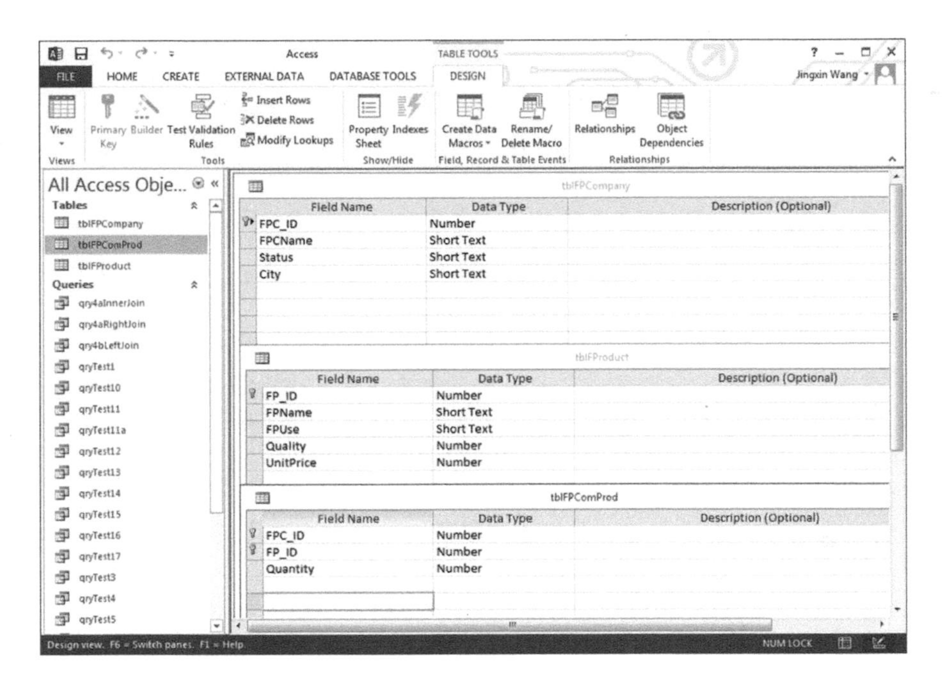

FIGURE 8.1
Schemas of three tables in `dbForProdSale` database.

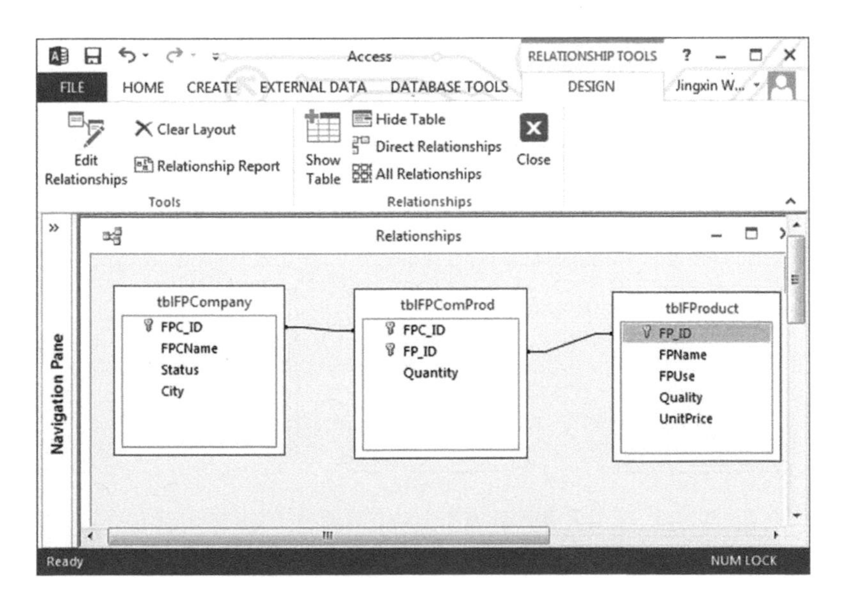

FIGURE 8.2
Building a many-to-many relationship between `tblFPCompany` and `tblFProduct`.

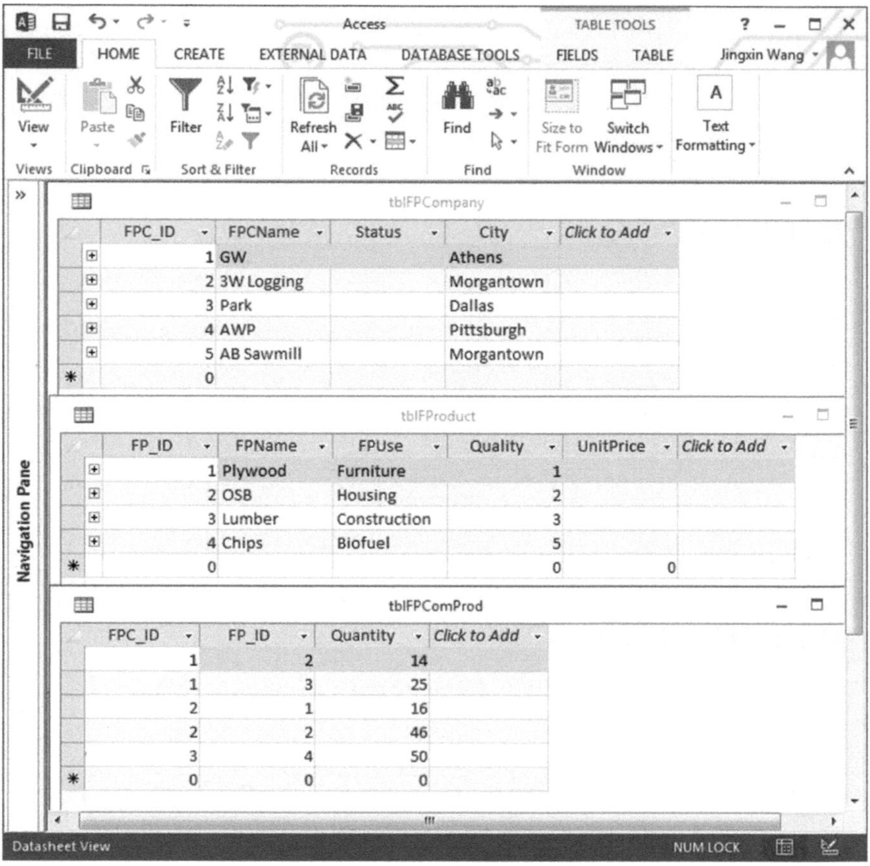

tblFPCompany

FPC_ID	FPCName	Status	City	Click to Add
1	GW		Athens	
2	3W Logging		Morgantown	
3	Park		Dallas	
4	AWP		Pittsburgh	
5	AB Sawmill		Morgantown	
0				

tblFPProduct

FP_ID	FPName	FPUse	Quality	UnitPrice	Click to Add
1	Plywood	Furniture	1		
2	OSB	Housing	2		
3	Lumber	Construction	3		
4	Chips	Biofuel	5		
0			0	0	

tblFPComProd

FPC_ID	FP_ID	Quantity	Click to Add
1	2	14	
1	3	25	
2	1	16	
2	2	46	
3	4	50	
0	0	0	

FIGURE 8.3
Data entries for three tables in the `dbForProdSale` database.

4. List the forest products company name, product ID, and supplied quantity; sort the result in ascending order of company name and descending order of quantity.

```
SELECT tblFPCompany.FPCName, tblFPComProd.FP_ID, Quantity
FROM tblFPCompany, tblFPComProd
WHERE tblFPCompany.FPC_ID = tblFPComProd.FPC_ID
ORDER BY tblFPCompany.FPCName, Quantity DESC;
```

5. List the name of the forest products companies that supply forest product #3 (lumber).

```
SELECT FPCName
FROM tblFPCompany
```

```
WHERE FPC_ID = any (SELECT FPC_ID FROM tblFPComProd WHERE
  FP_ID = 3);
```

Here, we use a nested query. The nested `SELECT` must be enclosed in parentheses `()` and the comparison "`= any`" may be replaced by "`in`."

6. List the IDs of forest companies that supply at least one type of forest products #2, #3, or #4.

```
SELECT FPC_ID
FROM tblFPComProd
WHERE FP_ID In (2,3,4);
```

7. List the names of forest products companies that currently supply no forest products.

```
SELECT FPCName
FROM tblFPCompany
WHERE not exists
(SELECT * FROM tblFPComProd WHERE FPC_ID = tblFPCompany.FPC_ID);
```

8. Find the ID of the company that supplies the maximum quantity of forest products.

```
SELECT FPC_ID
FROM tblFPComProd
WHERE Quantity >= all (SELECT Quantity FROM tblFPComProd);
```

We can use the MS Access built-in function, `MAX`, to do the same thing in this example. This query can be rewritten as:

```
SELECT FPC_ID
FROM tblFPComProd
WHERE Quantity >= (SELECT MAX(Quantity) FROM tblFPComProd);
```

9. Find the total number of forest products companies.

```
SELECT Count(*) AS No_Of_Companies
FROM tblFPCompany;
```

10. List the IDs of the forest products companies that supply at least one type of forest product.

```
SELECT DISTINCT FPC_ID
FROM tblFPComProd;
```

11. Find the total quantity of forest product #2 that is supplied.

```
SELECT SUM(Quantity) As TotalProducts
FROM tblFPComProd
WHERE FP_ID = 2;
```

12. List the names of forest companies that supply at least two different forest products.

```
SELECT FPCName
FROM tblFPCompany
WHERE 2<= (SELECT COUNT(FPC_ID) FROM tblFPComProd WHERE
  FPC_ID = tblFPCompany.FPC_ID);
```

13. List each forest products company ID with the average quantity it supplies.

```
SELECT FPC_ID, AVG(Quantity)
FROM tblFPComProd
GROUP BY FPC_ID
ORDER BY FPC_ID;
```

We logically group quantity together using an aggregate statistical function AVG.

14. List the names of the forest companies that supply exactly two different types of forest products.

```
SELECT FPCName
FROM tblFPCompany
WHERE FPC_ID IN (SELECT FPC_ID FROM tblFPComProd GROUP BY
  FPC_ID HAVING COUNT(*) = 2);
```

The HAVING clause complements the GROUP BY by providing a Boolean condition to decide if a group should or should not be displayed.

Class Exercises

1. What is SQL?
2. What is the basic structure of SQL?
3. How would we create simple queries with SQL and apply them in Natural Resource Management using ArcMap or other programming languages?

References

Atkins, J. 2001. *Database Management* (Lecture Notes). Department of Computer Science, West Virginia University, Morgantown, WV.

Microsoft Corporation. 2013. Introduction to Queries. http://office.microsoft.com/en-us/access-help/introduction-to-queries-HA102749599.aspx. Accessed on January 30, 2014.

SQLCourse.com. 2017. SQL Courses. http://www.sqlcourse.com/index.html. Accessed on February 3, 2017.

Section IV

Handheld Devices

9

Handheld Computers and Windows Mobile

To date, there are many types of handheld computers (HPCs) and versions of Windows Mobile operating systems (OSs) available. In this chapter, we will discuss some of the useful features of handheld PC and their OSs.

9.1 Handheld Terms and Features

Handheld PC—A portable computer, with a display and a built-in keyboard, that is small enough to be held in one hand (such as Windows Mobile personal digital assistant [PDA]).

Pocket PC—A computer that fits in the palm of your hand, runs the latest Windows Mobile OS, and is typically smaller than an HPC. It includes Pocket Word, Pocket Excel, Calendar, Contacts, and Tasks as well as other applications.

Personal Digital Assistant—A handheld or mobile device with personal information manager and basic computing functions such as web browsing, office applications, and data manipulations. It could be an HPC or a pocket PC (PPC).

Windows Mobile Device Center—Offers device management and data synchronization between a Windows Mobile-based device (HPC, PPC, or other PDAs) and a computer. It replaced MS ActiveSync.

CompactFlash card—A nonvolatile secondary storage card that is compatible with mobile devices. CompactFlash (CF) uses flash memory, so it does not require too much power to work, and no power is needed to maintain the information stored in the chip. With an adapter, it is compatible with a PC card slot.

Secure Digital card—Also a nonvolatile secondary storage card by SanDisk, the Secure Digital (SD) card can be used in just about every type of mobile device (digital cameras, video cameras, tablets, and smartphones).

Dynamic Host Configuration Protocol—Dynamic Host Configuration Protocol (DHCP) automatically configures the PC Companion with an Internet Protocol (IP) address, subnet, and gateway. DHCP is supported with Ethernet and wireless local area networks (WLANs).

File Transfer Protocol—A standard IP for transmitting files between computers on the Internet.

Transport Communications Protocol/Internet Protocol—The communications protocol that Windows Mobile uses to communicate with the Internet and for synchronization.

Universal Serial Bus—A connection standard used by computers and mobile devices. These are the types of cables and connectors used in a computer bus.

When selecting a handheld or PDA for applications in forest and natural resource management, a user must weigh the importance of size, performance, features, weight, and cost (Cnet.com 2014). Consider the following characteristics and uses before purchasing:

1. Size and Weight—Most current models of PDAs or HPCs are between 7.62 and 15.24 cm (3 and 6 in.) long, about 7.62 cm (3 in.) wide, and weigh between 113.40 and 198.45 g (4 and 7 ounces) (Cnet.com 2014). Ask yourself if a model is portable and light enough for you to carry it with you when you need it in the field.
2. Data Input—Data entry is essential to HPCs in field applications. Check the type of HPC keyboards your applications require. In addition to touch screen, do you need a touchpad for mouse functions?
3. Memory—PDAs typically have their OS stored in read-only memory (ROM) and use built-in random access memory (RAM) for processor memory and as file storage space. Thus, more attention needs to be paid to how much RAM is installed in the device. For example, NAUTIZ X7 and Juniper Archer have a RAM memory of 128 MB.
4. Secondary Storage—Do you need a CF, PC card, or SD slot? Do you have enough RAM for running programs and storage? A memory card is a good way to store more application data and is used as a backup of RAM data.
5. Processors—A fast processor is always preferred for an HPC, especially if it is used for playing music, video games, and some complex computing. For example, NAUTIZ X7 uses a Marvell 806 MHz processor, while Juniper Archer uses a PXA 270 processor.
6. Internet Connectivity—We can use Ethernet or wireless to connect HPCs to the Internet.
7. Battery—Battery life is critical for HPCs and PDAs. We would require a battery life longer than 8 hours under field conditions for natural resource applications.

8. GPS Capability—Oftentimes, your HPCs may be used to collect spatial data in natural resource applications. You need to ensure that your device has the hardware installed or the capability to support GPS.
9. Applications—Do you need MS Office Mobile including Excel Mobile, PowerPoint Mobile, and OneNote Mobile? We typically use HPCs for specially designed application programs in forestry and natural resources, such as timber cruising, field survey, and mapping programs. For these applications, we also need SQL Server CE for Windows Mobile Devices.

9.2 Handheld PCs and Windows Mobile

Several varieties of HPCs are available from a few major manufacturers (Table 9.1). These devices have similarities and differences that must be weighed when considering which device suits your forestry needs.

9.2.1 Hewlett-Packard Jornada Series

Although the HP Jornada 720 made its debut in the fall of 1999, the final production of the devices went to the end in 2001 (Fitch 2006). The HP Jornada 720 essentially marked the end of Hewlett-Packard's work in the HPC market.

Hewlett-Packard Jornada 720 Series HPC is a mobile device powered by Microsoft Windows for HPC 2000 (a version of Windows CE) OS (Hewlett-Packard Co. 2000). It uses the Intel SA-111-206 MHz processor with a memory of 32 MB. If you are familiar with Microsoft Windows products and notebook PCs, you will notice that your HP Jornada has many of the same characteristics, making it easy for you to quickly become proficient and productive. It has many useful features such as Pocket Office that includes Pocket Access.

9.2.2 Zebra Workabout Handheld Field PC

The Zebra Workabout Pro is designed for applications across a range of industries, including mobile field services, logistics, warehousing, transportation, and manufacturing. It used to be Psion Teklogix, and it is now owned by Zebra Technologies (https://www.zebra.com/us/en/products/mobile-computers/handheld.html). Its impressive flexibility enables us to apply it in forestry and natural resource management. There are a number of add-ins and software applications you can attach to the device, including a GPS module used to track and trace locations of trees or forest roads and landings. The device also provides a variety of features and benefits including mobility and rugged reliability. A few models of Workabout are available, including the Workabout Pro 3 or 4 (Figure 9.1).

TABLE 9.1

Comparisons of Features and Functions of Few HPCs

Device	HP Jornada 720	Zebra Workabout Pro3	Allegro MX	Juniper Archer	NAUTIZ X7	Trimble Geo XT
Operating system	Windows CE	Windows Mobile 6	Windows Mobile 6	Windows Mobile 6	Windows Mobile 6	Windows Mobile 6
Processor	Intel SA-111-206 MHz	PXA270 624 MHz	Intel Arm-XScale	PXA270 520 MHz	XScale 806 MHz	ARM920T PXA27x 520 MHz
RAM (MB)	32	256	128	128	128	128
Rugged	No	Yes	Yes	Yes	Yes	Yes
GPS	No	No	No	Yes	Yes	Yes
Bluetooth	No	Yes	Yes	Yes	Yes	Yes
Wireless	No	Yes	Yes	Yes	Yes	Yes

FIGURE 9.1
Zebra Workabout Pro 3. (From Zebra Technologies Corporation, Lincolnshire, IL, www.zebra.com/us/en.html, accessed March 15, 2017.)

Workabout runs on the Windows Mobile 6.1 OS and uses PXA270 624 MHz. It has 256 MB of RAM and one GB of flash memory storage. The USDA Forest Service plans to use it to run the Forest Service's FSCruiser and FSVeg inventory software, as well as other products, such as Fountains Forestry's Pocket Dog data collection and processing software (www.fountainsamerica.com/twodog/). The device comes with basic software, including Microsoft Office Mobile (Excel, PowerPoint, and Word), Internet Explorer, and calendar and contacts applications.

9.2.3 Allegro Field PC

The Allegro series Field PCs include Allegro CE, CX, MX, and Allegro 2 by the Juniper Systems. The ultra-rugged Allegro MX Field PC (Figure 9.2) is built to perform in the most demanding outdoor or industrial environments (www.junipersys.com). This device uses an Intel Arm-XScale processor with a RAM of 108 MB. It runs on the Windows Mobile 6.1. Other features include (a) integrated Bluetooth wireless technology and Wi-Fi 802.11b/g, (b) IP67 rating for rugged applications, (c) robust full alphanumeric keyboard with 62 large keys that are color-coordinated by function, 12 function keys that can be used as hot keys in application programs, and five Windows keys

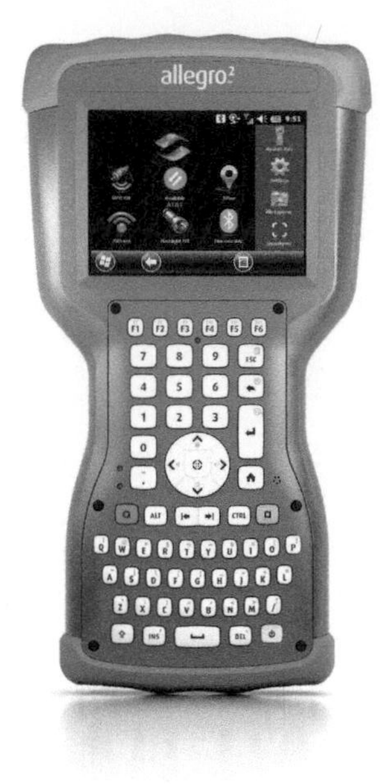

FIGURE 9.2
Allegro MX Rugged Handheld. (From Juniper Systems, Logan, UT, www.junipersys.com, accessed March 8, 2017.)

that provide enhanced use with Windows Mobile, (d) highly outdoor-visible display in color or monochrome, and (e) rechargeable battery that operates for more than 12 hours on one charge.

9.2.4 Archer Field PC

The Archer by Juniper Systems (Figure 9.3) is designed for the most demanding field applications (www.junipersys.com/Juniper-Systems-Rugged-Handheld-Computers/products/Archer-Field-PC). This rugged handheld PC can survive 1.52 m (5 ft) drops onto concrete, full immersion in water, and temperatures up to 60°C (140°F). The Archer is fully waterproof and dustproof, earning it an IP67 rating. It is also tested to MIL-STD-810F for water, humidity, sand and dust, vibration, altitude, shock, and temperature. It uses Windows Mobile 6.1 OS with a 520 MHz PXA270 processor and has a RAM of 128 MB. Its battery life can be up to 20 hours. A GPS module can be attached to the device. It comes with Microsoft Office Mobile.

9.2.5 NAUTIZ X7 Field PC

NAUTIZ X7 by the Handheld Group AB (Figure 9.4) exemplifies the evolution of handheld PC (https://www.handheldgroup.com/). NAUTIZ X7

FIGURE 9.3
Juniper Archer Field PC. (From Juniper Systems, Logan, UT, www.junipersys.com, accessed March 8, 2017.)

FIGURE 9.4
NAUTIZ X7 Field PC. (From HHCS Handheld USA, Inc. Lidköping, Sweden, https://www.handheldgroup.com/, accessed March 9, 2017.)

offers a lively 806 MHz XScale processor with 128 MB of onboard RAM and a generous 4 GB of flash storage. This field-ready device has a 5600 mAh Li-ion battery that will operate up to 12 hours on a single charge.

NAUTIZ X7 also delivers an unprecedented package of capability. It starts with integrated SiRF Star III GPS, Bluetooth 2.0, and 802.11b/g WLAN functionality, plus a built-in 3-megapixel camera with autofocus and an LED flash. NAUTIZ X7's innovations include 3G capability for GSM/UMTS phone and data transmission, an integrated compass and altimeter, and even a g-sensor/accelerometer that can measure speed, vibration, and rotation, opening the door to countless application possibilities. The Windows Mobile 6.1 OS, 8.89 cm (3.5 in.) VGA touch screen display, and numeric keypad make this handheld easy to operate in the field applications of forestry and natural resources.

9.2.6 Trimble GeoExplorer 3000 Series Handhelds

The GeoExplorer 3000 series (Figure 9.5) includes the GeoXH, GeoXM, and GeoXT handhelds (Trimble 2012). These handhelds combine a Trimble GPS

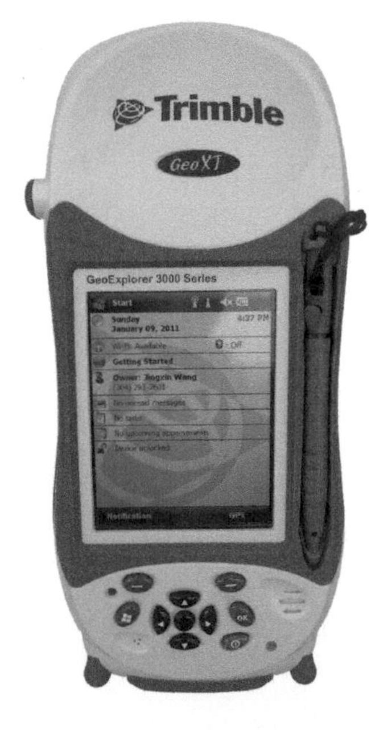

FIGURE 9.5
Trimble Geo XT. (From Trimble, *GeoExplorer 3000 Series User Guide*, Trimble Navigation Limited, Westminster, CO, 2012.)

receiver with a field computer powered by the Microsoft Windows Mobile 6 OS. The GeoXT handheld provides submeter accuracy. It uses an ARM920T PXA27x processor with a RAM of 104 MB and an expansion SD slot of up to 1 GB. It has built-in Bluetooth and WLAN connectivity options. This handheld device is rugged and resistant to heavy wind-driven rain and comes with an all-day battery.

9.3 Mobile Operating Systems and Data Communications

9.3.1 Mobile Operating Systems

There are four major mobile OSs for HPCs and smartphones: iOS, Android, Windows Mobile, and BlackBerry. All four of these OS platforms have their strengths and weaknesses, depending on what you are already using and what you want to get out of your computing experience (Colbert 2013). Android, iOS, and BlackBerry are specifically used for smartphones. For handheld field PCs, Windows Mobile is the most commonly used OS.

Windows Mobile is a mobile OS developed by Microsoft for smartphones, PPCs, and HPCs. It is supplied with a suite of basic applications developed with the Microsoft Windows API and is designed to have features and an appearance somewhat similar to desktop versions of Windows. Windows Mobile is composed of a variety of components, each providing specific functionalities and capabilities.

Windows Mobile 6, for example, was released on February 12, 2007. It is powered by Windows CE 5.0 (version 5.2) and comes in three different versions: Windows Mobile 6 Standard, Windows Mobile 6 Professional, and Windows Mobile Classic. Devices without an integrated phone are called Windows Mobile Classic devices instead of PPCs. Devices with an integrated phone and a touch screen are called Windows Mobile Professional devices, and devices without a touch screen are called Windows Mobile Standard devices (Hall 2007).

Table 9.2 compares four mobile OSs and lists the potential applications for Blackberry, iOS, Android, and Windows Mobile. Some of the differences among these OSs might be easily recognized, yet greater details of these OSs should be explored if we intend to use them for field computing and mapping applications. Their strengths and weakness should be compared in areas such as e-mail, GPS, multimedia, Microsoft Office compatibility, Internet connectivity, ease of use, and the quality of third-party applications.

TABLE 9.2
Mobile Operating Systems and Application Programs

Service	Android OS	Apple iOS	Blackberry OS	Windows Mobile OS
Web browser	Android (WebKit)	Safari	Blackberry	Internet Explorer
Flash playback	Yes	No	No	No
Outlook web access (web interface)	Yes	Yes	No	Yes
Adobe Acrobat	Yes	Yes	Yes	Yes
Excel	Yes	Yes	Yes	Yes
PowerPoint	Yes	Yes	Yes	Yes
Word	Yes	Yes	Yes	Yes

9.3.2 Data Communications between HPCs and PCs

We can connect a handheld device to a desktop PC using a serial connection, USB connection, infrared port, and wireless or network connection. However, a program called Windows Mobile Device Center (formerly called Microsoft ActiveSync) needs to be installed on the desktop PC. The Windows Mobile Device Center offers device management and data synchronization between a Windows Mobile–based device and a desktop PC. It provides a great synchronization experience with Windows-powered PCs and HPCs. The synchronization process offers advanced capabilities like autodetection of the serial port on which your PC companion is installed. You can configure the MS Windows Mobile Device Center including sync settings and rules.

One of the most convenient features of MS Windows for HPCs is the so-called "instant on." With instant on, there is no waiting for HPCs to start up or shut down; you can start working immediately by pressing the *Power* key. When you are finished, turn off your HPC by pressing the *Power* key.

9.4 Data Storage and Program Execution

One of the first things new mobile device users notice is that they can store information on their device right away. This information is stored in internal RAM. We can add a CF card, PC card, or SD card to store data and programs as well. The CF, PC, and SD cards are used as secondary storage in the device. Some of the HPC models might also allow for internal storage to the extra flash ROM.

9.4.1 Random Access Memory

For HPCs, RAM is used differently than on a desktop PC. The RAM on a mobile device is usually battery backed up. In a mobile device, the internal RAM is used to store both programs and data. RAM is typically divided into two parts: one for data storage and the other for program execution. It is compressed real time using two different methods to maximize the amount of information you can store. One method is to compress the whole file as it is received. This is good for programs and for data that are in ASCII format. The other method of compression is to split every other byte of data into two separate streams. Then the resulting streams of data are compressed.

9.4.2 External PC, CF, and SD Cards

PC, CF, and SD cards are separate external areas to store both programs and data. We can store much larger files on external storage than on internal RAM. There are two types of storage available: flash and hard disk. Reading and writing to flash storage is fairly fast and requires much lower power than reading and writing to hard disks. We can use external storage to install application programs.

The PC card, introduced in 1990, is a credit-card-sized memory or I/O device that fits into a personal computer (usually a notebook or laptop computer). Probably, the most common use of a PC card is the telecommunications modem for notebook computers.

There are two major differences between CF and SD memory cards. SD cards are significantly smaller than CF cards and are equipped with a nine-pin interface, as compared to the 50-pin interface of the CF. Another major difference between these two types of memory cards is the absence of a microcontroller in SD cards.

9.4.3 Program Execution

A portion of both the internal RAM and the external storage card is used to store programs only. When you choose to run a program stored in either place, the program is copied into RAM (the execution space) to execute (Hewlett-Packard Co. 2000). This means that there are two copies of a program if it is being executed.

In the Windows Mobile OS, virtual memory is used, meaning that memory is allocated on a page-by-page (usually 4k in size) basis for programs that are running. The system reclaims pages that are no longer being used so that they can be used by other programs. When data have been written to a page, that page does not get reclaimed since there is no swap file to write the data temporarily (Hewlett-Packard Co. 2000). This concept of virtual memory is radically different from the desktop. The Windows Mobile OS can also switch background applications to *suspended mode* to free up RAM.

Class Exercises

1. Compare/contrast synchronization vs. file copying.
2. Define HPC and PPC.
3. Where can we store data and programs on HPCs?
4. How can we execute a program on HPCs?
5. What is RAM and how is RAM used on an HPC?
6. What are the major OSs for HPCs and smartphones?

References

Cnet.com. 2014. Handhelds Buying Guide. http://reviews.cnet.com/4520-9580_7-5139854-1.htmltag=more. Accessed on February 3, 2014.

Colbert, D. 2013. Which is the superior mobile OS: iOS, Android, or Windows 8? http://www.techrepublic.com/blog/tablets-in-the-enterprise/which-is-the-superior-mobile-os-ios-android-or-windows-8/. Accessed on February 7, 2014.

Fitch, C. 2006. Flashback: Hewlett-Packard Jornada 720. http://www.hpcfactor.com/reviews/hardware/hp/jornada720/. Accessed on February 5, 2014.

Hall, R. 2007. New Windows Mobile 6 devices. *Smartphone & Pocket PC Magazine* http://mobile.smartphonemag.com/cms/_archives/Jun07/wmsix.aspx. Accessed on February 5, 2014.

Hewlett-Packard Co. 2000. *HP Jornada 720 Series Handheld PC—User's Guide*. Asia Pacific Personal Computer Division, Singapore.

Trimble. 2012. *GeoExplorer 3000 Series User Guide*. Trimble Navigation Limited, Westminster, CO.

10

Handheld PC Applications: An Integrated Computer-Based Cruising System

10.1 Introduction

Computer programs have been developed for many applications in forestry, from the development of species-specific optimal thinning schedules to stand generation and harvesting simulations (Farrar 1981, Brooks and Vodak 1986, Reisinger et al. 1988, Rose and Chen 1995). Computer simulation has proven to be sufficiently comprehensive to handle the various types of problems envisioned in forest operations and has been used for linking the variable components into production and cost analysis (Goulet et al. 1979, Stuart 1981). Simulation also provides an accepted method of evaluating a wide range of system configurations, operating environments, and forest utilizations. The use of computer programs in the forest industry can reduce costs, save time, and aid extensively in the practice of processing forest inventory data (Rennie 1991). Timber cruising and forest inventory are two important aspects of forest practices, and using a computer program to determine timber cruise and forest inventory design and plot layout can minimize time in the field and result in considerable savings in time and money (Wiant and Gambill 1985). A computer-based timber cruising and forest inventory system could help landowners and resource managers protect forest resources and assist in the management of timberlands (Blinn and Vandenberg-Daves 1993).

Several timber cruising programs have been developed, from simple calculator-assisted procedures for marking stands and rapid sawtimber/pulpwood estimates to relatively complicated handheld- or PC-based programs for yield curve design and cruising data collection and analysis (Moser and Raney 1990, Wiant 1990, MacLean et al. 1998). Wiant and Gambill (1985) developed a program to minimize timber cruise field time by determining the optimal basal area factor or plot size. This package was written with BASIC and run under a DOS environment.

Rennie (1991) showed new uses of Statistical Analysis System (SAS) for processing timber inventory data. The SAS routines would allow the forester to write a program that would enable the efficient design of a forest inventory to meet specified objectives.

The Private Lands Information System (PLIS) was developed to provide field foresters access to a map-based inventory of nonindustrial private timber lands in Minnesota (Blinn and Vandenberg-Daves 1993). The PLIS included three basic components: hardware, software, and a database. The database held not only data for individual timber stands but also spatial data for stand or property boundaries. A timber cruise and forest inventory program named CRUISE was developed by Dr. Harry Wiant to process cruise data into meaningful cruise statistics in the field (Wiant 1990). CRUISE was programmed with HotPaw™ Basic and is used on the Palm OS handheld. The CRUISE99 program was accordingly developed for the PC environment.

Reports need to be generated once cruising/inventory data are processed. Using a relational database management system such as MS Access, Oracle, or dBASE could fulfill the need of generating forest inventory reports (Belli et al. 1987, Wang et al. 2004). Belli et al. (1987) stored tabular data such as site descriptions, plot-level data, and tree-level data in a database. Reports generated by the database management system included stand/stock tables, land use summaries, and recalculation of individual tree volume estimates from specified input files (Belli et al. 1987).

Information on timber cruising programs is also available on the Internet. The USDA Forest Service provides four Windows-based generic programs—Check Cruise, Cruise Design, Timber Theft/Local Volume Table, and Traverse (http://www.fs.fed.us/fmsc/measure/cruising/). The Check Cruise program is used to compare measurements and volumes between an original cruise and a check cruise while the Cruise Design program was developed to help cruisers design timber cruises and meet predetermined sampling errors. The Timber Theft/Local Volume program is used not only to determine the volume of a tree but also to estimate the removed volume in a theft case. The Traverse program uses distances and compass bearings to construct maps and determine the acreage within the traversed area. Canal Forest Resources developed an advanced timberland inventory and investment management system. It offered silvicultural land management and investment information to timberland investors through secure, web-based technology. A forest inventory software program named TwoDog was initially developed and provided by Foresters Incorporated and is now operated and supported by Fountains America Inc. (http://www.fountainsamerica.com/twodog/). It is a comprehensive forest inventory application for field and desktop computers. Sampling options are also provided in the package.

In this chapter, we will discuss an integrated timber cruising system specifically for Appalachian hardwood forests (Wang et al. 2004), including object-oriented system design and entity-relationship (ER) data modeling techniques for handheld PC (HPCs).

10.2 System Structure

This integrated timber cruising system consists of three major components: a handheld data collection system, data transfer, and data analysis components (Figure 10.1). The handheld system is used to collect cruising data, including forest, plot, and tree measurements. Understory vegetation and wildlife data collection (referred to as transect data) are also implemented with the handheld system. Data transfer is an interface for communicating between the HPC and desktop PC and is used to synchronize data on the desktop PC. Data analysis is a component that is used to analyze and summarize the cruise data and generate desired reports. These components can be run as an integrated system, or on a stand-alone basis.

The handheld system was written with Microsoft VB CE, which is run under Microsoft Windows CE or Mobile environment. It contains two main modules: collect and edit (Figure 10.2). The collect module is designed to allow a user to collect plot, tree, and transect data associated with each cruise point, while the edit module is designed to permit field editing of tree data. Retrieval and storage of cruising data is accomplished with an MS Access database.

Both the data transfer and data analysis components were programmed with MS VB V6.0, which resides on the desktop PC under the MS Windows environment (Figure 10.2). The data transfer provides two major functionalities: it allows for the transfer of cruising data from HPC to desktop PC and accommodates updates of data tables on the desktop PC as well as readying the database for the next cruise transfer. The addition of species, cruise data

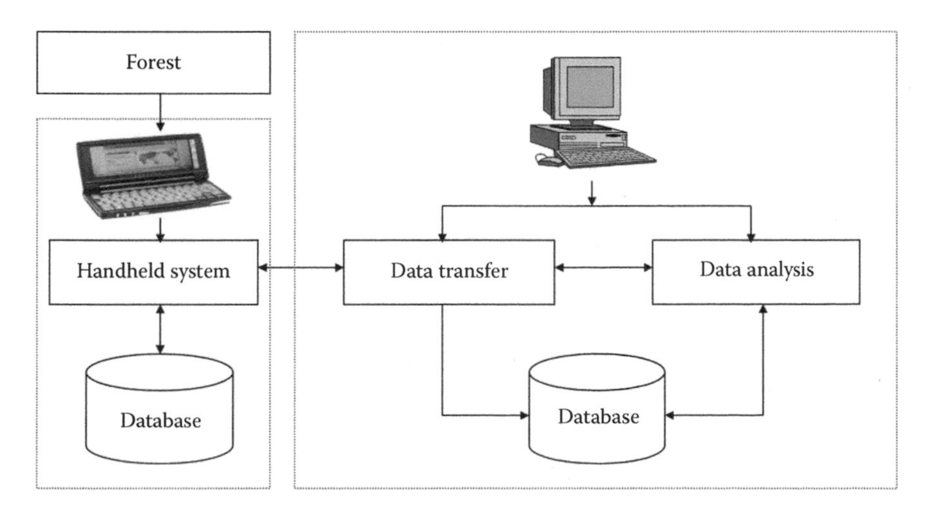

FIGURE 10.1
Structure of the integrated timber cruising system.

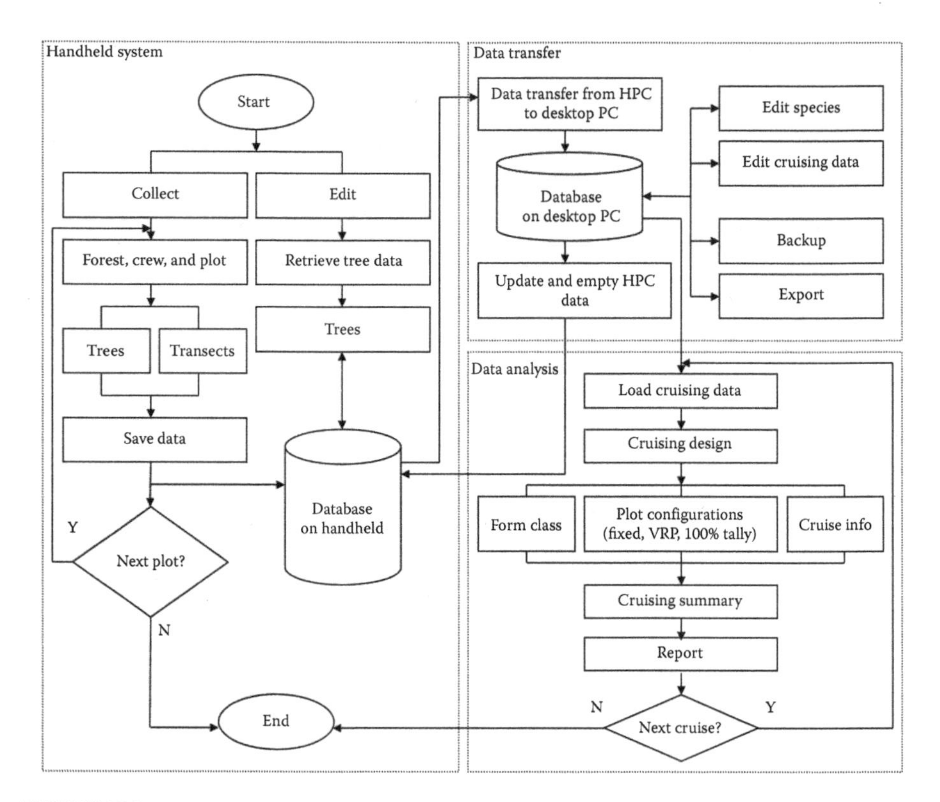

FIGURE 10.2
Flowchart of the integrated timber cruising system.

backups, and exportation of data (into ASCII text and MS Excel) are also provided in the data transfer component. ADO CE application programming interface was employed to conduct data transfer via a dynamic link library (DLL)—adofiltr.dll (Roof 1998). The adofiltr.dll, an MS Pocket Access file converter and synchronizer, is a part of the ADO 2.0 Solution Development Kit for Windows CE. It allows programmatic transfer of database tables between the host desktop computer and the mobile device. This DLL contains two functions, `DesktopToDevice()` and `DeviceToDesktop()`, which are used to transfer data or copy tables. It runs on the desktop PC, not the handheld. The desktop initiates and controls the transfer process. The key requirement for this transfer process is the presence of the same table schemas on both desktop and handheld. The ADO CE data transfer feature has a solid set of tools for transferring data. This feature allows the transfer of complete tables between devices rather than the synchronization of individual records.

The data analysis was organized in a modular way, in which modules were designed as independently as possible (Figure 10.2). A module, or self-contained unit of code, could be a single function or act as several functions. A data access object was used to load the cruising data from the database into this component. This allows tree-form class and cruise information to be edited for analysis summaries and reports. In the cruise summary module of this component, stand/stock table, point/plot, and cruising statistics are provided by business functions built into the component. All end-user reports were developed using an ActiveX component of an MS Access object, which is a reusable piece of programming code and data. It allows the user to call MS Access reports directly from a VB application without the use of complicated coding.

A relational database model was used for holding cruising data on both the handheld system and the data transfer component, which was implemented based on the ER model (Figure 6.5). The relational database model presents the data as a collection of tables. Instead of modeling the relationships in the data according to the way that they are physically stored, the structure of an ER model is defined by establishing relationships between data entity types. As described in Chapter 6, an entity type represents a data table in the ER model and has attributes (fields) that are the descriptive properties of the entity type. *Primary key* refers to one or more fields that make a record unique in a data table. The field (or fields) used to link to a primary key in another data table is known as a *foreign key*, and a foreign key is any field(s) used in a relationship.

There are four data entity types (*Plot, Tree, Transect,* and *Species*) in the model. Each entity type has its own attributes. For example, the *Plot* entity type has *forest, crew, plotNo, plotType,* and other attributes. Entity types are related using relationships such as *contains* and *associates* in the model. For a detailed description of this integrated timber cruising ER model, please refer to Section 6.3 (Figure 6.5).

10.3 System Implementation

10.3.1 Handheld System

The main method of user interface with the handheld system is accomplished through the touch screen using a stylus or fingertip. The touch screen is used in much the same way the user would use a mouse to navigate and select objects on a desktop computer screen. The user can also use a fingertip to tap the touch screen, but the stylus provides the greatest accuracy.

Two modules were implemented in the handheld system to collect cruising data (Figure 10.3). Forest and plot information are entered first, and the user then selects the *Add Plot* button. The tree data frame is then enabled accordingly (Figure 10.3a). There were three types of plots: *intensive, DBH only,* and *DBH and merchantable height*. Text boxes on the tree data frame are enabled or disabled depending on the type of plot being selected. A function was also

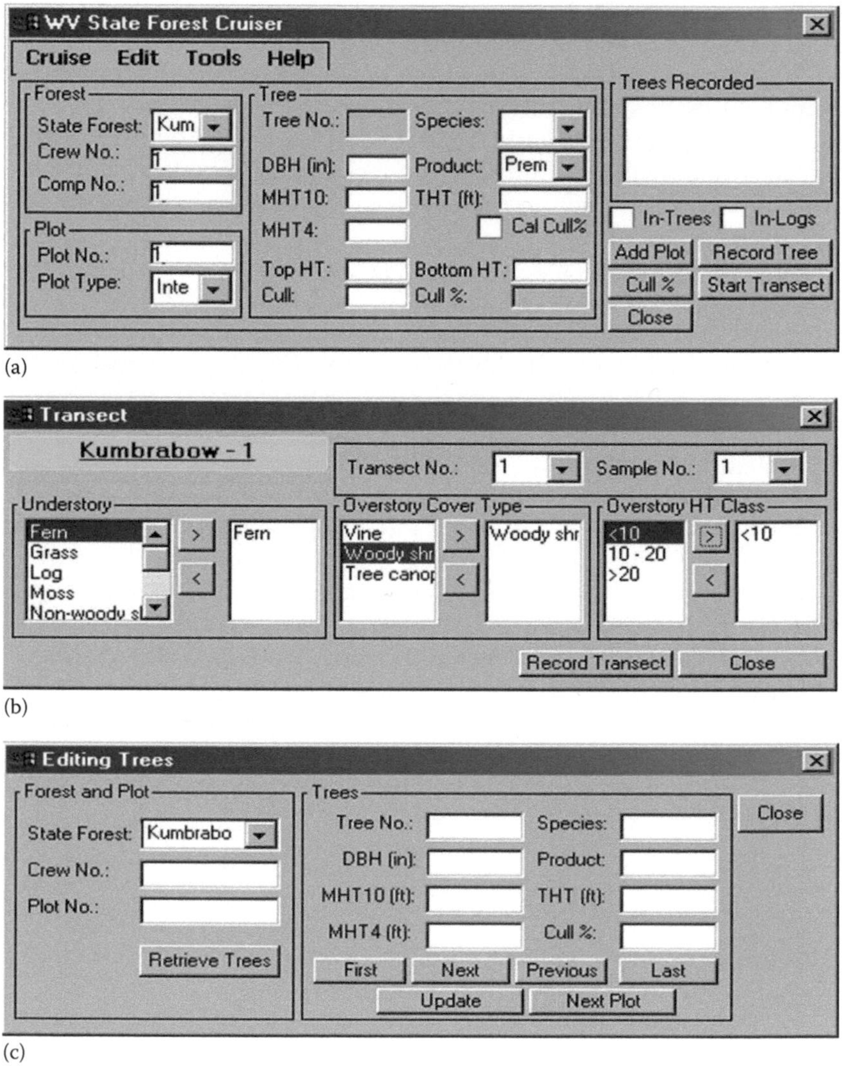

(a)

(b)

(c)

FIGURE 10.3
Main forms of data collection in the handheld system. (a) Data entry for forest plot and tree, (b) transect data entry, and (c) editing tree data.

implemented to calculate the cull percent of a tree. Once trees are entered, a list box is used to display the tree attributes found on a plot. Three elemental times can be collected at each plot: *IP_Time*, *DP_Time,* and *HP_Time*. *IP_Time* is the time interval from when the *Add Plot* button is pressed to when the *Start Transect* (understory vegetation data collection) button is pressed. *DP_Time* is the time interval between when the *Add Plot* button is pressed and the *In-Trees* box is checked, and *HP_Time* is the time between when the *Add Plot* button is pressed and the *In-Logs* box is checked. These time values can be used to examine the cruising efficiency of the field users. A list of convenient list boxes were implemented for adding or removing single or multiple items for transect data collection (Figure 10.3b).

Field transcription mistakes are common; therefore, the handheld system provides functionality that allows the user to edit tree data in the field (Figure 10.3c). Trees sampled on a plot can be retrieved based on *forest*, *crew number*, and/or *plot number*. Once trees are retrieved for a plot, the user can navigate the selected record set and modify the data for that tree and move to the next tree or to another plot.

10.3.2 Data Transfer

The data transfer component was implemented on the desktop PC and provides two basic functions: transferring cruise data from the HPC to the PC and updating and clearing data tables on the HPC (Figure 10.4a). For the sake of data security, this component was designed to run these events separately. The system is designed to first update the species table, and then empty temporary tables on the HPC before the cruise data can be entered in the field. Once users collect the cruising data, the data tables on HPCs are copied and synchronized to a temporary database on the PC. The tables in the temporary database are then appended to the related tables in the main database. Finally, the system again needs to empty the data tables on handhelds in preparation for the next timber cruise.

Another major function of the data transfer component is editing cruise data (Figure 10.4b). A VB tab strip control was used to implement three tabs: *Plot*, *Tree*, and *Transect*. A VB DB gird on each tab is associated with a record set defined in the database. The *Plot* tab is the first displayed. Structured Query Language was used for retrieving trees and transects associated with the plot the user selects. Once the user selects a record from the plot tab, they can select either the *Tree* tab or the *Transect* tab to continue the editing process.

10.3.3 Data Analysis

The data access object was implemented to connect the database, and a VB flex grid control was used to display the data. Once the cruising data are loaded into computer memory, a *Cruise Design* window pops up (Figure 10.5). The user can browse the data loaded and invoke another dialog box to view species and grade codes used in the system. In this *Cruise Design* window, the user can design and save cruising information for a report header. The user simply enters data into the required text boxes and then selects the *Save Cruise Info* button. To modify the form class for a specific species, the user simply clicks the *Girard Form Class* button. A form is displayed using a DB grid control associated with a form class table in the database. This allows the user to modify the form class data.

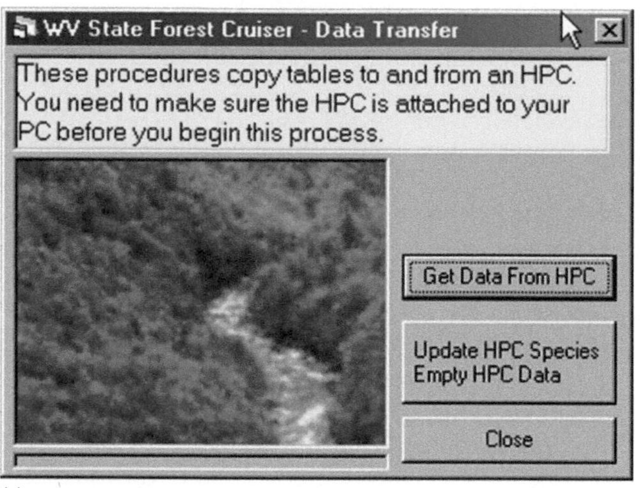

(a)

Edit Cruising Data

Plot | Tree | Transect

Date	Time	PlotNo	IPTime	DPTime	HPTime	PlotType
		1	3	0	0	Intensive
5/15/2002		2	144	0	0	Intensive
5/15/2002		1	88	58	0	Intensive
5/15/2002		2	144	0	0	Intensive
5/15/2002		1	88	58	0	Intensive
5/15/2002		2	144	0	0	Intensive
5/15/2002		1	88	58	0	Intensive
5/15/2002		1	53	0	0	Intensive
5/15/2002		1	53	0	0	Intensive
5/15/2002	5:20:44 PM	1	53	0	0	Intensive

Delete Close

(b)

FIGURE 10.4
Major functions of data transfer component. (a) Transfer of cruise data and (b) editing of cruise data.

Cruise type selections can also be made during the cruise design process. The user can choose *Fixed Area*, *Variable Radius Plot* (*VRP*), or *100% Tally* by checking VB radio buttons. If a plot type of *Fixed Area* is chosen, *Plot Size* must be entered. *Basal Area Factor* should be given if *VRP* is checked. Four types of summaries are provided depending on the user's requirements. *Volume* summaries can be done on both per acre and per tract basis. *Stand/Stock Tables* are further summarized by *DBH Class*, *Species & DBH*, or by

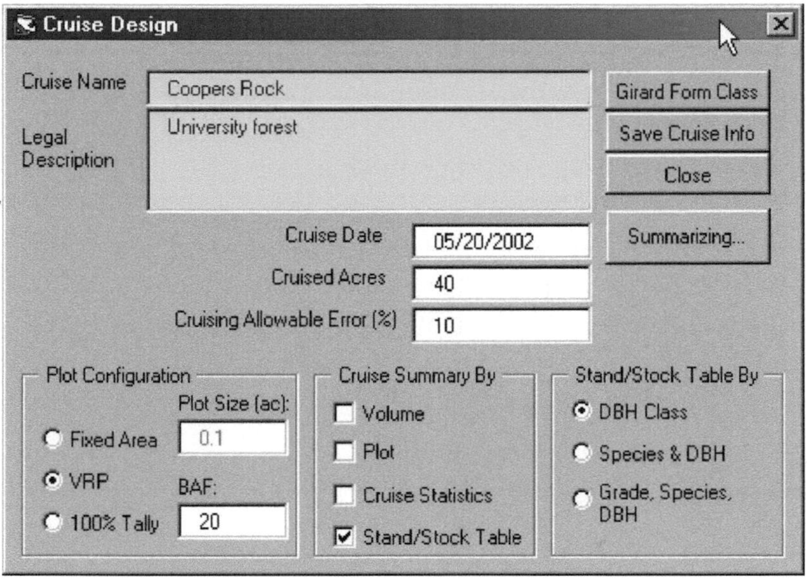

FIGURE 10.5
A major form in the data analysis component.

Grade, Species, DBH. Cruise statistics are provided for volume per acre in cubic feet, International board feet (1/4) (IBFV), Doyle (DBFV), trees per acre, and basal area per acre. Statistics include the mean, standard deviation, standard error, variance, coefficient of variance, confidence interval at 95% level, percent of error, and sample size.

Once the cruising data are summarized, the results are stored in the database for report generation. A total of seven reports can be generated in the data analysis component of this system.

Class Exercises

Using a Handheld PC to Collect Timber Cruising Data

In this lab, we are going to use a Windows Mobile handheld computer to collect timber cruising data. You may not be required to conduct a real timber cruise in this lab, depending on your class instruction. Simulated cruise data will be provided for entry into handheld computers.

1. Using the 240TimbInv or a similar program on a handheld computer (such as Nautiz X7), enter plot and tree data.

2. Import raw data from the handheld to the desktop computer using the Windows Mobile Device Center.
3. Download and Save the Microsoft Access database file *Handheld.accdb*. This file has the tables set up for the importation of the raw data.
4. Import raw data into Microsoft Access.
5. Create queries to summarize:
   i. Basal area and volume by plot AND by diameter class (DBH).
   ii. Basal area and volume by species AND by diameter class (DBH).
6. Create reports based on the two queries created. Sort reports by plot or species and by DBH.

References

Belli, K.L., A.R. Ek, M.H. Hansen, and J.T. Hahn. 1987. Statewide forestry databases for microcomputers. *Northern Journal of Applied Forestry* 4: 117–118, 165.

Blinn, C.R. and J. Vandenberg-Daves. 1993. Evaluation of a computerized timber inventory system for nonindustrial private landowners. *Northern Journal of Applied Forestry* 10(3): 123–127.

Brooks, D.G. and M.C. Vodak. 1986. YPOP: A microcomputer program for evaluating thinning alternatives in natural stands of yellow-poplar. *Northern Journal of Applied Forestry* 3: 3–5.

Farrar, K.D. 1981. In situ stand generator for use in harvesting machine simulations. MSc thesis, Virginia Polytechnic and State University, Blacksburg, VA, 211pp.

Goulet, D.V., R.H. Iff, and D.L. Sirois. 1979. Tree-to-mill forest harvesting simulation models: Where are we? *Forest Products Journal* 29(10): 50–55.

MacLean, D.A., K.B. Porter, and J. Kerr. 1998. Forester's yield curve designer software. *Northern Journal of Applied Forestry* 15(1): 23–27.

Moser, J.W. and J.D. Raney. 1990. A programmable calculator-assisted procedure for marking unevenaged stands. *Northern Journal of Applied Forestry* 7: 140–142.

Reisinger, T., W.D. Greene, and J.F. McNeel. 1988. Microcomputer-based software for analyzing harvesting systems. *Southern Journal of Applied Forestry* 12: 37–41.

Rennie, J.C. 1991. Forest inventory processing with statistical software. *Northern Journal of Applied Forestry* 8: 41–44.

Roof, L. 1998. *Professional Visual Basic Windows CE Programming*. Wrox Press Ltd., Birmingham, U.K., 447pp.

Rose, D.W. and C.M. Chen. 1995. An interactive thinning simulation for red pine stands. *Northern Journal of Applied Forestry* 12(1): 43–48.

Stuart, W.B. 1981. Harvesting analysis technique: A computer simulation system for timber harvesting. *Forest Products Journal* 31(11): 45–53.

Wang, J., S. Grushecky, and J. Brooks. 2004. An integrated computer-based cruising system for central Appalachian hardwoods. *Computers and Electronics in Agriculture* 45(2004): 133–138.

Wiant, H.V. 1990. An inexpensive computer system for rapid sawtimber estimates. *Northern Journal of Applied Forestry* 7: 142–145.

Wiant, H.V., Jr. and C.W. Gambill. 1985. Minimize field time when cruising Appalachian hardwoods. *Northern Journal of Applied Forestry* 2: 70.

11

Introduction to Geospatial Technology Applications in Forest Management

Geospatial technology refers to technology used for visualization, measurement, and analysis of features or phenomena that occur on earth. Geospatial technology includes three different technologies: global positioning systems (GPS), geographical information systems (GIS), and remote sensing (RS). In this chapter, we will introduce some basic applications of GPS, GIS, and RS in forest resource management.

11.1 GPS Applications

11.1.1 What Is GPS?

GPS is a location system based on a constellation of about 24 satellites orbiting the earth at altitudes of approximately 17,702.78 km (11,000 miles) (Corvallis Microtechnology Inc. 1996). Each of these satellites makes two circuits around the earth every 24 h. GPS was conceived in 1960 and developed in the 1970s by the U.S. Department of Defense for its tremendous application as a military locating utility. The first satellites were launched into space in 1978. The system was declared fully operational in April 1995. Over the past several years, GPS has proven to be a useful tool in nonmilitary mapping applications such as in forest and natural resource management. Uncorrected positions determined from GPS satellite signals produce accuracies in the range of 50–100 m. When using a technique called differential correction, users can get positions accurate to within 5 m (Corvallis Microtechnology Inc. 1996). GPS needs four satellites to provide a three-dimensional (3D) position.

As GPS devices are becoming smaller and less expensive, there are an expanding number of applications for GPS, such as GPS-based piloting, driving navigations, precision agriculture and forestry, and navigation tools for foresters, hikers, and hunters. There are three classes of GPS receivers:

1. Geodetic
   a. Capable of subcentimeter accuracy
   b. Bulky, expensive

   c. High-precision mapping applications such as surveying, geodetics
2. Mapping
   a. Capable of less than 3 m accuracy
   b. Portable, less expensive
   c. Accurate mapping for integration with GIS
3. Navigation
   a. Capable of less than 3 m accuracy
   b. Light-weight, cheap
   c. Basic navigation, limited data storage

Global navigation satellite system (GNSS) is a satellite system that is used to pinpoint the geographic location of a user's receiver anywhere in the world. There are four GNSSs that are currently in either full or partial operation: the U.S. GPS, the Russian Federation's global orbiting navigation satellite system, the European Union's Galileo, and China's Beidou. Each of the GNSSs employs a constellation of orbiting satellites working in conjunction with a network of ground stations in many applications.

11.1.2 GPS in Forestry and Natural Resource Management

GPS can be used to gather spatial data in forestry and natural resources. This data is collected in the form of features (points, lines, or polygons). When using GPS, it is best to know what types of features you would like to collect. Point features can be used to obtain spatial locations of trees or water bars in forest operations. Line features can be used to determine forest road layout or to collect stream data for best management practices (BMPs). A polygon can be used to determine a landing site or a tract boundary. To increase accuracy, it is best to collect as many points for a feature as possible. A minimum of 10 points is recommended for accurate positions of spatial features. You can set the GPS unit to collect points at different intervals such as 1, 5, or 10 seconds. Furthermore, in order to improve accuracy, you will need to be using a minimum of 4 satellites. An increased number of satellites will speed up the collection process and make your features more accurate.

11.1.2.1 Data Collection

Depending on the handheld GPS device you are using, you may begin collecting data just by powering on the GPS field software installed on the device. The GPS field software product could be Trimble TerraSync, ESRI ArcPad, or another third-party product. You will need to select the type of information you want to collect, which can simply be *General*. Using this application, you can collect and name features. You will want to create a new file and give it a proper name (such as the tract or lab you are working with).

Now you are ready to collect new features. Select the type of feature you want to collect and name the individual feature. While naming the feature you will already be collecting data. After collecting your desired number of points, store the feature. To collect a new feature, simply select *Collect Data* and begin with a new feature.

11.1.2.2 Data Transfer

If using a Tremble GeoXT you will need to transfer your data from the device to a computer using the GPS Pathfinder Office. Other types of GPS units may not require this step due to different software.

To begin transferring data using the GPS Pathfinder Office, connect the device to your computer via a cable or using a wireless connection. Find the toolbar on the far left of the computer's screen, where you will see three tools that you will use to transfer, collect, and export data. You will first need to transfer the data from the GPS device to the computer. Click the transfer button and select your file in order to transfer it to a premade folder on the desktop. Select the data file you wish to transfer and proceed to transfer the data.

Real-time differential correction for GPS (real-time DGPS) has had a very positive effect on navigation and the verification of spatial data (Trimble Navigation Limited 2004). The US-based Wide Area Augmentation System (WAAS) is one example of real-time DGPS. It is also called satellite-based augmentation system (SBAS). To receive WAAS corrections, a GPS receiver needs to be SBAS-capable. However, in some places, there are no DGPS services, and some applications need more accurate data than from DGPS. GPS data are typically corrected after transferring to achieve better accuracy. This process is called postprocessing. To postprocess data, you select a base station or tower near you and download the necessary files to correct your data. You will then need to correct these data and save them in the same folder as a corrected file. Finally, you will need to export the data into the folder and save it as a shapefile. After this step, you are ready to import your shapefiles into ArcMap.

If you use a Garmin Etrex Legend GPS unit, you need to follow the steps below to transfer your data from a GPS unit to a computer:

a. Connect the GPS data transfer cable to a serial port of the personal computer (PC). Connect the other end to the data connector on the back of the GPS unit.
b. Open ArcGIS on your desktop and turn on your GPS unit. Go to the menu and click *Tools* → *Extensions*, check *MxGPS 9x* in the pop-up dialog. Click *View* → *Toolbars*, check *MxGPS 9x*. Now the tool is ready to use in ArcMap.

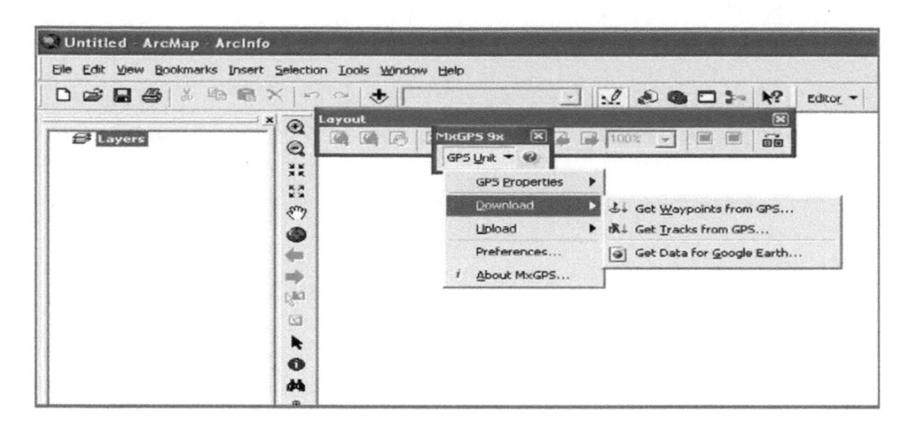

FIGURE 11.1
MxGPS9x in ArcMap.

c. Click *GPS Unit* (Figure 11.1) → *Download*. Select *Get Waypoints from GPS*. Click *OK*.
d. To get all the points from the GPS unit, click the top-left button (Figure 11.2a). Select all the points and then click *Save* (Figure 11.2b). You can select to save them in a designated directory. The file will also be added into ArcMap automatically.
e. To download tracks from a GPS unit, follow steps (c) and (d).

11.2 GIS Applications

11.2.1 What Is GIS?

Geographic information system (GIS) is a powerful tool for spatial data analysis. Although many different definitions of GIS exist (Clarke 2001), the common concept is that GIS is a computer system that includes software and hardware designed to provide users with the capability to visualize and analyze spatial information.

GIS has emerged as an essential tool for urban and natural resource planning and management. The capacity to store, retrieve, analyze, model, and map large areas with huge volumes of spatial data has led to an extraordinary proliferation of applications. Geographic information systems are now used for land use planning, utilities management, ecosystems modeling, landscape assessment and planning, transportation and infrastructure planning, market analysis, visual impact analysis, facilities management, and real estate analysis, as well as many other applications in forestry and natural resources.

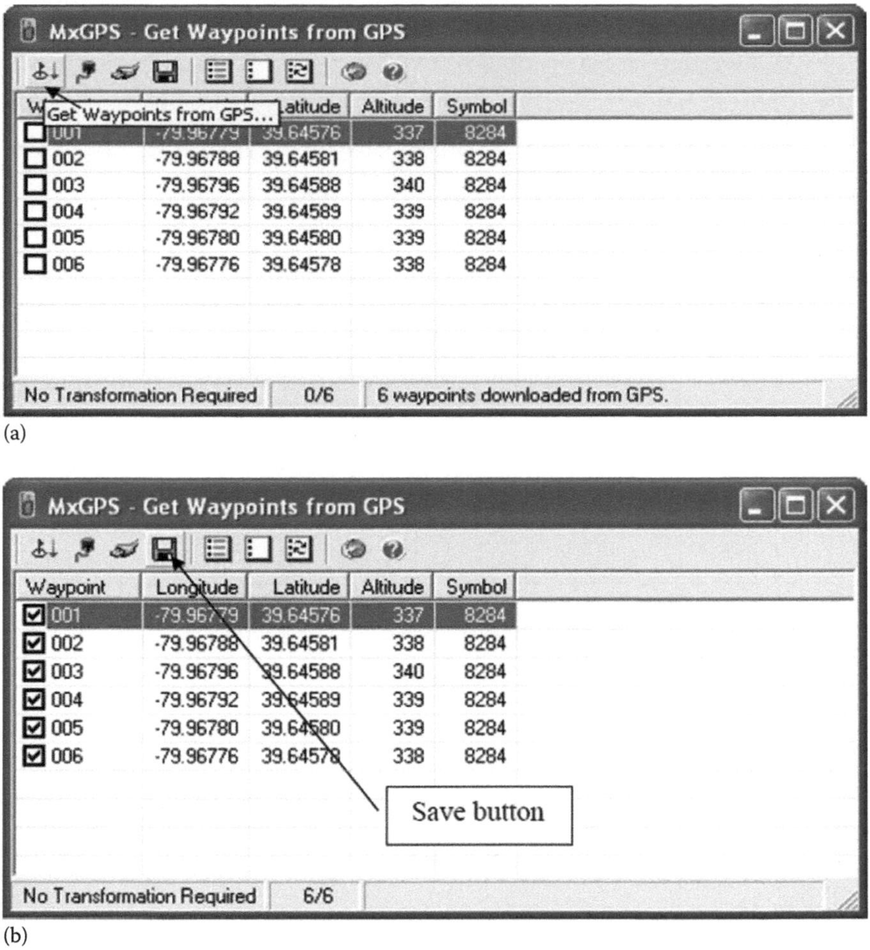

MxGPS - Get Waypoints from GPS

Get Waypoints from GPS...

[Waypoint]	[Longitude]	Latitude	Altitude	Symbol
001	-79.96779	39.64576	337	8284
002	-79.96788	39.64581	338	8284
003	-79.96796	39.64588	340	8284
004	-79.96792	39.64589	339	8284
005	-79.96780	39.64580	339	8284
006	-79.96776	39.64578	338	8284

No Transformation Required | 0/6 | 6 waypoints downloaded from GPS.

(a)

MxGPS - Get Waypoints from GPS

Waypoint	Longitude	Latitude	Altitude	Symbol
001	-79.96779	39.64576	337	8284
002	-79.96788	39.64581	338	8284
003	-79.96796	39.64588	340	8284
004	-79.96792	39.64589	339	8284
005	-79.96780	39.64580	339	8284
006	-79.96776	39.64578	338	8284

Save button

No Transformation Required | 6/6

(b)

FIGURE 11.2
Get points from a GPS unit. (a) Download GPS data and (b) save GPS data to computer.

A GIS map is made up of layers (Figure 11.3), or collections of geographic objects that are alike (ESRI 2001). To make a map, you can add as many layers as you want. Layers may contain **features** or **surfaces**. Geographic objects (features) have a variety of shapes. All of them can be represented as one of three geometric forms: point, line, or polygon. Polygons usually represent large objects, such as a harvested forest site, a tract of forest, tracts of other lands, and countries. Lines represent things too narrow to be polygons, such as forest roads, pipelines of natural gas, or recreation trails. Points are used to represent locations, such as trees, landings, sawmills, cities, schools, or other plant locations. Polygons, lines, and points collectively are called **vector** data.

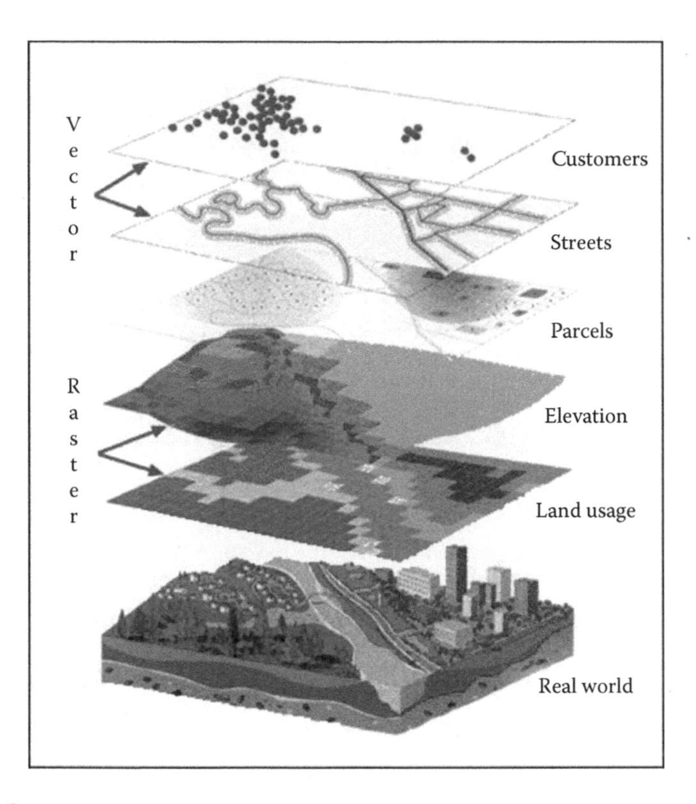

FIGURE 11.3
The concept of GIS map layers. (From ESRI, *Getting to Know ArcGIS Desktop*, ESRI Press, Redlands, CA, 2001.)

Surfaces have numeric values rather than shapes. For example, we can measure values for elevation, slope, rainfall, temperature, and wind speed for any particular location on the earth's surface and generate a raster surface using these values. A **raster** is a matrix of identically sized square cells. For example, a 30 m elevation map for the state of West Virginia is made up of 30×30 m^2 cells.

11.2.2 GIS in Forestry and Natural Resource Management

The management of forests and natural resources has become progressively more complex since there are multiple objectives to achieve, as well as multiple criteria and constraints to consider. This makes GIS an important tool in decision-making during planning, policy-making, and management (Upadhyay 2009). GIS can be established to provide crucial information about resources and can simplify planning and management of resources (e.g., recording and updating resource inventories, harvest estimation, scheduling and planning, ecosystem management, and landscape and habitat planning [ESRI 2003]).

As technology has advanced, GIS has become increasingly popular in natural resource management. GIS applications can be grouped into the following four broad functional categories (with examples and descriptions related to forestry and natural resources):

- Location—Locate forest resources, including the property location, name, boundary, and other geographic references (Figure 11.4).
- Analysis—For example, what has changed in the forest (landscape, land use, or vegetation) in the last 20 years? Overlaying several map layers of spatial data is one common technique used in GIS. These layers can be used to analyze different spatial features available in GIS.
- Modeling and simulation—GIS is very powerful in spatial modeling due primarily to its ability to deal with large quantities of spatial data efficiently and effectively. Examples include impacts of forest BMPs on harvested sites and phenology and vegetation changes due to climate change.
- Visualization—Visualization is one powerful framework in GIS that allows the user to see spatial data in the format of a simple map or of a map with complex multidimensions. These simulations coupled with animation techniques can be used to present spatio-temporal data effectively. GIS can be used to generate 3D visual models of

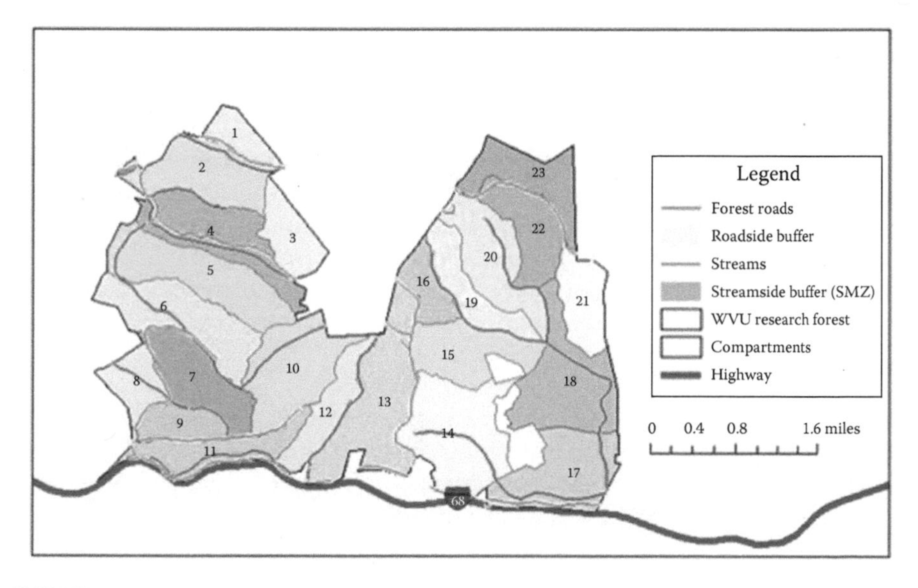

FIGURE 11.4
Overlaying compartments and other spatial attributes for West Virginia University (WVU) Research Forest. Map was prepared by Dr. Benktesh Sharma using several data sources available from the WVU Division of Forestry and Natural Resources' archive.

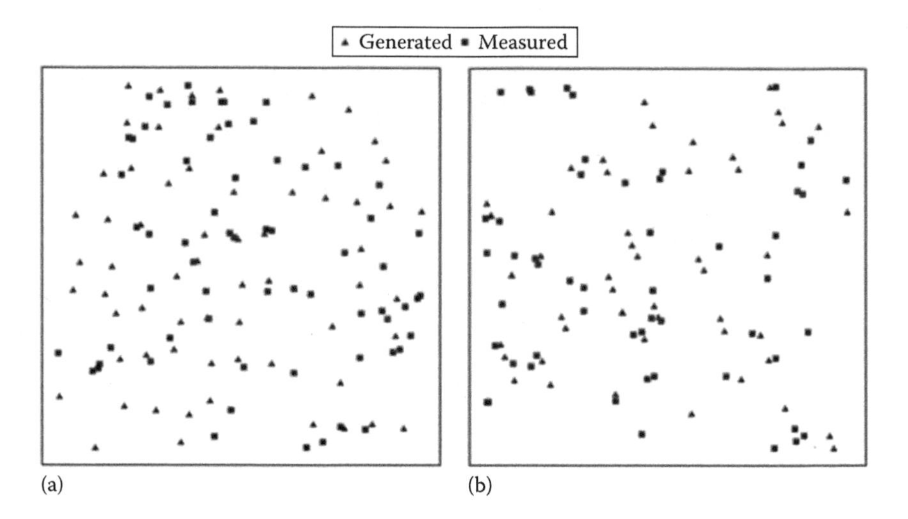

FIGURE 11.5
Spatial pattern of generated and measured stands. (a) Generated and measured stand with random spatial pattern for a plot with 400 trees per ha; (b) Generated and measured stand with clustered spatial pattern for a plot with 330 trees per ha.

forests and different forest management activities. For example, the effect of different cutting patterns can be visually assessed in three dimensions. Similarly, 3D visual models can be used to study spatial patterns of trees in a forest (Wang et al. 2009b) (Figure 11.5). In general, ArcScene in Esri's ArcGIS can be used for making a 3D visualization using digital elevation models (Figure 11.6).

GIS applications in forest and natural resource management can range from simple analysis (such as overlaying different layers, measurement of area and volume of forest resources, zonation and calculations of statistics) to complex modeling and simulation. Overlaying different spatial attributes such as roads, streams, and forest boundaries is one common task implemented in GIS. Examples of modeling and simulation-related functions can include spatial patterns of trees or landscapes (Wang et al. 2009b), spatial regression analyses of forest resources (Overmars et al. 2003), geo-statistical approaches to predictions (Grushecky and Fajvan 1999), and simulations of resources over different spatial and temporal horizons.

GIS has been widely applied in the major areas of forest and natural resource management. Some examples include:

Forest resource assessment and monitoring: Assessing deforestation, degradation, and land use/cover change; examining forest types, age classes, and succession stages; evaluating forest resources at various spatial scales.

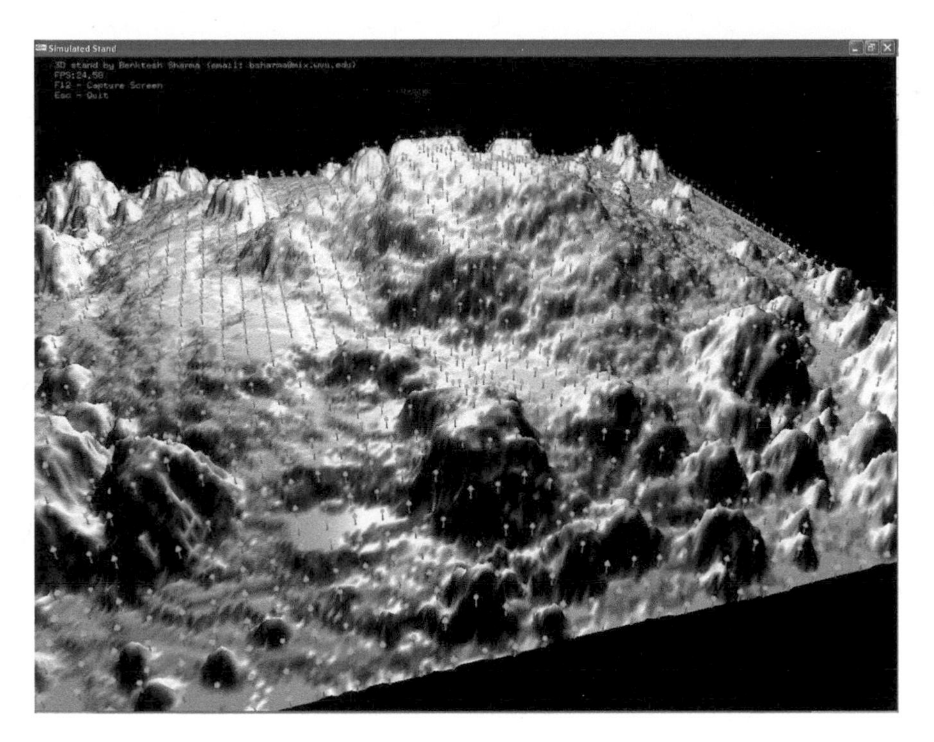

FIGURE 11.6
A forested landscape simulation model using digital elevation models to generate terrain and trees of central Appalachian hardwood forest.

Forest protection: Forest protection, whether against pests, fire, disease, or human actions, needs spatial data to improve the process of forest planning and implementation (VDF 2008). For example, with fire-related information gathered (such as location, direction, rate of spread and intensity of forest fire, vegetation types, topography, and historical records), fire occurrence can be predicted using GIS.

Forest and biomass harvest planning and scheduling: Spatial forest modeling via GIS can substantially enhance the planning of forests and biomass harvesting strategies. Spatial modeling could help forest managers and landowners understand the economic, environmental, and social impacts of the proposed harvest. Sharma (2010) developed a computer-based forest planning/scheduling system to study forest harvest strategies. The system could generate and visualize a spatio-temporal forest plan for different management objectives. The system was used to optimize different harvest schedules with different objectives ranging from maximization of timber production, to maximization of timber production and stand carbon stock, to maximization of carbon stock only under clear-cut and selection-cut methods applicable for both long and short rotations.

Forest best management practices: Forest BMPs are guidelines for controlling sediment and protecting water quality during forest operations. Spatial data such as soil, stream type, and population density are collected for the sites to identify how these spatial attributes affect BMP application, effectiveness, and compliance. Using spatial analysis, Wang et al. (2009a) reported that higher levels of application, effectiveness, and compliance were found on sites with either intermittent or ephemeral streams, wider streamside management zones (SMZs), or low soil moisture index, and no significant differences were presented in BMP application, effectiveness, and compliance among stream type, SMZ width, soil series, moisture index, and population category. Road and landing layouts located outside of high water cumulative flow areas also contributed to higher rates of BMP application, effectiveness, and compliance, which substantiated the importance of preharvest planning.

Forest conservation and biodiversity: GIS can help in the preparation of conservation policies and plans, particularly in support of legislation via the functions of identification, selection, design, and management of protected areas and nature reserves (SIC 2009).

Climate change: Application of GIS in climate change is still in its early stages. However, the strength of GIS in spatial and modeling analysis and in organizing digital spatial data sets of different themes indicates that it will play a big role in future studies, especially in phenology, CO_2 flux, and carbon- and water-cycling-related research.

Wildlife habitat conservation and planning: GIS is often used in habitat and vegetation mapping, monitoring, assessment, and analysis of the progression of conservation activities, ecological patterns, and encroachment upon protected wildlife conservation areas by comparing images from different time periods.

Recreation and park management: GIS allows us to map recreation trails and assess the flow of visitors to a park. With GPS data, we can accurately map and analyze locations, times, and returns of visitors to some specific spots in a park.

Soil and watersheds: Watershed spatial databases from local to national scales are being maintained in many countries to serve the interests of multiple stakeholders in watershed management (Musinguzi et al. 2008). Hamons (2007) investigated the amount of sediment delivered to stream channels and the determination of the topographical attributes responsible for the origination and transfer of sediment using a spatial analysis method. The sample data were collected in a central Appalachian mixed hardwood forest from 2002 through 2005 and were analyzed spatially and statistically to determine the magnitude of effect topographical attributes, road construction, and harvesting operations had on sediment delivery to the stream channel.

11.2.3 GIS Software and Data

The many GIS packages currently on the market range from free software products to expensive commercial software. ArcGIS, Smallworld, Manifold System, Mapinfo, Autodesk, ILWIS, Erdas Imagine, GRASS, and MapServer are some common software products that dominate the GIS system market.

GIS relies on spatial data for its application. In several ways, GIS provides the basic functionality of a spatial data management system in which queries can be made using structured query language. Therefore, an added advantage of GIS is spatial information management.

Spatial data have some reference to geographic locations. Trees, for example, are linked to a geographic domain (a forest in a county, a path in some forest, etc.) yet several nonspatial analyses can be made on trees (such as tree diameter and height). If the question we are considering does not require the analysis of spatial attributes of our data, then GIS is not necessary. Spatial information can be in the format of coordinates in latitude and longitude or Universal Transverse Mercator coordinates. These coordinates are linked with the spatial attributes. For example, if we prepare a map of counties in a state, we usually have location information of these counties in terms of X and Y coordinates in GIS. If we want to store elevation, then a third coordinate "Z" is used.

GIS can be grouped into vector or raster categories based on the data type and software used to represent spatial features. In vector GIS, spatial features are represented by points (for point-based features such as locations of trees), lines (for linear features such as rivers and forest roads), and polygons (for area-based features such as forest stands, buffer zones, and forest tracts). In raster GIS, spatial features are represented as pixels (the smallest picture elements). In most systems, both data types are available and they are interconvertible. Suitability of one type over another really depends on the analysis in question and on the scale of use (Figure 11.7).

Forest resource inventory and analysis was one of the earliest uses of GIS. GIS is used to inventory forest types at the landscape level, delineate different land covers, and delineate different management zones in a forest management plan. GIS can also be used to plan for several harvest-related activities. For example, it can be used to estimate the cost and plan the location and design of forest access roads, skid trails, landing locations, SMZs, and stream crossings. In such applications, it is customary to use the following analyses:

- Terrain and slope stability analyses
- Cut and fill estimates

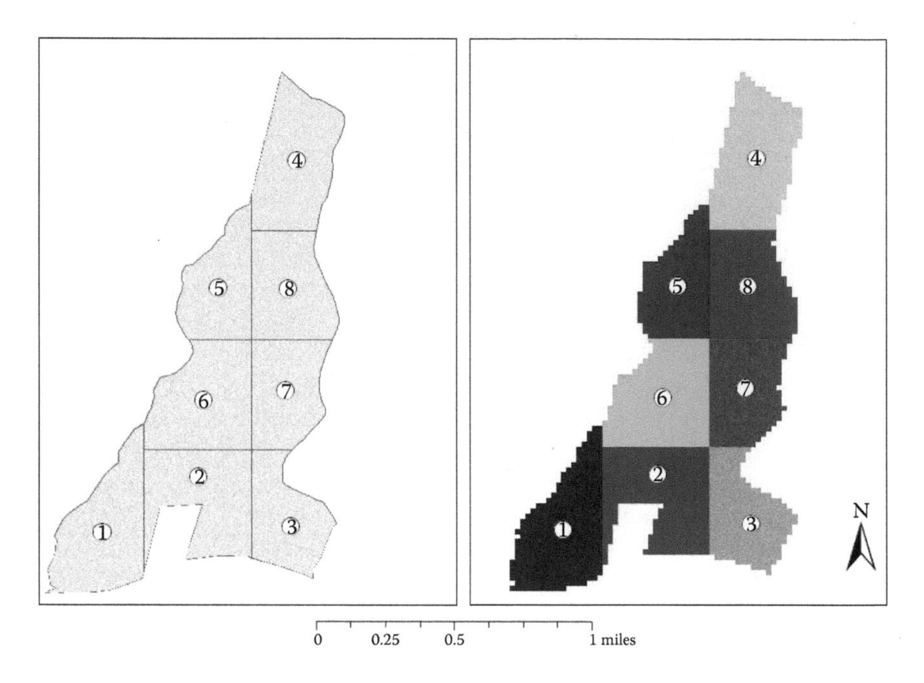

FIGURE 11.7
Vector representation of forest stand (as polygon) on left and raster representation of the same stand on right. This map shows how stand boundary can be delineated. These stands can also be displayed with attributes such as growing stocks, density, or other related forest stand properties. (Maps courtesy of Dr. Benktesh Sharma.)

- Site suitability analyses
- Alignment and grade calculations
- Rights-of-way corridor studies
- Environmental impact assessments
- Integration of survey data
- Cost and material flow analyses

Zoning and buffering can be used in forest type classification (vegetation stratification, slope/elevation zoning, growth zoning), SMZs, road and trail buffers, recreational areas, protected forests, historical sites/area designation, water conservation areas, wetland areas, and wildlife habitat.

Mobile GIS is one of the recent field technologies that has been used in forest inventory and design. This technology allows mapping and real-time GPS tracking to be integrated when conducting forest inventory. Using mobile GIS techniques, tree locations and their attributes can be stored in a database. This technique essentially requires a portable computer such as a handheld PC, smart phone, or tablet PC loaded with GIS software (i.e., ArcPad, ArcGIS for Windows Mobile and Tablet).

While GIS has great utility in forest and natural resource management planning, it also presents some challenges. GIS technology changes rapidly, and its costs can be significant. There are also several functional limitations of GIS such as how to show temporal constraints in a spatial representation, which is an important topic for forest and natural resource management planning. Furthermore, while general-purpose GIS software is very efficient in 2D spatial problems, multidimensional problems such as 3D modeling with terrain features (which are important in ecological studies) have limitations in terms of GIS applications.

11.3 Remote Sensing Applications

11.3.1 What Is Remote Sensing?

Remote sensing is the technique for obtaining information about an object or phenomenon without making physical contact with the object. RS generally refers to the use of aerial sensor technologies to detect and classify objects on earth by means of propagated signals (Schowengerdt 2007).

Based on the source of the signal obtained, RS may be categorized as either passive or active. Passive RS occurs when information emitted passively, like sunlight, is merely recorded. Active RS occurs when a signal is first actively emitted from its source, such as an aircraft or satellite, and then recorded. (Schott 2007, Schowengerdt 2007, Liu and Mason 2009).

In active RS, the device emits energy in order to scan objects and areas, and then its sensor detects and measures the radiation that is reflected or backscattered from the target (Natural Resources Canada 2016). RADAR and LiDAR use active RS. A simple example of active versus passive RS is taking a picture using a camera with and without a flash (Natural Resources Canada 2016). As described, "when we use a camera with a flash to take a picture, the camera sends light to the target and the light reflects off the target back to the camera lens. The reflected light will then be measured by the camera. If the flash is not used, then the camera is a passive sensor as the light measured is from a source other than the camera (sensor)" (Natural Resources Canada 2016).

11.3.2 RS in Forestry and Natural Resource Management

RS has been extensively applied in agriculture, forestry, and natural resources. Some examples include how RS is used in forestry and natural resources management:

- Forest land use and land cover changes, for example, deforestation and fire impacts
- Forest inventory and vegetation growth and yield

- Greenhouse gas emission of biomass utilization monitoring
- Watershed protection, for example, changes of watershed due to forest operations
- Forest health and sustainable management, for example, pest infestation and management

11.4 GIS Mapping and Analysis Examples

11.4.1 Identifying and Mapping Vegetation Phenology

Phenological records are a useful model in the study of climate change, because the seasonal pattern of vegetation is sensitive to small variations in climate (Tan et al. 2011). Vegetation phenologies derived from remote-sensing methods are unequal to phenology records of species levels from a ground-measured approach. The onset of growing seasons at a species level can hardly be detected by sensors on a landscape scale. However, the large-scale land-surface phenology, which refers to aggregated information from the spatial resolution of satellite sensors, can be observed from remote-sensing platforms (Tan et al. 2011). Rather than attempting to identify a vegetation-specific developmental stage, the growing season derived from RS refers to the stage of great intensity in large-scale plant activities (Yu et al. 2013a).

The normalized difference vegetation index (NDVI) time-series data set is widely used because it has an advantage in presenting vegetation information in broad spatial coverage, especially in quantifying vegetation variations at regional, continental, and global scales (Reed et al. 1994, Tucker et al. 2001). Biweekly NDVI data can be obtained from the Global Inventory Modeling and Mapping System at NASA's Goddard Space Flight Center (Figure 11.8). The NDVI data were derived from the advanced very-high-resolution radiometer (AVHRR) sensor aboard National Oceanic and Atmospheric Administration (NOAA) polar orbiting satellites and were calculated from AVHRR bands as:

$$NDVI = \frac{R_{nir} - R_r}{R_{nir} + R_r}$$

where

R_r is the spectral reflectance in visible bands (550–700 nm)

R_{nir} is *the* spectral reflectance in near-infrared bands (730–1000 nm) (Sun et al. 2012)

A generalized NDVI temporal profile is continuous and smooth because vegetation canopy changes are small with respect to time (Ma and Veroustraete 2006). However, there are frequent fluctuations because of data transmission

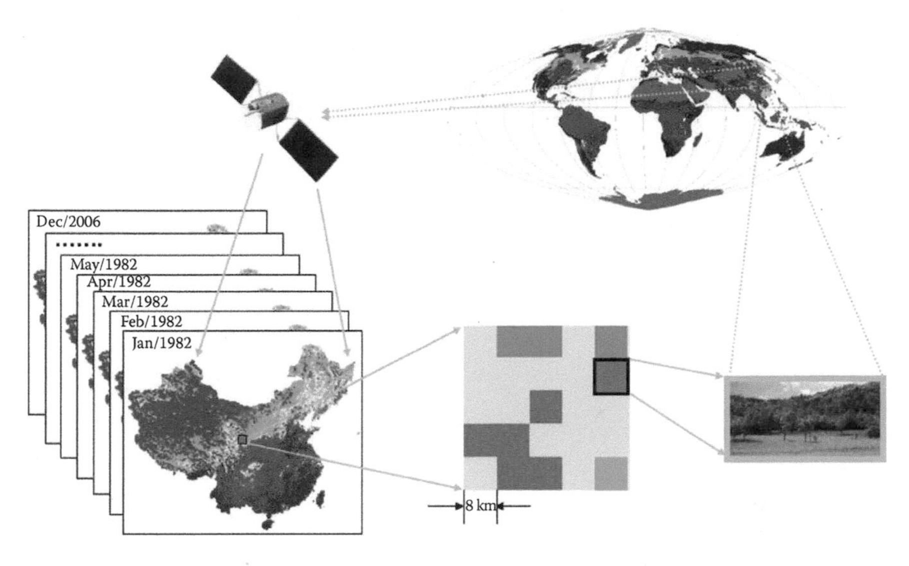

FIGURE 11.8
NDVI time-series data retrieved from the RS technique. (Courtesy of Dr. Zhen Yu.)

errors, incomplete or inconsistent atmospheric corrections, variations in cloudiness, and bidirectional effects in the NDVI data sets (Ma and Veroustraete 2006). The smoothing spline and double logistic algorithms are two of the most commonly used methods to identify the phenology.

11.4.1.1 Data Preparation

The downloaded biweekly NDVI data are organized in images (2D spatial arrays shown in Figure 11.8). The images can be read by different software, such as ArcGIS, Erdas, ENVI, R, and Matlab. To extract the time-series NDVI data for a specific location, the images should be read and stored as a matrix in the computer memory by the software. Then the user can save the NDVI time-series data based on the row and column of the location in the matrix. As an example, Figure 11.9 shows the biweekly NDVI time series of a location (black "x") and the way to smooth the wave (red dots).

11.4.1.2 Identifying Phenology

Spline-fitting analysis can be used to build the relationship between the biweekly NDVI and the corresponding days of the year (DOY), and to represent the daily changes in NDVI as a function of DOY (Figure 11.10).

This can be accomplished in Matlab. Import NDVI time-series data and use the spline function to fit the NDVI data into smooth time-series data (Figure 11.10). The smoothing spline is constructed for the specified

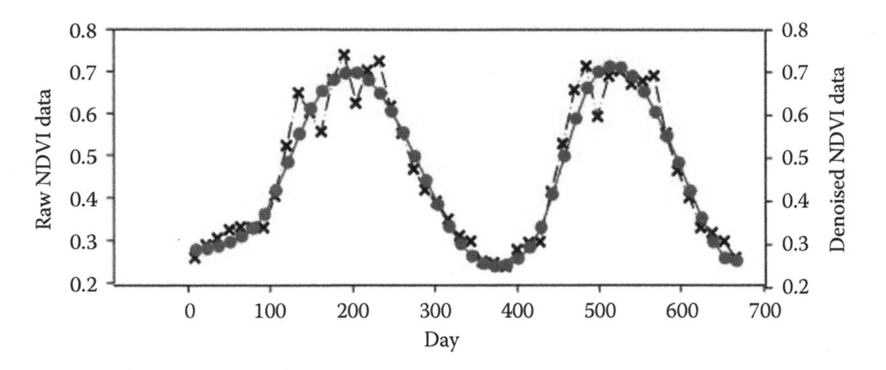

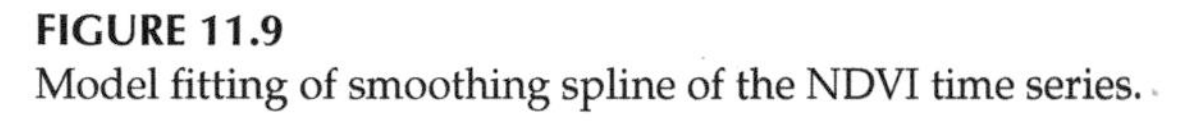
FIGURE 11.9
Model fitting of smoothing spline of the NDVI time series.

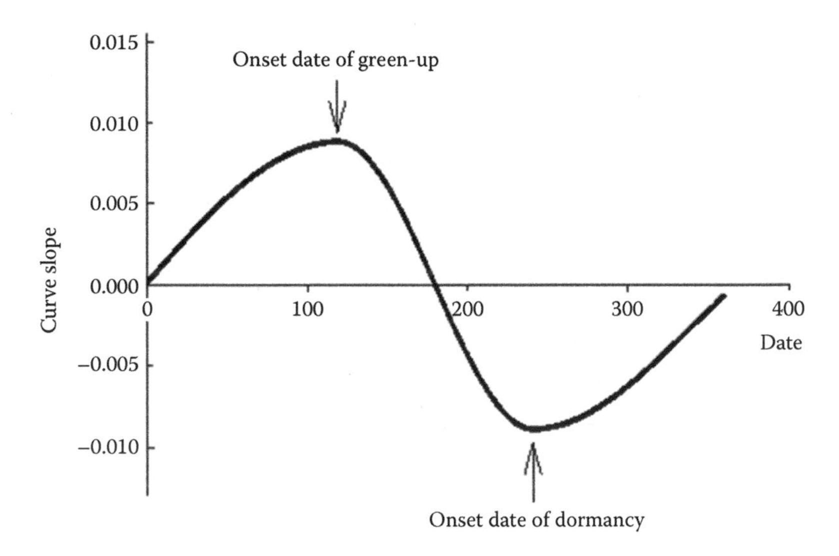

FIGURE 11.10
Identify the phenology using slope of smoothed NDVI curve.

smoothing parameter t (Yu et al. 2010) and minimizes the following two terms for the solution:

$$t\sum_{i}\left(y_i - s(x_i)\right)^2 + (1-t)\int\left(\frac{d^2 s}{dx^2}\right)^2 dx$$

where t is defined between 0 and 1. The first term is the sum of the squares of all the entries, which represents the deviation from input data points. The second term denotes the second derivative of the function, which represents the quantification of roughness of the fitted curve. As t moves from 0 to 1, the **smoothing spline** changes from a least-square straight-line fit to a **cubic spline interpolant** (Mathworks 2012).

After the smooth NDVI data have been created, the phenology can be identified by the slope of the smoothed NDVI time series for each year individually (Figure 11.10).

Timesat software (v3.1) can be applied to smooth and extract the phenology information from NDVI time-series images (2D spatial arrays). Here is the basic function of the double logistic algorithm used in Timesat:

$$g(t;x_1,\ldots,x_4)=\frac{1}{1+\exp\left(\frac{x_1-t}{x_2}\right)}-\frac{1}{1+\exp\left(\frac{x_3-t}{x_4}\right)}$$

where x_1 determines the position of the left inflection point while x_2 gives the rate of change. Similarly, x_3 determines the position of the right inflection point while x_4 gives the rate of change at this point (Eklundh and Jönsson 2010). Figure 11.11 shows the spatial distribution of the start of the growing season in temperate China using the double logistic approach of Timesat software (Yu et al. 2013b).

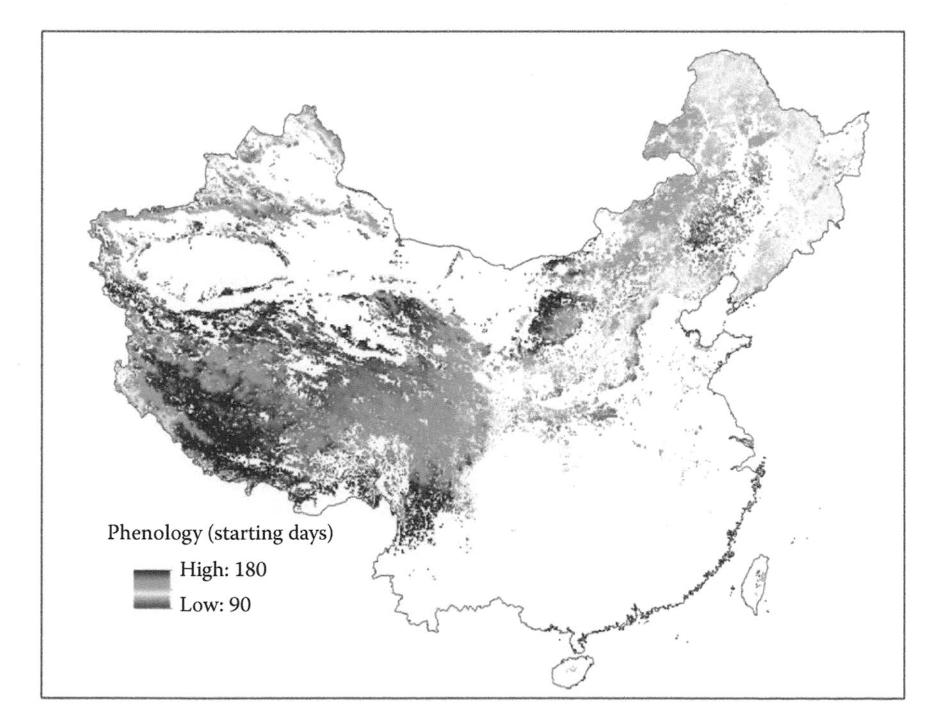

FIGURE 11.11
Spatial distribution of start of growing season in temperate China using double logistic approach. (From Yu, Z. et al., *Global Change Biol.*, 19, 2182, 2013b.)

11.4.1.3 Mapping the Phenology Using ArcGIS

For this mapping example, we need the phenology image file and the country boundary of China file. Here are the mapping procedures:

a. Launch ArcMap. From the Windows taskbar, click *Start* → *All Programs* → *ArcGIS* → *ArcMap*.
b. Add data. Click the *Add Data* button, browse to and click the phenology and boundary data, and click *OK*.
c. Right click the data layer, and click *Properties*. Click the *Symbology* tab. In the *Show* box, click *Stretched*. Choose a color type from the *Color Ramp*. Set *Stretched Type* as *Minimum-Maximum*. Check the checkbox of *Display Background Value* and click *OK*.
d. Click the *Layout view*. Click *Insert-Legend* to add a legend for the map (Figure 11.11).
e. Export the map from the *File-Export* map menu, and *Save As* an image file.

11.4.2 Mapping and Analyzing Winter CO_2 Efflux in a Conifer Forest Area of North America

In this example, we use flux data from twelve flux stations to calculate and map the winter CO_2 flux in a conifer forest area of North America. For forest ecosystems, the winter net ecosystem exchange (NEE = CO_2 flux between ecosystem and atmosphere) can be expressed as:

$$NEE = RE - GPP$$

where
RE is the ecosystem respiration
GPP is the gross primary productivity

During winter, RE and GPP mainly come from soil respiration and photosynthesis, respectively. Negative NEE indicates uptake of CO_2 by the ecosystem, and positive RE indicates loss of CO_2 from the ecosystem. Winter soil respiration is mainly controlled by soil temperature, which in turn, is affected by the snow depth. We can build the relationship between winter snow depth and CO_2 flux. Then, the regression equation can be applied to the entire study area to retrieve the spatial distribution of carbon efflux.

11.4.2.1 Data Preparation

The Canadian Meteorological Centre daily snow depth analysis data from August 1998 to December 2012 are used to explore the relationship with carbon flux at the site level (http://nsidc.org/data/docs). The flux data of twelve tower sites can be downloaded from FLUXNET websites (http://fluxnet.ornl.gov/). The downloaded level 3 flux data used in this example has been standardized and gap-filled for the user. The forest map used in this example is the 2005 North American Land Cover that can be downloaded from the Commission for Environmental Cooperation of the U.S. Geological Survey (http://landcover.usgs.gov/nalcms.php).

11.4.2.2 Data Analysis

The downloaded flux data are in ".csv" format. Open the data using MS Excel and summarize the winter CO_2 flux and snow depth data for each station by year. Save the results as an Excel file for further use.

Then use MS Excel to find and fit the relationships between snow depth and CO_2 flux. First, draw an x–y scattered chart to check the relationship between these two variables visually (x is snow depth and y is the CO_2 flux from RE and NEE).

Right click the scatter points and click *Add Trend Line*. Choose *Polynomial* in the *Trendline Options* box. Check *Display Equation on Chart* and *Display R-squared Value on Chart* at the bottom, then click *Close*. The results are shown in Figure 11.12.

The quadratic equation is:

$$y = a + bx + cx^2$$

where x is the snow depth, and y is the CO_2 flux (RE or NEE).

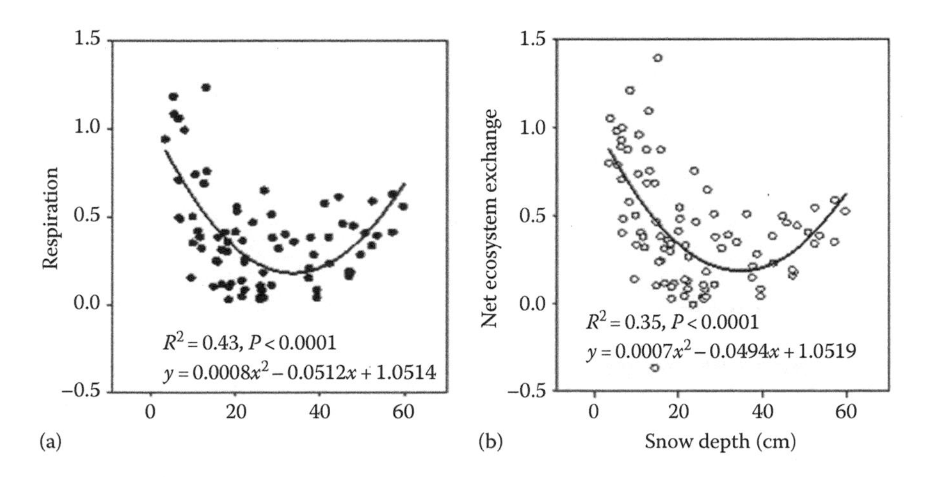

FIGURE 11.12
Correlation between snow depth, respiration, and net ecosystem exchange: (a) snow depth and respiration; (b) snow depth and NEE.

Apply the equation to the entire study area. Calculate NEE based on the winter snow depth and the equation from the scatter plot.

To extract the NEE of the conifer forest area, we need to create a conifer forest map first. Launch ArcMap. Click the *Arctoolbox* window. Browse to and double-click *equal to* under *Logical* of *Math Toolbox*. For the *Input raster or constant value* box, choose the downloaded forest map and input value 4, respectively (as the conifer forest value is 4 in the map). Choose the output path and click *OK* to create the conifer forest map. Then open the *Times* tool under *Math toolbox* and input the NEE data and the conifer forest map. Choose the output path and click *OK* to create the NEE map of the conifer forest area.

11.4.2.3 Mapping CO_2 Efflux Using ArcGIS

For this mapping example, we need the conifer forest NEE image file and the country boundary of the U.S. file. The mapping steps are similar to the previous example:

a. Launch ArcMap. From the Windows taskbar, click *Start* → *All Programs* → *ArcGIS* → *ArcMap.*

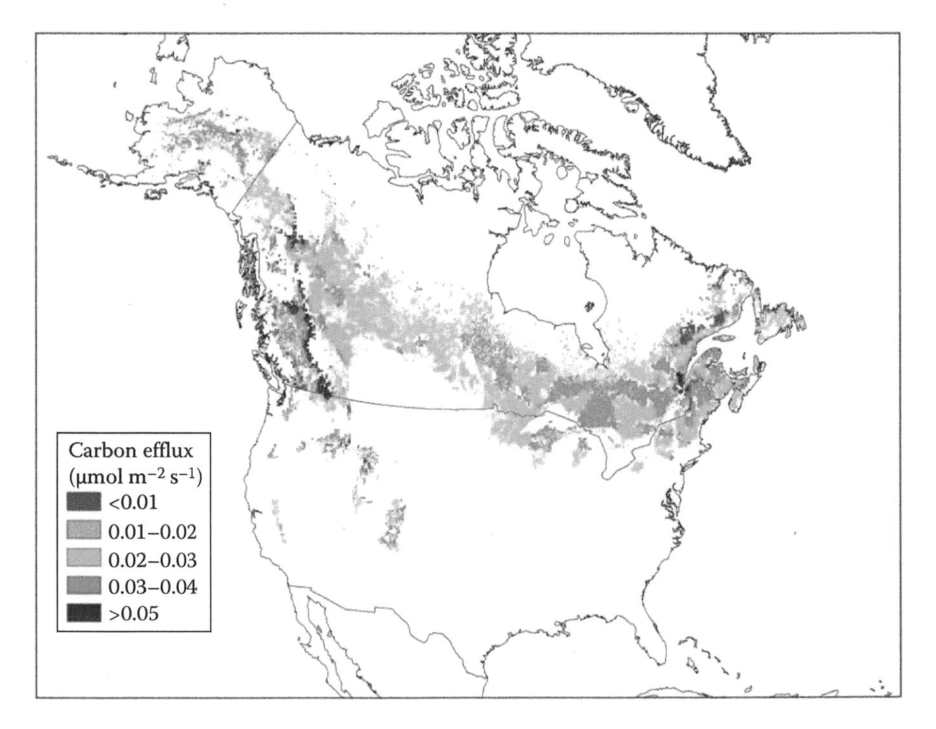

FIGURE 11.13
Spatial distribution of NEE in conifer forest of North America. (From Yu, Z. et al., *Global Ecol. Biogeogr.*, 25(5), 586, 2016.)

b. Add data. Click the *Add Data* button, browse to and click the *conifer forest NEE data* and *boundary data*, and click *OK*.
c. Right click the data layer and click *Properties*. Click the *Symbology* tab. In the *Show* box, click *Stretched*. Choose a color type from the *Color Ramp*. Set *Stretched Type* as *Minimum-Maximum*. Check the checkbox of *Display Background Value* and click *OK*.
d. Click the *Layout view*. Click *Insert-Legend* to add a legend for the map (Figure 11.13).
e. Export the map from the *File-Export* map menu, and *Save As* an image file.

Class Exercises

1. Use GPS to Collect Spatial Data of Forest Management
   In this lab, we are going to use GPS to collect timber cruising data. You are not required to conduct a real timber cruising. However, you do need to select one plot on or close to campus and each plot should contain at least three trees. We need to use a Nautiz X7 handheld unit and ArcMap software to complete the following:
   a. Identify a plot with three or more trees.
   b. Mark the boundary of a plot using a GPS unit.
      i. In *START* menu, choose *POCKET NAVIGATOR*
      ii. *MENU* → *OVERLAY* → *DELETE* → *DELETE ALL*
      iii. *MENU* → *OVERLAY* → *CREATE* → *ROUTE* → walk along the perimeter of the plot until you return to the starting point → *MENU* → *OVERLAY* → *SAVE AS* → name as "boundary," and hit SAVE
   c. Create a centerline of a skidding trail (route) by connecting a point outside the plot and a point within the plot.
      i. *MENU* → *OVERLAY* → *DELETE* → *DELETE ALL*
      ii. *MENU* → *OVERLAY* → *CREATE* → *ROUTE* → start to walk along the skidding trail you set. The starting point is somewhere outside the plot boundary, and the ending point is somewhere inside the plot → *MENU* → *OVERLAY* → *SAVE AS* → name as "SkiddingTrail/Route," and hit *SAVE*
   d. Mark locations of trees.
      i. *MENU* → *OVERLAY* → *DELETE* → *DELETE ALL*
      ii. *MENU* → *OVERLAY* → *CREATE* → *MARK* → walk to where the tree is and hit the point where you are, a flag will show on the

map; repeat this set of steps for two or more trees → MENU → OVERLAY → SAVE AS → name as "Trees," and hit SAVE

e. Move GPS data to your computer and convert to GPX files.
   i. Go to Convert Link: http://www.gpsvisualizer.com/ → choose the original file from handheld system (mmo files) → choose output format as Google Earth → hit MAP IT → Download and rename *kmz* files to your new folder
f. Generate a map with tree locations, a skidding trail, and a plot boundary. **This map should be completed individually**.
   ii. In Arctool box → Coversion tool → From KML: KML to layer
g. Generate a map. Open ArcMap and click *Add* button (which looks like a plus sign on the top menu). Navigate to the folder where you saved your file and add all the data into ArcMap.
   i. Click *View* on the top menu and then *Layout View*
   ii. Click Insert on the top menu and add all the necessary elements of a map such as *Title, Legend, North Arrow*, and *Scale Bar*. Position the elements attractively. Figure 14.9 is an example of a map.
   iii. When you finish your map, click *File* → *Export Map*. Export map as a PDF file for report.

2. Use GIS to Analyze Spatial Data
   We will use ArcGIS to manipulate spatial data. This lab consists of two parts. In Part I, you need to manipulate spatial data in ArcMap (similar to what you collected in Exercise 1), perform data analysis, and create a map. In Part II, you are required to perform similar data manipulation and create a map based on different spatial data. Please follow the instructions below to finish this lab.

Part 1

1. Create a folder on your desktop, download the related shape files from Exercise 1.
2. Add the data to a new ArcMap document.
3. Based on the sample data, what is the total area of the polygon? What is the total distance of the road? How many points are within your polygon? How many points are within 7.62 m (25 ft) of your trail? (1 acre = 0.4 ha) To get the answers, complete the following steps:

   Calculate polygon area and road length:

   a. Right click the *Boundary_poly* data layer, select *Open Attribute Table.*
   b. Click *Table Options*

c. Add a field to the Attribute Table
d. Enter *Acres* in the Name field, change type to *Float*, and then click *OK*.
e. A new field named *Area* has been added in the data table.
f. Now go back to the attribute table, right click the field *Acres* and select *Calculate Geometry*. When you see a pop-up message, hit *Yes*. In the *Calculate Geometry* window, change the Property to *Area* and Units to *Acres*. Click *OK*. Now, your area has been computed.
g. Similarly, to get the total road distance, right click the *Road_line* layer, add a new field *Length*, and calculate the length of the road in feet (1 ft = 0.305 m).
h. To get the total points within a polygon, select *Selection* on the top menu bar and then select *Select by Location*. In the pop-up window, select the *tree_points* layer, change the source layer to the *Boundary_poly* layer, and change the Spatial Selection method to target layer features that *are within the source layer feature.*
i. Similarly, to get the points within 7.62 m (25 ft.) of the road, select *Selection* on the top menu bar and then select *Select by Location*. In the pop-up window, select the *tree_points* layer, change the Source layer to the *Road_line* layer, and change the Spatial Selection method to target layer features that *are within a distance of the source layer feature*. Change the distance in the box to *25.0* and units to *feet* (1 ft = 0.305 m).

4. Create a map that includes a Title, Legend, North Arrow, and Scale Bar. In addition to those elements, insert a textbox that contains the answers to the following questions:
    a. How many acres (1 acre = 0.4 ha) are enclosed by the boundary polygon?
    b. What is the total length of the road?
    c. How many trees are inside the boundary?
    d. How many trees are within 7.62 m (25 ft) of the road?
5. Export the map to a PDF.

Part 2
In this part, you will use real spatial data to create a map.

1. Download and extract the zipped file of spatial data from eCampus that is in the *Lab_9_p2* folder. In the downloaded folder, you will see:
    a. *Landing.shp*—Existing landing area; *Haul_road.shp*—Existing haul road;
    b. *Preston_Roads.shp*—County road;

   c. *wvu_for_topo.jp2*—Topographic map
   d. Press the *Add Data* button in ArcMap to import all the spatial data.
2. Right click on the *haul_road shapefile → zoom to layer.*
3. Calculate the length of the haul road and the landing area based on the instructions in Part 1.
4. Create a map to clearly display the Preston County road, haul road, and landing, with the topographic map as the base map. In addition, the map should have a North Arrow, Scale Bar, Legend, Title, and a Textbox containing the results of your analysis (length of the haul road in feet (1 ft = 0.305 m) and area of the landing in acres).

Two maps are required for this lab, one for Part 1 and one for Part 2. Each map should include all the attributes listed previously.

References

Clarke, K.C. 2001. *Getting Started with Geographic Information Systems* (3rd Edition). Prentice Hall, Upper Saddle River, NJ.

Corvallis Microtechnology, Inc. 1996. Introduction to the global positioning system for GIS and TRAVERSE. Available online at http://www.cmtinc.com/gpsbook/index.htm. Accessed on February 10, 2014.

Eklundh, L. and P. Jönsson. 2010. *TIMESAT 3.0: Software Manual.* Lund University, Sweden, p. 74.

ESRI. 2001. *Getting to Know ArcGIS Desktop.* ESRI Press, Redlands, CA.

ESRI. 2003. *Geography Matter to Forestry.* Environment System Research Institute, Redlands, CA. Available online at http://www.esri.com/industries/forestry/. Accessed on October 4, 2016.

Grushecky, S.T. and M.A. Fajvan. 1999. A geostatistical comparison of forest spatial structure immediately following shelterwood and diameter-limit harvesting in West Virginia. *Forest Ecology and Management* 114(1999): 421–432.

Hamons, G. 2007. Modeling sediment movement in forested watersheds using hillslope attributes. Master thesis, West Virginia University, Morgantown, WV.

Liu, J.G. and P.J. Mason. 2009. *Essential Image Processing for GIS and Remote Sensing.* Wiley-Blackwell, Chichester, U.K., p. 4.

Ma, M.G. and F. Veroustraete. 2006. Reconstructing pathfinder AVHRR land NDVI time-series data for the Northwest of China. *Advances in Space Research* 37: 835–840.

Mathworks. 2012. Curve Fitting Toolbox: Smoothing splines. Mathworks, Natick, MA. Available online at http://www.mathworks.com/help/curvefit/smoothing-splines.html. Accessed on December 21, 2012.

Musinguzi, M., G. Bax, and S.T. Togboa. 2008. A methodology for coding wetlands for identification in a GIS based wetlands database. Available online at GISdevelopment.net, from http://www.gisdevelopment.net/application/environment/wetland/maf06_20abs.htm. Accessed on October 4, 2016.

Natural Resources Canada. 2016. Tutorial: Fundamentals of remote sensing. Available online at http://www.nrcan.gc.ca/earth-sciences/geomatics/satellite-imagery-air-photos/satellite-imagery-products/educational-resources/14639. Accessed on October 14, 2016.

Overmars, K.P., G.H.J. Koning de, and A. Veldkamp. 2003. Spatial autocorrelation in multi-scale land use models. *Ecological Modeling* 164(2–3): 227–270.

Reed, B.C., J.F. Brown, D. Vanderzee, T.R. Loveland, J.W. Merchant, and D.O. Ohlen. 1994. Measuring phenological variability from satellite imagery. *Journal of Vegetation Science* 5: 703–714.

Schott, J.R. 2007. *Remote Sensing: The Image Chain Approach* (2nd Edition). Oxford University Press, New York, p. 1.

Schowengerdt, R.A. 2007. *Remote Sensing: Models and Methods for Image Processing* (3rd Edition). Academic Press, London, U.K., p. 2.

Sharma, B. 2010. Modeling of forest harvest scheduling and terrestrial carbon sequestration. PhD Dissertation, West Virginia University, Morgantown, WV.

SIC. 2009. Wildlife and marine conservation. Satellite Imaging Service, Satellite Imaging Corporation, Houston, TX. Available online at http://www.satimagingcorp.com/svc/wildlife_and_marine_conservation.html. Accessed on October 4, 2016.

Sun, P.S., Z. Yu, S.R. Liu, X.H. Wei, J.X. Wang, and N. Zegre. 2012. Climate change, growing season water deficit and vegetation activity along the north-south transect of eastern China from 1982 through 2006. *Hydrology and Earth System Sciences Discussions* 9: 6649–6688.

Tan, B., J.T. Morisette, R.E. Wolfe, F. Gao, G.A. Ederer, J. Nightingale, and J.A. Pedelty. 2011. An enhanced TIMESAT algorithm for estimating vegetation phenology metrics from MODIS data. *IEEE Journal of Selected Topics in Applied Earth Observations and Remote Sensing* 4: 361–371.

Trimble Navigation Limited. 2004. Why post process GPS data? Westminster, CO. Available online at www.trimble.com. Accessed on October 14, 2016.

Tucker, C.J., D.A. Slayback, J.E. Pinzon, S.O. Los, R.B. Myneni, and M.G. Taylor. 2001. Higher northern latitude normalized difference vegetation index and growing season trends from 1982 to 1999. *International Journal of Biometeorology* 45: 184–190.

Upadhyay, M. 2009. Making GIS work in forest management. Available online at http://www.forestrynepal.org/images/GIS%20and%20Forest%20Management_0.pdf. Accessed on October 14, 2016.

VDF. 2008. Forest protection and fire prevention. Virginia Department of Forestry, Charlottesville, VA. Available online at http://www.dof.virginia.gov/index.shtml. Accessed on October 10, 2016.

Wang, J., T. Goff, and M. Strager. 2009a. Using spatial features to review application, effectiveness, and compliance of forestry best management practices in West Virginia. *International Journal of Forest Engineering* 20(2): 36–46.

Wang, J., B.D. Sharma, Y. Li, and G. Miller. 2009b. Modeling and validating spatial patterns of a 3D stand generator for central Appalachian hardwood forests. *Computers and Electronics in Agriculture* 68(2009): 141–149.

Yu, Z., S.R. Liu, J.X. Wang, P.S. Sun, W.G. Liu, and D.S. Hartley. 2013b. Effects of seasonal snow on the growing season of temperate vegetation in China. *Global Change Biology* 19: 2182–2195.

Yu, Z., P.S. Sun, and S.R. Liu. 2010. Phenological change of main vegetation types along a North-South Transect of Eastern China. *Chinese Journal of Plant Ecology* 34: 316–329.

Yu, Z., P.S. Sun, S.R. Liu, J.X. Wang, and A. Everman. 2013a. Sensitivity of large-scale vegetation greenup and dormancy dates to climate change in the north–south transect of eastern China. *International Journal of Remote Sensing* 34: 7312–7328.

Yu, Z., J. Wang, S. Liu, S. Piao, P. Ciais, S.W. Running, B. Poulter, J.S. Rentch, and P. Sun. 2016. Decrease in winter respiration explains 25% of the annual northern forest carbon sink enhancement over the last 30 years. *Global Ecology and Biogeography* 25(5): 586–595.

Section V

Visual Basic .NET Programming

12

Introduction to Visual Basic .NET Programming

Visual Basic (VB) is a third-generation event-driven programming language that was first released in 1991 for Microsoft (Microsoft Corporation 1998). It is possibly the fastest and easiest way to create applications for MS Windows. Designed to be both easy for beginners to learn and powerful for experts, VB uses an English-like syntax that promotes clarity and readability in combination with an extensive set of advanced features and a philosophy that emphasizes flexibility, productivity, and great tooling (Microsoft Corporation 2014). VB 6.0 was released in 1998 and was a popular programming language with many new features including web-based applications.

Succeeding VB 6.0, the first version of VB.NET was released in 2002. By 2016, eight versions of VB.NET had been released as components of Microsoft Visual Studio. VB.NET introduced many exciting new features to VB developers, though these enhancements have caused some minor compatibility issues with legacy code (Wakefield et al. 2001, McKeown 2010). The integrated development environment (IDE) (including code editing, debugging, and execution) incorporates some of the best ideas of VB 6.0 and InterDev (a user-friendly program development software) to make it easier and more intuitive to quickly create applications using a wider variety of development resources.

Possibly the most valuable addition to VB.NET is object-oriented programming (OOP). Although approximations of object orientation have been available in earlier versions of VB, only in VB.NET do developers gain the advantages of true object inheritance, which allows business logic to be more easily and reliably programmed through an object or a block of code (Wakefield et al. 2001).

12.1 What Is Visual Basic?

According to Microsoft's definition (Microsoft Corporation 1998), VB consists of two parts:

1. The "Visual" part refers to the method used to create the graphical user interface (GUI). Rather than writing numerous lines of code to

describe the appearance and location of interface elements, you simply put prebuilt objects into place on screen.

2. The "Basic" part refers to the Beginners All-Purpose Symbolic Instruction Code (BASIC) language. Notice that VB has evolved from its original BASIC language and now contains several hundred statements, functions, and key words, many of which relate directly to the Windows GUI.

There are programming languages derived from VB. Visual Basic for Applications (VBA) in MS Word, Excel, and Access, and VB Script for web-based programming are applications or programming languages that use VB.

12.1.1 Visual Basic and VB.NET Concepts

12.1.1.1 Windows, Events, and Messages

It is necessary for us to have a better understanding of some of the key concepts VB has built. For any Windows application programs, VB employs three key concepts: windows, events, and messages (Microsoft Corporation 1998).

A window is a rectangular object with its own boundaries. There are many types of windows in MS Windows applications, such as file explorer windows, document windows, or dialog boxes. Other windows include command buttons, text boxes, option buttons, and menu bars. The MS Windows operating system manages these windows by assigning each one a unique identification number (window handle) (Microsoft Corporation 1998).

An event is a significant occurrence in a program. Events are typically invoked by users' actions (such as a mouse click or a key press) through programmatic control, or even as a result of another event.

Each time an event occurs, it causes a message to be sent to the Windows operating system. The system processes the message and delivers it to the other related windows.

12.1.1.2 Procedural Programming

In traditional or procedural programming, the code is organized into small "procedures" that use and change the data. Program execution starts with the first line of code and follows a predefined path through the application, calling defined procedures as needed (Microsoft Corporation 1998).

12.1.1.3 Event-Driven Programming

In event-driven programming, the flow of the program does not follow a predetermined path and is determined by events such as user actions, messages from other programs, or events from the same program.

12.1.1.4 Object-Oriented Programming

In OOP, the data and related functions are bundled together into an object. An object is a code-based abstraction of a real-world entity or relationship (Sheldon et al. 2010). For example, a tree object can represent a real-world tree in a forest growth and yield application. As we learned in Section 2.2, OOP supports three foundational concepts of encapsulation, inheritance, and polymorphism.

12.1.2 VB Integrated Development Environment

VB IDE integrates many different functions (such as design, editing, compiling, and debugging) within a common environment, while each of these tools would operate as a separate program in a traditional development environment (Microsoft Corporation 1998).

To demonstrate the steps to start a VB IDE, we'll use the following example of an IDE of VB.NET in Visual Studio 2013 (or a later version):

a. Click *Start* and choose *All Programs* → *Visual Studio 2013* → *Visual Studio 2013.*
b. Select *Visual Basic Development Settings* as your default environment settings when you first use the Visual Studio. You can always change it to other programming languages, such as Visual C++ or Visual C#.
c. In the *New Project* dialog box, select *Windows Forms Application* and click *OK.*
d. The VB.NET IDE is displayed (Figure 12.1).

The VB.NET IDE consists of these key elements: menu bar, toolbars, tool box, solution explorer window, properties window, form designer, code editor window, object browser, error list, and other windows. By clicking the *View* tab on the menu bar, you may view them.

12.2 VB.NET Programming Examples

12.2.1 First Application

There are three main steps to creating an application in VB (Microsoft Corporation 1998): (1) Creating the interface, (2) Setting properties, and (3) Writing code. For this application, we will use the classic programming example "Hello World!" found in many programming textbooks. In this exercise, we want to create an application that will display "Hello World!" in a text box once the user clicks a command button.

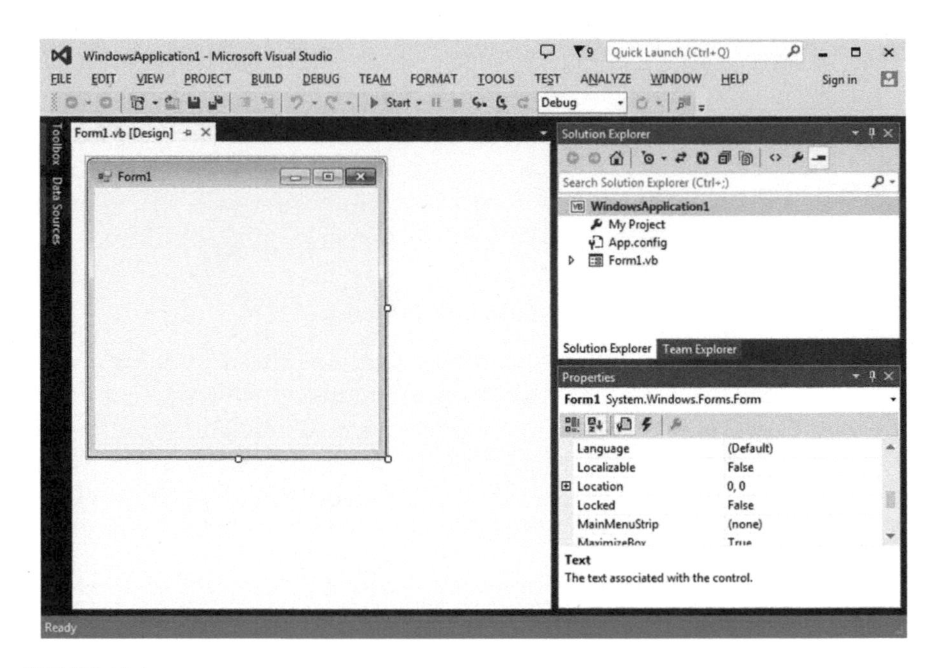

FIGURE 12.1
Visual Basic .NET IDE.

Creating the interface: Using the default Form1, click *text box* and *button controls* in the toolbox and then add them consecutively to Form1.

Setting properties: Select *Form1* or other objects on it, and then in the properties windows, you can set the properties of the selected object. For example, change the TEXT property of Button1 to "Hello World."

Writing code: Double click the *command* button, then the Code Editor window will be displayed. Type `TextBox1.Text = "Hello World!"`

Running the application: From the menu bar, click the *Start Debugging* button, then run the program. To end the program, you need to click the *Stop Debugging* button on the menu bar.

12.2.2 Example for Calculating Basal Area of Trees

In this example, we would like to calculate the basal area (BA) of a tree based on its diameter at breast height (DBH). Once a user enters the DBH for a tree, the BA for that tree should be calculated and displayed in a text box by clicking a *Calculate* button. The cumulative results of several trees also should be listed in a list box. The BA is calculated by using the following equation:

$$BA = 0.005454154 * DBH^2$$

where
BA is basal area in ft^2 (1 ft^2 = 0.0929 m^2)
DBH is the tree's diameter at breast height in inches (1 in. = 2.54 cm)

The cumulative results displayed in a list box could be used for comparison among trees. Here are the controls or objects we need to create the interface for this project: a text box, a list box, two buttons, and three labels.

12.2.2.1 Creating the Interface and Setting Properties

To start this project, from the *Start* menu, click *All Programs → Visual Studio 2013 | Visual Studio 2013*. Start a new project by choosing *New Project* from the *File* menu and then selecting *Windows Forms Application* in the *New Project* dialog box (when you first start VB, the *New Project* dialog box is presented). VB creates a new project and displays a new form for you. To design the interface:

- Create a directory like C:\For240\VBApps\CalBA\
- Start VB.NET
- Put the required VB.NET controls on the form (Figure 12.2)
- Name the controls (Table 12.1)
- Clear the default text in the boxes
- Change the form text to "Calculate BA"
- Save the project as "prjCalBA"

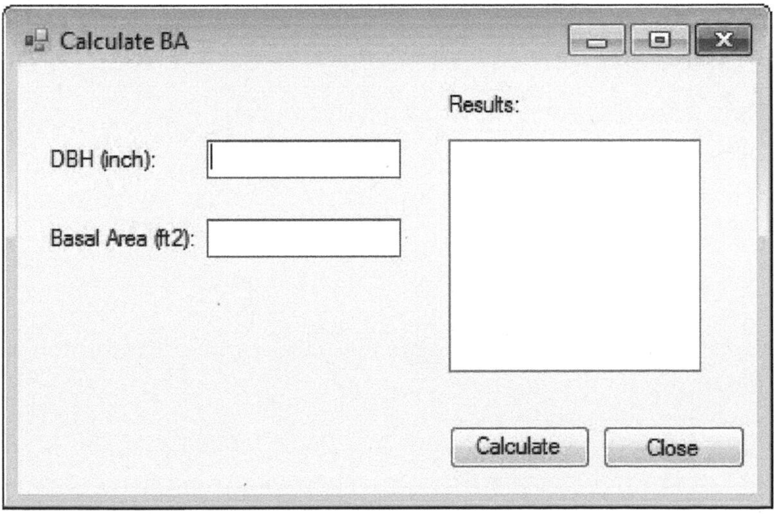

FIGURE 12.2
Interface for calculating BA of trees.

TABLE 12.1
Property Settings of the Objects

Object	Property	Setting
Form1	Name	frmCalBA
	Text	Calculate BA
TextBox1	Name	txtDBH
	Text	
TextBox2	Name	txtBA
	Text	
ListBox1	Name	LstResult
Button1	Name	cmdCalBA
	Text	Calculate
Button2	Name	cmdClose
	Text	Close

12.2.2.2 Writing Code

Double-click the *command* button and a *code-editing* box will pop up. Type the following lines under the Button 1 (cmdCalBA):

```
Private Sub cmdCalBA_Click(ByVal sender As System.
  Object, ByVal e As System.EventArgs) Handles
  cmdCalBA.Click
        Dim DBH, BA
        DBH = txtDBH.Text
        BA = 0.005454154 * DBH * DBH
        txtBA.Text = BA
        lstResult.Items.Add(DBH & "," & BA)
 End Sub
```

Similarly, double-click the Button 2 (cmdClose) and type:

```
Private Sub cmdClose_Click(ByVal sender As System.
  Object, ByVal e As System.EventArgs) Handles
  cmdClose.Click
        End
End Sub
```

Remember to save the project again by clicking the *Save* button on the menu bar.

12.2.2.3 Running the Application

Click the *Start Debugging* button on the menu bar to run the project. Enter "12" in the DBH box, then click *Calculate,* and you will add the first result to

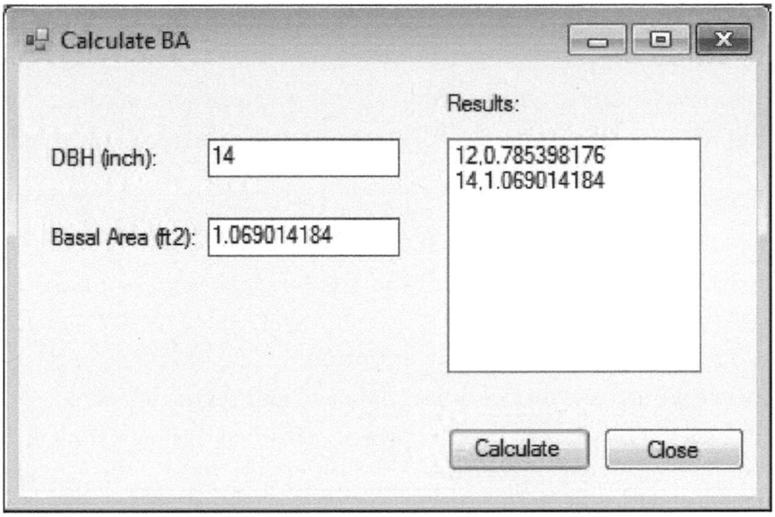

FIGURE 12.3
Output of BA calculations.

the list box. If you change the DBH from 12 to 14, then click *Calculate,* you will add the second result to the list box (Figure 12.3). You can repeat the above procedures to calculate the BA for any other trees you wish.

12.2.3 Database Connection and Data Display Application

This application demonstrates how a DataGridView control can be used to display a table of information from a database. VB makes it easy to access database information from within your application. The data control provides the ability to navigate through the database record set, synchronizing the display of records in the grid control with the position in the record set. This application needs two controls: a DataGridView control and a command button.

The database we are going to use is TimberCruising.accdb created in the database application section (Chapter 7 "Class Exercise"). The record set is the table *tblTrees* in the database.

12.2.3.1 Creating the Interface

You begin creating the application by choosing *New Project* from the *File* menu, then selecting *Windows Forms Application* in the *New Project* dialog box (when you first start VB, the *New Project* dialog box is presented). VB creates a new project and displays a new form. Now, we need to design the interface. Follow the steps below to add DataGridView and data source:

a. Select *DataGridView* under the group *Data* in the Toolbox, and drag it to the form interface. By default, the control will be named "DataGridView1."
b. Highlight *DataGridView1* and click the right arrow on top of the control to add the data source to it.
c. In the pop-up dialog, expand the *Choose Data Source* options and click *Add Project Data Source*.
d. Select *Database* and click the *Next* button.
e. Choose a database model, select data set, and then click *Next*.
f. Click *New Connection...*, select *Microsoft Access Data File*, click the *Continue* button, and browse the folders to locate the database that you will use in the project (Figure 12.4a). In this case, select *Timbercruising.accdb*, which is saved in the same folder as the project. Click *OK*. Go ahead and save the connection (Figure 12.4b).
g. As in Figure 12.5, select the table that you want to display using DataGridView. In this case, choose *tblTrees* and then click the *Finish* button.

Once everything is set up, your design interface should be completed. You also need to add a command button to the form (as demonstrated in the two previous VB.NET examples). Now save your project with the name "prjDBCDApp."

12.2.3.2 Setting Properties

In the Properties window, set properties for the objects according to Table 12.2. Use the default settings for all other properties.

12.2.3.3 Writing Code

Double-click the form or control to display the *Code* window and then type the following code for each event procedure. Add the code to the *cmdClose_Click* event procedure to end the application when you click the *Close* button.

```
Private Sub cmdClose_Click(sender As System.Object,
  e As System.EventArgs) Handles cmdClose.Click
              Me.TblTreesTableAdapter.Update(Me.
                TimberCrusisingDataSet.tblTree)
        End

End Sub
```

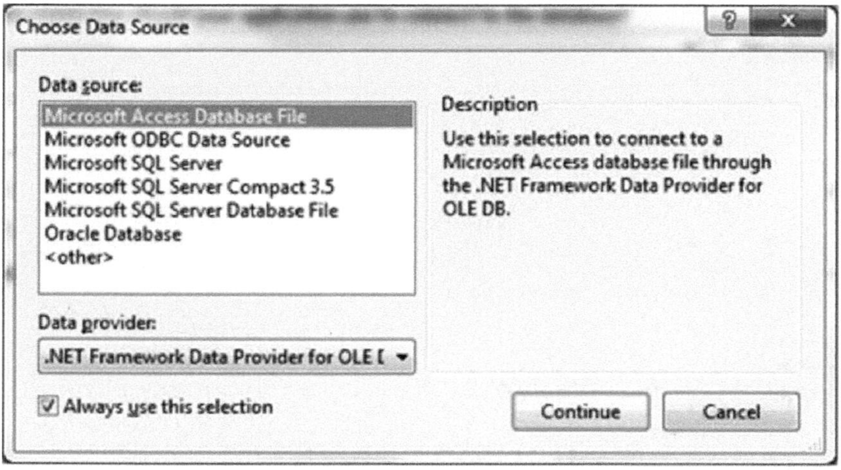

Choose Data Source

Data source:
Microsoft Access Database File
Microsoft ODBC Data Source
Microsoft SQL Server
Microsoft SQL Server Compact 3.5
Microsoft SQL Server Database File
Oracle Database
<other>

Description
Use this selection to connect to a Microsoft Access database file through the .NET Framework Data Provider for OLE DB.

Data provider:
.NET Framework Data Provider for OLE [

Always use this selection

Continue Cancel

(a)

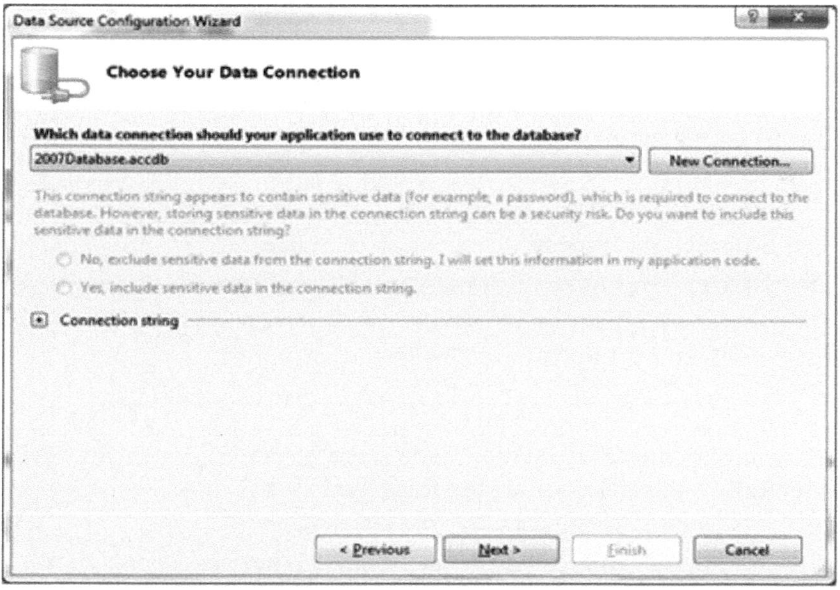

Data Source Configuration Wizard

Choose Your Data Connection

Which data connection should your application use to connect to the database?
2007Database.accdb
New Connection...

This connection string appears to contain sensitive data (for example, a password), which is required to connect to the database. However, storing sensitive data in the connection string can be a security risk. Do you want to include this sensitive data in the connection string?

No, exclude sensitive data from the connection string. I will set this information in my application code.

Yes, include sensitive data in the connection string.

Connection string

< Previous Next > Finish Cancel

(b)

FIGURE 12.4
(a) Select data source and (b) define data connection.

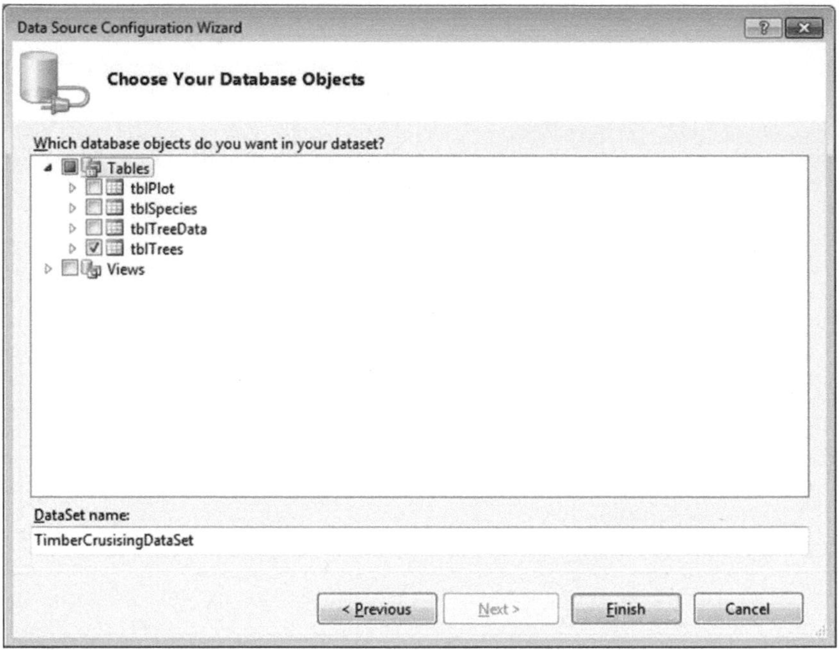

FIGURE 12.5
Choose data source table.

TABLE 12.2

Property Settings of the Objects

Object	Property	Setting
Form	Name	frmDBCDApp
	Text	Database Connection and Data Display
Button1	Name	cmdClose
	Text	Close

Since we have defined the data source as DataGridView1, Visual Studio 2013 automatically generates code for the form load event.

```
Private Sub frmDBCDApp_Load(sender As System.Object,
  e As System.EventArgs) Handles MyBase.Load
        'TODO: This line of code loads data into the
          'TimberCrusisingDataSet.tblTrees' table.
          You can move, or remove 'it, as needed.
          Me.TblTreesTableAdapter.Fill(Me.
          TimberCrusisingDataSet.tblTrees)

End Sub
```

Again, save your project at this point.

Database Connection and Data Display

TreeID	PlotID	Species	DBH	MHT	THT
1	1	RM	34	1.5	117
2	1	YB	21	0	0
3	1	YB	13	0	0
5	1	SM	12	0.5	83
6	1	OH	2	0	0

Close

FIGURE 12.6
Running the application.

12.2.3.4 Running the Application

There are two ways that you can run the application: from the menu bar or toolbar. Click the *Start Debugging* button, and your application will enter the running mode (Figure 12.6).

You may need to test your program. You can add data for new trees, edit existing trees, delete trees, or close your application.

Class Exercises

1. What is Visual Basic?
2. Compare and contrast VB and VBA.
3. Compare and contrast event-driven programming, object-oriented programming, and procedural programming.
4. What is VB.NET IDE?
5. What are the three main steps to create a VB.NET application?

References

McKeown, J. 2010. *Programming in Visual Basic 2010—The Very Beginner's Guide*. Cambridge University Press, Cambridge, U.K., p. 693.

Microsoft Corporation. 1998. *Visual Basic 6.0—Programmer's Guide*. Microsoft Press, Redmond, WA, p. 959.

Microsoft Corporation. 2014. Visual Basic resources. Available online at http://msdn.microsoft.com/en-us/vstudio/hh388573.aspx. Accessed on February 14, 2014.

Sheldon, B., B. Hollis, K. Sharkey, J. Marbutt, R. Windsor, and G. Hillar. 2010. *Professional Visual Basic 2010 and .NET 4*. Wiley Publishing, Inc., Indianapolis, IN, p. 1276.

Wakefield, C., H.E. Sonder, and W.M. Lee. 2001. *VB.NET Developer's Guide*. Syngress Publishing, Inc., Rockland, MA, p. 785.

13

VB.NET Controls, Project, and Menu Design

13.1 Types of Controls

Visual Basic .NET (VB.NET) controls make programming easier and faster. There are properties and events associated with each control we add/use. VB.NET has several categories of controls including:

1. Common controls, such as the command button and textbox controls
2. Containers, such as GroupBox and TabControl
3. Menus and toolbars, including MenuStrip and ToolStrip
4. Data, such as Chart, DataSet, and DataGridView
5. Components, including .COM objects, which exist as separate objects/files

In the previous chapter, we used DataGridView, textbox, and command button controls. There are many other built-in VB.NET controls. When you begin a VB.NET project, the toolbox shows different categories of all the controls (Figure 13.1). However, other controls can be added into the toolbox by right clicking *All Windows Forms* in the toolbox and choosing *Choose Items*.

VB.NET forms and controls are objects that expose their own properties, methods, and events. Properties can be thought of as an object's attributes, methods as its actions, and events as its responses. The following exercises demonstrate how to use VB.NET controls.

13.2 Using Timer, Option Button, Groupbox, and Checkbox Controls

In this example, you will use several VB.NET controls to build an interface and application to allow a user to form a timber harvesting system by selecting one felling machine and one skidding machine using radio buttons. The selected

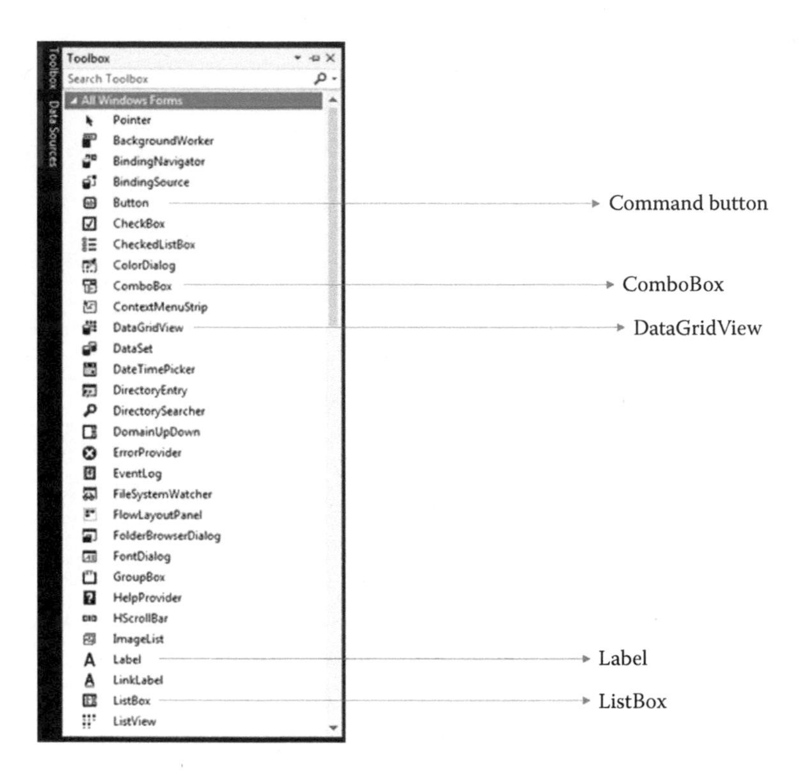

FIGURE 13.1
Default toolbox of VB.NET.

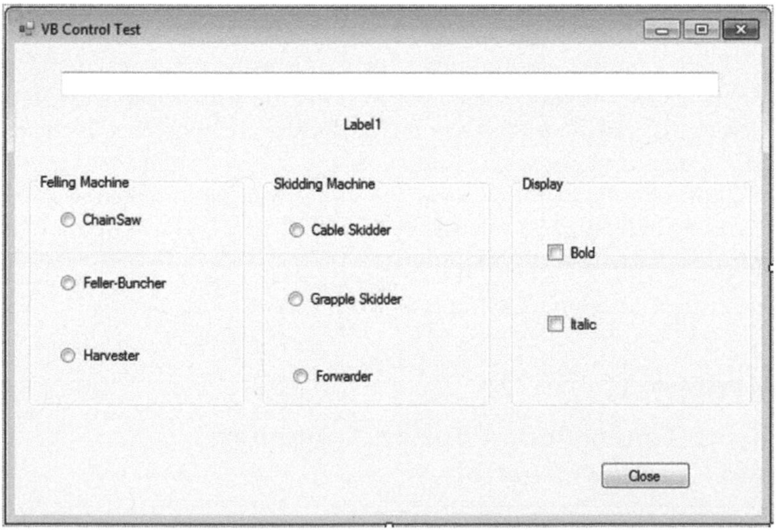

FIGURE 13.2
Interface of control test.

harvesting system is displayed in a textbox. Meanwhile, the font of the text in this textbox can be changed to italic, bold, or both. A label control is used to display time while a command button control is employed to end the application.

13.2.1 Creating the Interface and Setting Properties

Put the following controls on the form:

- A textbox to display the harvesting system
- A label to show the clock
- Three groupboxes to hold radio buttons and checkboxes
- A command button to end the application

The interface of your application will look like what is displayed in Figure 13.2. Table 13.1 lists the property settings of these controls.

TABLE 13.1

Property Setting of Controls

Control	Property	Setting
Form1	Name	frmVBControl
	Text	VB control test
Text1	Name	txtDisplay
Label1	Name	lblTime
GroupBox1	Text	Felling machine
GroupBox2	Text	Skidding machine
GroupBox3	Text	Display
Radio Button1	Name	optCS
	Text	Chainsaw
Radio Button2	Name	optFB
	Text	Feller-buncher
Radio Button3	Name	optHV
	Text	Harvester
Radio Button4	Name	optCD
	Text	Cable skidder
Radio Button5	Name	optGD
	Text	Grapple skidder
Radio Button6	Name	optFD
	Text	Forwarder
Checkbox1	Name	chkBold
	Text	Bold
Checkbox2	Name	chkItalic
	Text	Italic
Button1	Name	cmdClose
	Text	Close

13.2.2 Writing Code

Double-click the related controls and type the event codes there.

```
Public Class frmVBControl
        Dim strfeller As String
        Dim strskidder As String

        Private Sub DisplayCaption()
           'concatenate the caption with the two string
             variables
            txtDisplay.Text = "You selected a " & strfeller &
              " and " & strskidder & " harvesting system."
        End Sub

        'Add the code under command button -cmdClose:
        Private Sub cmdClose_Click(ByVal sender As System.
          Object, ByVal e As System.EventArgs) Handles
          cmdClose.Click
                End
        End Sub

        'Double-click frmVBControl and add the code:
        Private Sub frmVBControl_Load(ByVal sender As System.
          Object, ByVal e As System.EventArgs) Handles MyBase.
          Load
                'Load system time
                lblTime.Text = Format(Now, "short time")
        End Sub

        'Double-click each radio button and add the following
          codes there:
        Private Sub optcs_CheckedChanged(ByVal sender As
          System.Object, ByVal e As System.EventArgs) Handles
          optcs.CheckedChanged
                'assign a value to the first string variable
                  - strfeller
                strfeller = "Chainsaw"
                'call the subroutine DisplayCaption
                Call DisplayCaption()
        End Sub

        Private Sub optfb_CheckedChanged(ByVal sender As
          System.Object, ByVal e As System.EventArgs) Handles
          optfb.CheckedChanged
                strfeller = "Feller-buncher"
                Call DisplayCaption()
        End Sub
```

```
Private Sub opthv_CheckedChanged(ByVal sender As
  System.Object, ByVal e As System.EventArgs) Handles
  opthv.CheckedChanged
        strfeller = "Harvester"
        Call DisplayCaption()
End Sub


Private Sub optcd_CheckedChanged(ByVal sender As
  System.Object, ByVal e As System.EventArgs) Handles
  optcd.CheckedChanged
        'assign a value to the second string variable
          - strskidder
        strskidder = "Cable Skidder"
        'call the subroutine DisplayCaption
        Call DisplayCaption()
End Sub


Private Sub optgd_CheckedChanged(ByVal sender As
  System.Object, ByVal e As System.EventArgs) Handles
  optgd.CheckedChanged
        strskidder = "Grapple Skidder"
        Call DisplayCaption()
End Sub


Private Sub optfd_CheckedChanged(ByVal sender As
  System.Object, ByVal e As System.EventArgs) Handles
  optfd.CheckedChanged
        strskidder = "Forwarder"
        Call DisplayCaption()
End Sub


Private Sub chkbold_CheckedChanged(ByVal sender As
  System.Object, ByVal e As System.EventArgs) Handles
  chkbold.CheckedChanged
        If chkbold.Checked = True Then
            txtDisplay.Font = New Font(txtDisplay.
              Font, FontStyle.Bold)
        Else
            txtDisplay.Font = New Font(txtDisplay.
              Font, Not FontStyle.Bold)
        End If

End Sub
```

```
        Private Sub chkitalic_CheckedChanged(ByVal sender As
          System.Object, ByVal e As System.EventArgs) Handles
          chkitalic.CheckedChanged
                If chkitalic.Checked = True Then
                    txtDisplay.Font = New Font(txtDisplay.
                      Font, FontStyle.Italic)
                Else
                    txtDisplay.Font = New Font(txtDisplay.
                      Font, Not FontStyle.Italic)
                End If

         End Sub

End Class
```

13.2.3 Running the Application

Start the program; click your preferred radio buttons and checkboxes. The running application should look like Figure 13.3.

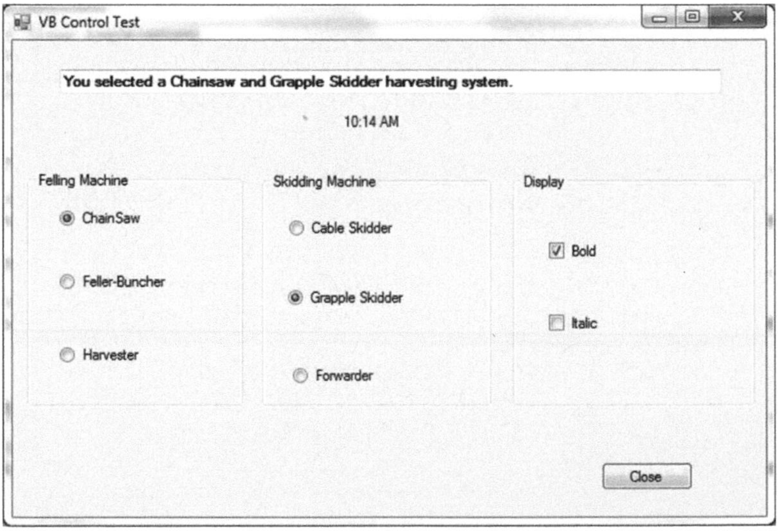

FIGURE 13.3
Running the application.

13.3 Using Drive, Dir, File, Combo, List, Frame Controls

In this exercise, you will use the Drive, Directory, and File controls to read a table in a database from the secondary storage. Then, you will click a button to populate the data from this table into a Combo Box. Finally, you will add the selected item in the Combo Box to a ListBox by clicking another button.

13.3.1 Creating the Interface and Setting Properties

We need to add the following controls to the form:

- GroupBox1 control
- DriveListBox1 control
- DirListBox1 control
- FileListBox1 control
- Label1
- Label2
- Command button—cmdClose
- Command button—cmdGetFile

If Drive, Dir, and File controls are not listed in the toolbox, we need to add them, so we can use them. To add them, we need to use Microsoft.VisualBasic.Compatibility. To reference Microsoft.VisualBasic.Compatibility:

a. Select *Add Reference…* from the *Project* menu, then the *Reference Manager* dialog box will be displayed.
b. Find and check/select *Microsoft.VisualBasic.Compatibility* in the listbox under *Assemblies/Framework* (Figure 13.4).
c. Click *OK*.
d. Click the *ToolBox* tab, then right click *All Windows Forms* and select *Choose Items…* under the *.NET Framework Components* tab.
e. In the *Choose Toolbox Items* dialog box, check *DriveListBox*, *DirListBox*, and *FileListBox* controls (Figure 13.5).

To save time, we will use the default settings for most of the controls. The properties of controls that need to be reset are listed in Table 13.2. The interface should look like Figure 13.6.

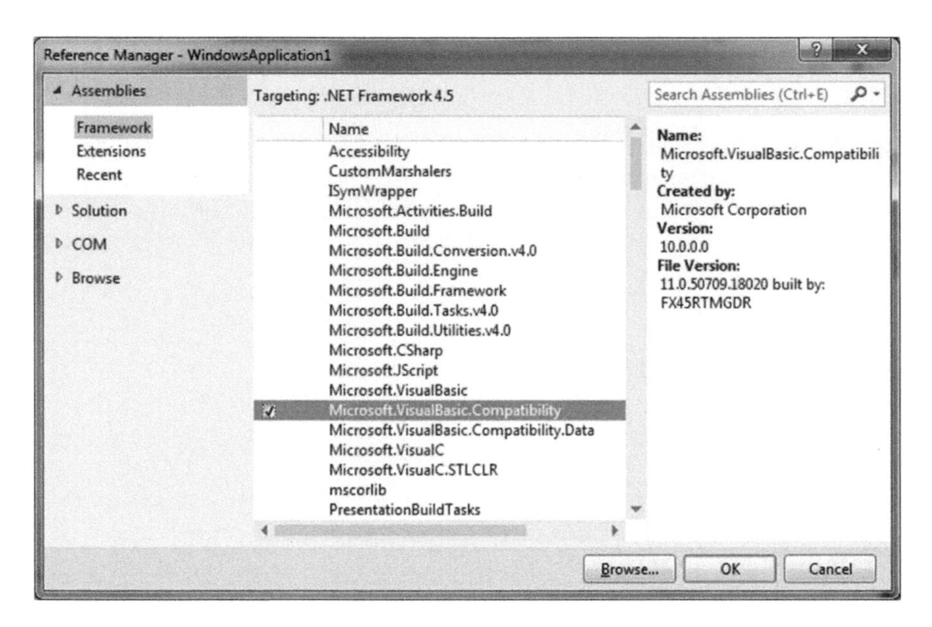

FIGURE 13.4
Add reference.

Choose Toolbox Items

Silverlight Components | Windows Phone Components | WPF Components
.NET Framework Components | COM Components | System.Activities Components

	Name	Namespace	Assembly Name
☐	DiscoveryClientProtoc	System.Web.Services.Discovery	System.Web.Services
☑	DomainUpDown	System.Windows.Forms	System.Windows.Forms
☐	DoubleBufferedListVie	Microsoft.TeamFoundation.VersionControl.Cor	Microsoft.TeamFoundation.Versic
☑	DriveListBox	Microsoft.VisualBasic.Compatibility.VB6	Microsoft.VisualBasic.Compatibili
☐	DriveListBoxArray	Microsoft.VisualBasic.Compatibility.VB6	Microsoft.VisualBasic.Compatibili
☑	DropDownList	System.Web.UI.WebControls	System.Web
☐	DropDownListConten	Microsoft.Office.Tools.Word	Microsoft.Office.Tools.Word.v9.0
☑	DynamicControl	System.Web.DynamicData	System.Web.DynamicData
☑	DynamicDataManage	System.Web.DynamicData	System.Web.DynamicData

Filter: Clear

DriveListBox
Language: Invariant Language (Invariant Country)
Version: 10.0.0.0

Browse...

OK Cancel Reset

FIGURE 13.5
Select toolbox items.

TABLE 13.2
Property Settings

Control	Property	Setting
Form1	Name	frmDrive
	Text	VB controls
GroupBox1	Text	GroupBox1
Button1	Name	cmdClose
	Text	Close
Button2	Name	cmdGetFile
	Text	Get file
Label 1	Text	File name
Label 2	Text	Label2

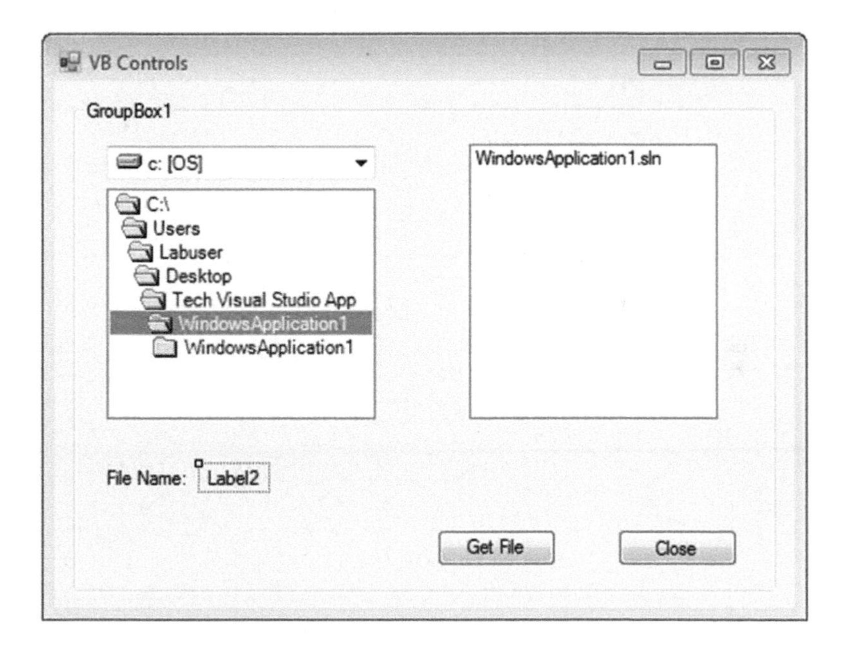

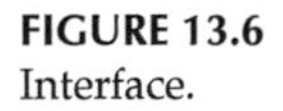
FIGURE 13.6
Interface.

13.3.2 Writing Code

Double-click *DriveListBox1*, *DirListBox1*, *FileListBox1*, and the two command buttons, respectively, to type the code for the following five procedures:

```
Public Class frmDrive
    Dim filename As String
```

```
        Private Sub DriveListBox1_SelectedIndexChanged(ByVal
          sender As System.Object, ByVal e As System.
          EventArgs) Handles DriveListBox1.
          SelectedIndexChanged

                DirListBox1.Path = Mid(DriveListBox1.Drive,
                  1, 1) & ":\"
        End Sub

        Private Sub DirListBox1_SelectedIndexChanged(ByVal
          sender As System.Object, ByVal e As System.
          EventArgs) Handles DirListBox1.SelectedIndexChanged
              FileListBox1.Path = DirListBox1.Path
        End Sub

        Private Sub FileListBox1_SelectedIndexChanged(ByVal
          sender As System.Object, ByVal e As System.
          EventArgs) Handles FileListBox1.SelectedIndexChanged
            filename = FileListBox1.Path
            If Microsoft.VisualBasic.Right(filename, 1) <> "\"
              Then
                filename = filename & "\" & FileListBox1.
                  FileName
            Else
                filename = filename & FileListBox1.FileName
            End If
        End Sub

        Private Sub cmdClose_Click(ByVal sender As System.
          Object, ByVal e As System.EventArgs) Handles
          cmdClose.Click


              End
        End Sub

        Private Sub cmdGetFile_Click(ByVal sender As System.
          Object, ByVal e As System.EventArgs) Handles
          cmdGetFile.Click
              Label2.Text = filename
        End Sub

End Class
```

13.3.3 Running the Application

1. Start the program.
2. Find an MS Access database file.
3. Click the *Get File* button.

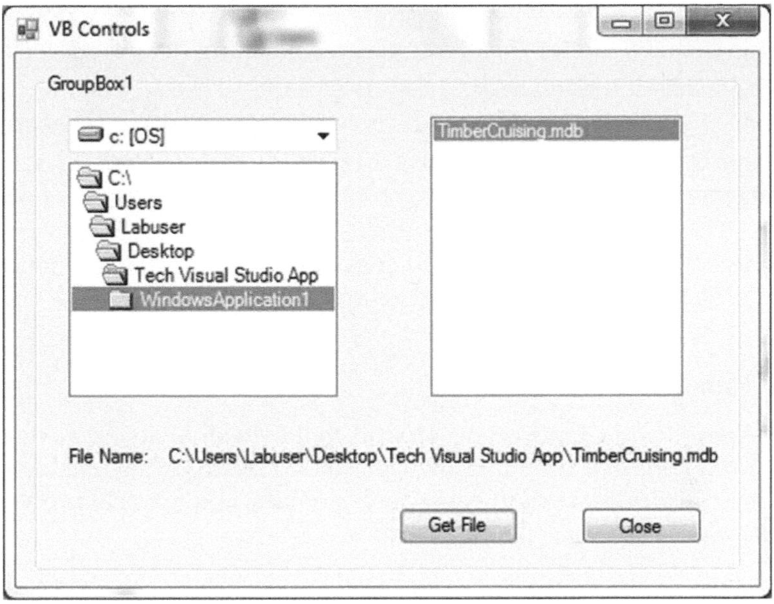

FIGURE 13.7
Running the application.

4. The path and name of the selected file are displayed.
5. Click the *Close* button to end this program.

The interface of your running application will look like Figure 13.7.

13.4 Working with a Project

As you develop a VB.NET application, you will work with a project to manage all the different files that make up the application (Microsoft Corporation 2015). Visual Studio creates many files and several folders when you write a program. It is best to leave the files and folders where they are. If you move them, save them in another location, delete them, or rename them, your project may not work. You should continue to work with the same folder for the solution when moving or making a backup of a project. Visual Studio expects the files to be in a specific location, and any changes could destroy a project. Although you can save your projects anywhere, you need to remember where you saved them. Open the *Solution* folder for your project and double-click on the file that ends with ".sln". That is the Solution file and it will open your project in Visual Studio. A VB.NET project typically consists of:

- One project file that keeps track of all the components (.sln)
- One folder that contains all the other files and folders for the project

- One file for each form (.vb)
- One resource file for each form (.resx) containing data properties of controls on the form, such as Picture or Icon
- Some files (.xsc, xsd, xss) if your project is connected to a database
- Some optional files such as standard module, class module, and data objects

A typical VB.NET project should consist of some or all of the following objects, including class, module, and control.

13.4.1 Form Class

Form classes (.vb file name extension) can contain textual descriptions of the form and its controls, including their property settings. They can also contain form-level declarations of constants, variables, and external procedures, event procedures, and general procedures.

13.4.2 User-Defined Class

Class modules (.vb file name extension) are similar to form modules, except that they have no visible interface. You can use class modules to create your own objects, including both properties and methods of the objects.

13.4.3 Standard Modules

Standard modules (.vb file name extension) can contain public or module-level declarations of constraints, variables, and procedures.

13.4.4 Standard Controls

Standard controls are supplied by VBNET. Standard controls, such as the command button or frame control, are always included in the toolbox.

13.5 Menu Design of VB Project

So far, you might notice that all the applications in this book consist of only one form and a few controls. However, for a larger VB.NET project, you should always use Menu Editor to create new menus and menu bars. From there, you can invoke other forms and procedures included in the project.

What follows is a simple exercise to show you how to create menus. Suppose we need to create the following menus of three levels:

- File
  - Close
  - Exit
- Edit
  - Species
  - Stand
    - Natural
    - Planted
- Help
  - About
  - Content

In this exercise, we are required to create an application that will do the following things: (1) allow the user to close all window items in the multiple document interface (MDI) form if she/he clicks the *Close* menu, (2) exit the program if the user clicks the *Exit* menu, (3) invoke the Species form if the user clicks the *Species* menu, and (4) invoke the About form if the user clicks the *About* menu.

To complete this exercise:

1. Start a new VB.NET project by clicking *Microsoft Visual Studio.*
2. Click *New Project…*
3. Select *Add Windows Form* under *Project* menu, choose *MDI Parent Form*, and then click the *Add* button (Figure 13.8).
4. Create menus.
   By default, Microsoft Visual Studio (2013 or later version) has provided several built-in menus, including File, Edit, View, Tools, and others. However, you can edit these menus manually to fit your own needs. There are two ways that you can create menus for a VB.NET project. You can right click any existing menu, then simply add, change, or delete it. Alternatively, you can right click the *Edit* menu and select *Edit DropDownItems…* to display the menu item editor (Figure 13.9). From there, you can perform the following menu editing actions:
   a. Add submenu—Click the *Add* button and change the text to "Species" (Figure 13.10).

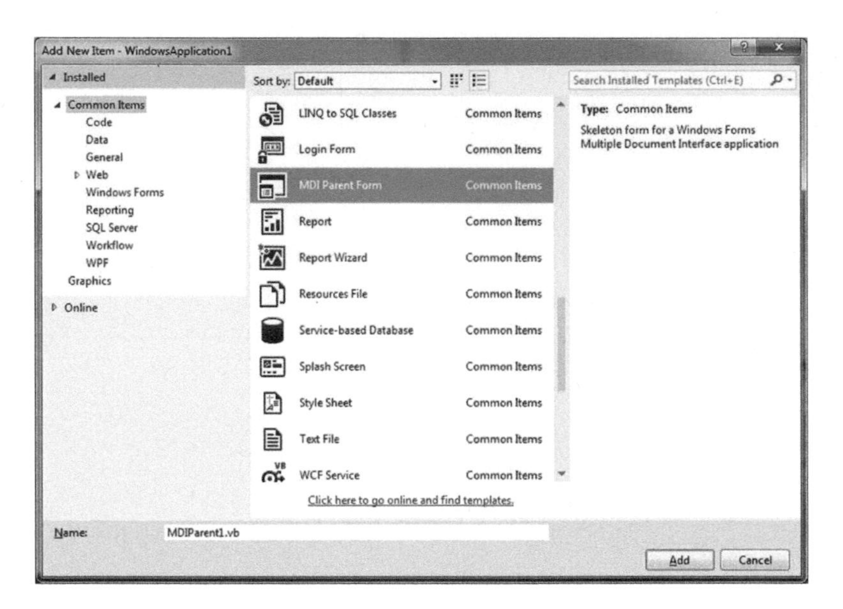

FIGURE 13.8
Addition of a MDI parent form to a VB.NET project.

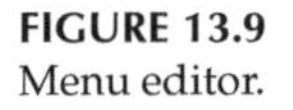

FIGURE 13.9
Menu editor.

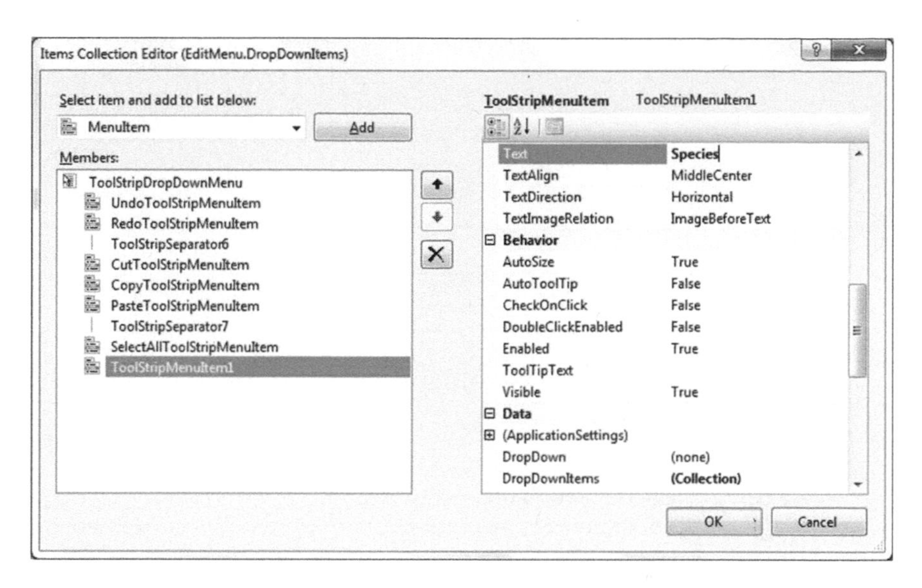

FIGURE 13.10
Edit submenu.

b. Add collection items—As is visible in the listbox on the right side of Figure 13.10. You can continue to add collection items for a submenu by clicking *DropDownItems* (Figure 13.11).

c. Delete submenu—Use the up and down arrow to select any existing menu and then hit the *Delete* button to delete it.

d. Edit submenu—Select any existing submenu and then go to the right-hand listbox to edit its properties (such as name, text, font color, etc.)

e. Click the *OK* button to finish editing. Remember, the edit process can be accomplished for each menu individually.

Follow the menu requirements to complete the menu editing (Figure 13.12).

5. Now you need to add a *Windows Form* and an *About* box. From the *Project menu*, select *Add Windows Form*, then repeat the procedure to add a second form. The two forms will be added and used to edit *Species* and display *About* information.
6. Double-click the menu items and add related code accordingly.

```
Imports System.Windows.Forms

Public Class MDIParent1

    'Add the codes for the menu Exit.
    Private Sub ExitToolsStripMenuItem_Click(ByVal sender
      As Object,
```

```
        ByVal e As EventArgs) Handles ExitToolStripMenuItem.
          Click
              End
        End Sub

        'Add the codes for the menu Species.
        Private Sub SpeciesToolStripMenuItem_Click(ByVal
          sender As
        System.Object, ByVal e As System.EventArgs) Handles
        SpeciesToolStripMenuItem.Click
              Form1.Show()
              Form1.Text = "Edit Species"

        End Sub

        'Add the codes for the menu About.
        Private Sub AboutToolStripMenuItem_Click(ByVal sender As
        System.Object, ByVal e As System.EventArgs) Handles
        AboutToolStripMenuItem.Click
              AboutBox1.Show()
              AboutBox1.Text = "About My Program"

        End Sub

        'Add the codes for the menu Close.
        Private Sub CloseToolStripMenuItem_Click(ByVal sender As
        System.Object, ByVal e As System.EventArgs) Handles
        CloseToolStripMenuItem.Click

        For Each ChildForm As Form In Me.MdiChildren
            ChildForm.Close()
        Next

        End Sub

        'Add the codes for the load event of the MDIParent1.
        Private Sub MDIParent1_Load(ByVal sender As System.
          Object, ByVal
        e As System.EventArgs) Handles MyBase.Load

              Form1.MdiParent = Me
              AboutBox1.MdiParent = Me

        End Sub

End Class
```

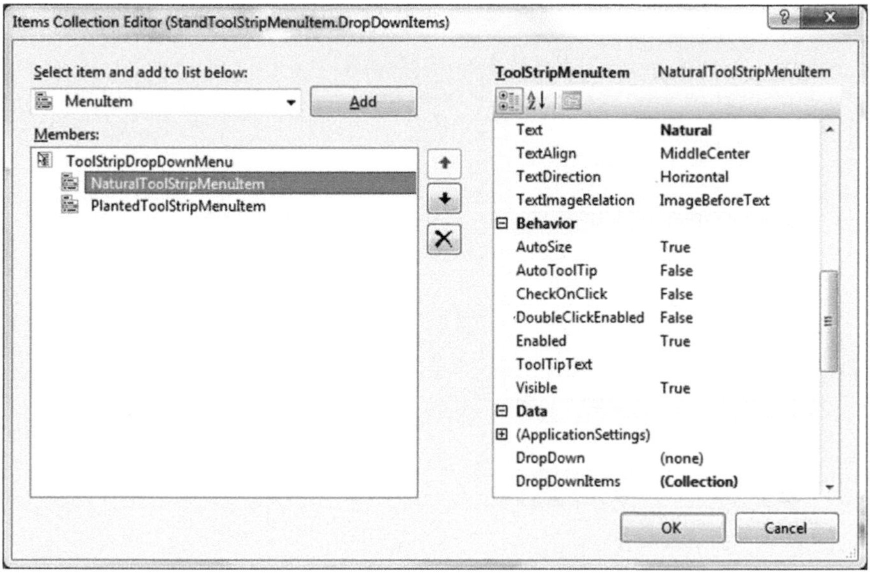

FIGURE 13.11
Add collections for submenu.

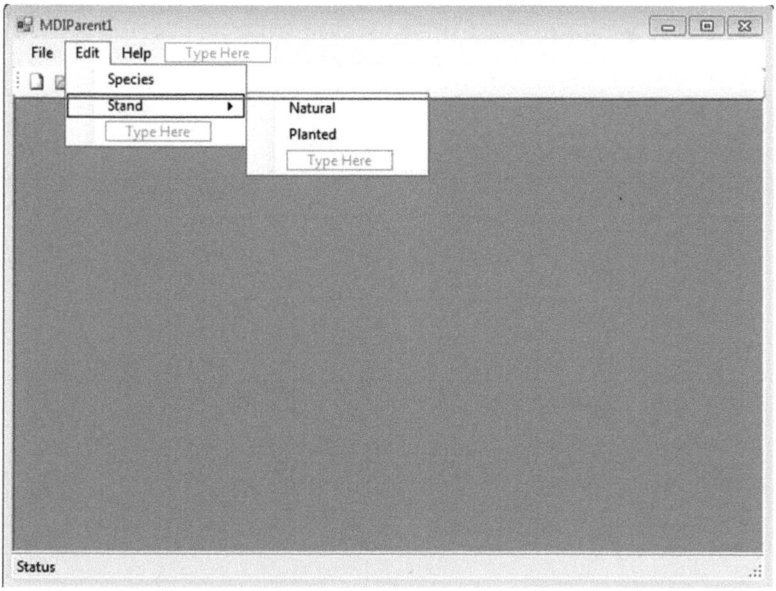

FIGURE 13.12
Menu design of a VB.NET project.

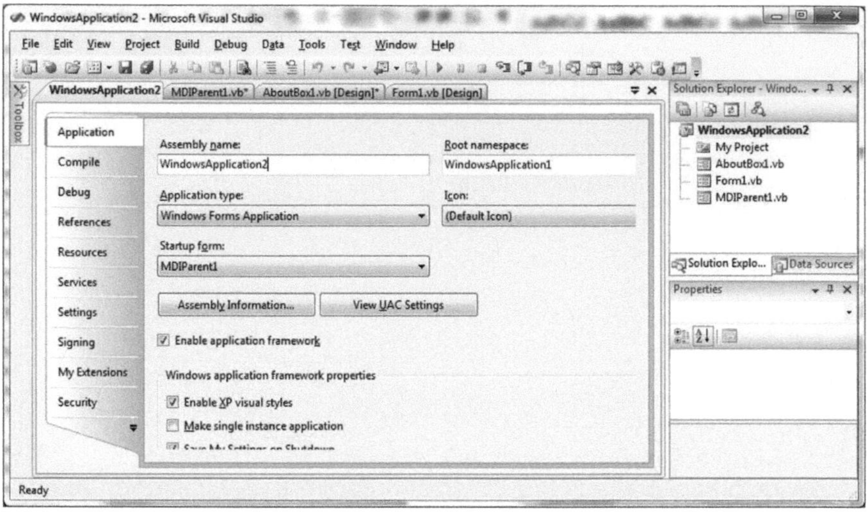

FIGURE 13.13
Project properties' dialog box.

7. In the *Solution Explorer Window*, right click your project name and choose *Property* and then change the start-up form of your application to *MDIParent1* (Figure 13.13).
8. Before you start to run the project, you need to set the form properties as listed in Table 13.3.
9. Run the project (Figure 13.14).
   a. Click *Edit | Species*
   b. Click *Help | About*
   c. Click *Close*
   d. Click *Exit*
10. *Save* your project.

TABLE 13.3
Property Settings of Forms

Form	Property	Setting
MDIParent1	WindowState	Maximized
	IsMdiContainer	True
Form1	IsMdiContainer	False
About Box1	Name	AboutBox1

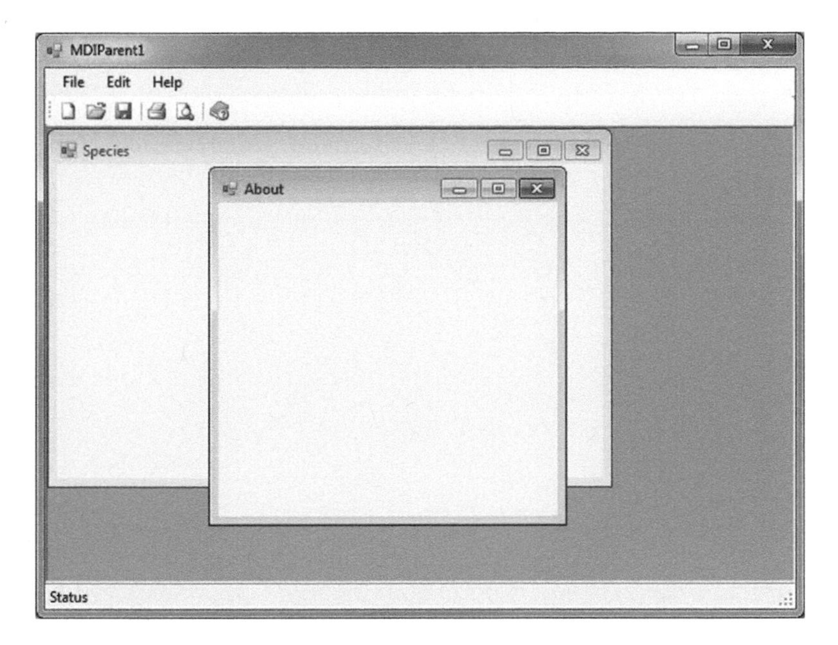

FIGURE 13.14
Running mode of a VB.NET MDI project.

Class Exercises

1. What does a VB.NET project usually consist of?
2. Compare and contrast form class, user-defined class, and standard module.
3. What are the types of VB.NET controls?

Reference

Microsoft Corporation. 2015. Managing Application Resources (.NET). Available online at https://msdn.microsoft.com/en-us/library/9za7fxc7.aspx. Accessed on February 14, 2017.

14

VB.NET Programming Fundamentals

14.1 Variables and Constants

Visual Basic .NET (VB.NET), like other programming languages, uses variables for storing values. Each variable has a name and a data type.

14.1.1 Declaring Variables

Users have to be more specific with data types while programming in VB.NET. Variables are declared with the *Dim* statement, supplying a name for the variable:

```
Dim variablename [as data type]
```

A variable name:

- Must begin with a letter.
- Can't contain an embedded period or embedded type-declaration character.
- Must not exceed 255 characters.
- Must be unique within the same scope, which is the range from which the variable can be referenced.

14.1.2 Scope of Variables

The scope is the lifetime of a variable. Depending on how it is declared, a variable is scoped as either a procedure-level (local), class-level, or module-level variable (Table 14.1).

14.1.3 Private versus Public

Private variables (or procedures) are **local** for a procedure or a module or a class. Public variables (or procedures) are available to all modules or classes in a project and are considered **global**.

TABLE 14.1

Scope of Variables

Scope	Private	Public
Procedure-level	Variables are private to the procedure in which they are declared and appear.	Not applicable. You cannot declare public variables within a procedure.
Class-level	Variables are private to the class in which they are declared and appear.	Variables are available outside the class.
Module-level	Variables are private to the module in which they are declared and appear.	Variables are available to all modules.

14.1.4 Static versus Dim

Values in local variables declared with *Static* exist the entire time your application/project is running, while variables declared with *Dim* exist only as long as the procedure is executing.

14.1.5 Constants

Constants also store values, but as the name implies, those values remain constant through the execution of an application. Using constants can make your code more readable by providing meaningful names instead of numbers.

```
[Public|Private] Const constantname[As Type] = expression

Public Const conBACal As Single = 0.005454154
```

14.2 Data Types

Data types control the internal storage of data in VB.NET. The data type determines how much memory is allocated to store the data (Table 14.2). While you should always specify the data type for a variable, if you don't, it defaults to an object data type. VB.NET has a number of built-in data types, with the most commonly used being:

1. *Numeric data types*
   - Integer

     `Dim NoTrees as Integer` *(Stores a value of integer)*

   - Single

     `Dim DBH as Single` *(Stores a value with single precision floating point)*

TABLE 14.2
Major Data Types in VB.NET

Data Type	Size in Bytes	Range
Boolean	4	True or false
Date	8	1/1/1 to 12/31/9999
Double	8	–1.79769313486231E308 to –4.94065645841247E-324 for negative values; 4.94065645841247E-324 to 1.79769313486232E308 for positive values
Integer	4	–2,147,483,648–2,147,483,647
Object	4 (32-bit) or 8 bytes (64-bit)	Any object type
Single	4	–3.402823E-38 to –1.401298E-45 for negative values; 1.401298E-45 to 3.402823E38 for positive values
String	10	0 to approximately 2 billion Unicode characters

Sources: Barwell, F. et al., *Professional VB.NET*, 2nd edn., Wiley Publishing, Inc., Indianapolis, IN, 2003, 985pp; Microsoft Corporation, Data type summary (Visual Basic), 2017, Available online at https://msdn.microsoft.com/en-us/library/47zceaw7.aspx, Accessed on February 9, 2017.

- Double

```
Dim DBH as Double
```
(*Stores a value with double precision floating point*)

2. *String data type*

```
Private S As String
Private S As String * 10
```
(*Variable-length string*)
(*Fixed-length string*)

If you assign a string of fewer than 10 characters to the above-defined fixed-length variable S, S is padded with enough trailing spaces to total 10 characters. If you assign a string that is too long for the fixed-length string, VB.NET simply truncates the characters.

3. *Boolean data type*

```
Dim blFlag As Boolean
```
(*Stores a value indicating True or False*)

4. *Object data type*

```
Dim objDB As Object
```
(*Stores any type of data*)

14.3 Arrays

Arrays can be used to store indexed collections of related variables. If you have experience in programming with other languages (such as C/C++ or Java), you will be familiar with the concept of arrays. Arrays allow you to use a series of variables of the same name, with different numbers (indices) to distinguish them. This helps create smaller and simpler code in many situations, because by using the index number the programmer can set up loops that deal efficiently with any number of cases. Always try to avoid declaring a larger array than necessary.

In VB.NET, there are two types of arrays: (1) a fixed-size array that always remains the same size and (2) a dynamic array whose size can change at run-time.

14.3.1 Declaring Arrays

```
Dim arrayValue(20) as Single          (for fixed size array of 21 elements)

Dim arrayValue() as Single            (for dynamic array)
...
Redim arrayValue(ComputedSize)        (for assigning dynamic array size)
```

14.3.2 Multidimensional Arrays

With VB.NET, you can declare arrays of multiple dimensions. For example, the following statement declares a two-dimensional 10-by-10 array within a procedure:

```
Dim StandStock(9, 9) as Single

Dim StandStock(1 to 10, 1 to 10) as Single
```

14.4 Classes and Procedures

14.4.1 Class

A **class** is a template to specify the properties and methods of an object. Class is a data structure that can contain data members such as constants, variables, and events and function members such as methods and properties. Let's start by looking at the basic syntax for declaring a class:

```
Class name

        [statements]

End Class
```

The *Class* keyword starts the class definition. The *name* is the name to be used to create instances of this class. The *statements* primarily comprise two parts: data and functions including the methods, properties, variables, and events of the class. Within a class, access to each member can be specified. A member declared as *Private* is available only from within the class. A member declared as *Public* is available inside the class as well as outside the class. *Public* is the default declaration if not specified. In object-oriented programming, a program is made up of various objects that interact with each other. Objects of the same type belong to the same class. An object is an instance of a class and its state is created by the values assigned to its instance variable.

Let's create a user-defined tree class to illustrate the application of a class or an object in VB.NET. As we know, a tree typically has attributes/data of height, species, and diameter at breast height (DBH). Depending on specific applications, its function could include calculation of the basal area (BA) and volume. Here is the code listing for a tree class:

```
Public Class Tree
    'data
    Private dbh As Single    'inch
    Private height As Single '# of 16 ft logs
    Private spp As String

    'methods
     'get tree data
    Public Sub TreeData()
        dbh = 12.2
        height = 2.5
        spp = "oak"
    End Sub

    'calculating basal area
    Public Function CalBA() As Single
        CalBA = 0.005454154 * dbh * dbh
    End Function

    'calculating volume
    Public Function CalVol() As Single
```

```
        CalVol = Math.Round(((0.55743 * height ^ 2 + 41.51275
          * height - 29.37337) + (2.78043 - 0.04516 * height ^
          2 - 8.77272 * height) * dbh + (0.04177 - 0.01578 *
          height ^ 2 + 0.59042 * height) * dbh ^ 2), 2)
    End Function

    'method to display tree data
    Public Sub DisplayTree()
        Console.WriteLine("This tree is: {0}", spp)
        Console.WriteLine("Tree DBH in inches: {0}", dbh)
        Console.WriteLine("Tree Height in # of 16-ft Logs: {0}
          ", height)
        Console.WriteLine("Basal Area in square feet: {0} ",
          CalBA())
        Console.WriteLine("Volume in Doyle Board Feet: {0}",
          CalVol())
    End Sub

    Shared Sub main()
        Dim t As New Tree()'an instance of Tree class or
          object
        t.TreeData()
        t.DisplayTree()
        Console.ReadLine()
    End Sub
End Class
```

You can group programming tasks by breaking programs into smaller logical components. These components, called **procedures**, can then become building blocks that let you enhance and extend VB. Procedures are useful for condensing repeated or shared tasks, such as frequently used calculations, text and control manipulations, and database operations. For example, the procedure "DisplayCaption()" in Section 13.2 has been used 8 times.

There are two major benefits of programming with procedures:

1. Debug more easily
2. Reuse the procedures for other programs with little or no modification

There are three basic types of procedures used in VB.NET:

1. Sub procedures that do not return a value.
2. Function procedures that return a value.
3. Property procedures that can return and assign values and set references to objects.

14.4.2 Sub Procedures

A *Sub* procedure is a block of code that is extended in response to an event. There are two types of Sub procedures: general and event procedures. The syntax for a Sub procedure is:

```
[Private][Public][Static] Sub procedurename (arguments)
  handles event name

        statements

End Sub
```

14.4.2.1 Event Procedures

When an object in VB recognizes that an event has occurred, it automatically invokes the event procedure using the name corresponding to the event. Because the name establishes an association between the object and the code, event procedures are said to be attached to forms and controls.

An event procedure for a control combines the control's name, underscore "_", and the event name. For example, if you are using a command button to invoke actions by clicking the button, this event procedure will be:

```
Private Sub Button1_Click(ByVal sender As System.
  Object, ByVal e
As System.EventArgs) Handles Button1.Click

        statements

End Sub
```

An event procedure for a form combines form name, underscore, and event. For example, if you are using Form1, you can have the following event procedures for this form:

```
Private Sub Form1_Click(ByVal sender As Object, ByVal
  e As
System.EventArgs) Handles Me.Click

        statements

End Sub

Private Sub Form1_Load(ByVal sender As System.Object,
  ByVal e As
System.EventArgs) Handles MyBase.Load

        statements

End Sub
```

14.4.2.2 General Procedures

A general procedure tells the application how to perform a specific task. Once a general procedure is defined, it must be specifically invoked by the application. By contrast, an event procedure remains idle until called upon to respond to events caused by the user or triggered by the system.

Why use general procedures?

- Several different event procedures might need the same actions performed.
- General procedures eliminate the need to duplicate the code.
- General procedures can enhance modular programming of your applications.

14.4.3 Function Procedures

Like MS Excel and Access, VB.NET includes built-in functions, like *Sqr, Cos,* or *Sin*. In addition, we can build our own functions. The syntax for a function procedure is:

```
[Private][Public][Static] Function functionname (args)
  [As type]

            statements

End Function
```

Like a Sub procedure, a function is a separate procedure and can perform a series of statements. However, there are two basic differences between a Sub procedure and a function:

- A function can return a value. While calling a function, we need to use an expression such as return_value = functionname(arguments).
- A function has a data type.

14.4.4 Sample Exercise

Here is an example to show you how to use procedures. Suppose we would like to calculate the BA of a tree and display it in a list box.

a. Start a new VB project.
b. Create the interface.

   We need to add the following controls on Form2 (Figure 14.1):

   - Two labels
   - A text box

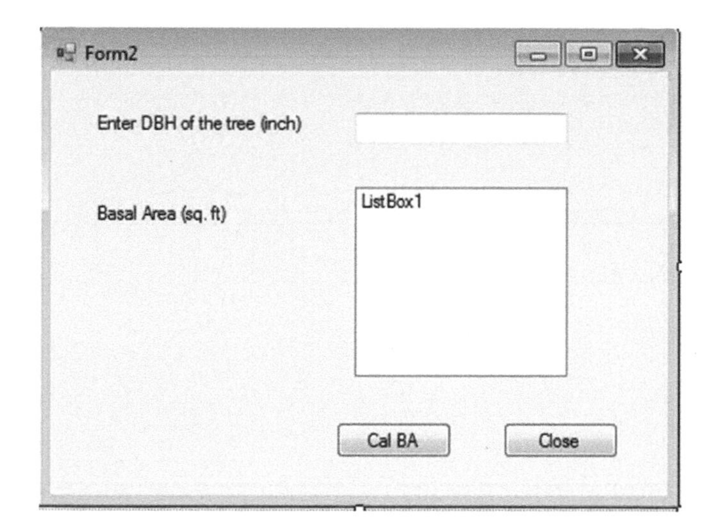

FIGURE 14.1
Interface of your application.

- A list box
- Two command buttons

c. Code the procedures.
   Using a function to calculate the BA and a Sub procedure to display the BA in the list box might be an efficient way to program this exercise.

```
Public Class Form2
    'In General Declaration
    Dim DBH As Single
    Dim BA As Single

    'A Procedure for displaying basal area
    Private Sub DisplayBA(ByVal basal_area As String)
        ListBox1.Items.Add(basal_area)
    End Sub

    'A Function for calculating basal area
    Private Function CalBA(ByVal d As Single) As Single
        CalBA = 0.005454154 * d * d
    End Function

    'Event procedure of clicking command1
    Private Sub Button1_Click(ByVal sender As System.Object,
      ByVal e As
    System.EventArgs) Handles Button1.Click
```

```
        DBH = CSng(TextBox1.Text)
        BA = CalBA(DBH)
        Call DisplayBA(CStr(BA))
    End Sub

    'Event procedure of clicking command2
    Private Sub Button2_Click(ByVal sender As System.Object,
      ByVal e As
    System.EventArgs) Handles Button2.Click

          End

    End Sub
End Class
```

You should notice that two VB.NET built-in functions were used in the program:

- *CSng* (Convert to single)
- *CStr* (Convert to string)

d. Run the project.

- Enter a DBH in the text box.
- Click the *Cal BA* button.
- Repeat the above procedures.

14.4.4.1 Passing by Value versus Passing by Reference

Only a copy of a variable is passed when an argument is passed by value. If the procedure changes the value, the change affects only the copy but not the variable itself. Using the ByVal key word indicates an argument passed by value.

Passing arguments by reference gives the procedure access to the actual variable content in its memory address location. As a result, the variable's value can be changed permanently. Passing by reference is the default in VB.NET.

14.5 Control Structures

Control structures allow you to control the flow of your program's execution. In this section, we will discuss decision structures and loop structures.

14.5.1 Decision Structures

VB.NET procedures can test conditions and then, depending on the results of the test, perform different operations. The decision structures that VB.NET supports include:

1. If … Then

```
If condition Then statement

 If condition Then

       Statements
End If
```

2. If … Then … Else

```
If condition1 Then
       Statement block 1
ElseIf condition2 Then
       Statement block 2
...
Else
       Statement block n
End If
```

3. Select Case

```
Select Case testexpression
       Case expressionlist1
               Statementblock1
       Case expressionlist2
               Statementblock2
       ...

       Case Else
               Statementblockn
End Select
```

14.5.2 Loop Structures

Loop structures allow you to execute one or more lines of code repetitively. The loop structures that VB supports include:

1. While … End While

```
While conditions

       Statements

End While
```

2. Do While … Loop

```
Do While condition
   Statements
Loop
```

3. For … Next

```
For counter = start To end [Step increment]

   Statements

Next counter
```

4. For Each … Next

```
For Each element In group

   Statements

Next element
```

14.6 Input and Output Files

We already learned how to input data from, and output results to, a VB.NET form. In this section, we will demonstrate how to read the data from a text file and write your data to a file. The ability to open up a text file and read its contents can be very useful in your programming life.

14.6.1 Direct File Access

Ever since the first version of VB, files have been processed using Open statements and other related functions. These mechanisms are fully supported in VB 6.0. VB.NET supports a new object called System.IO for the direct file functions. Direct file access is convenient and easy to use.

To open up a text file in VB.NET, you need to create an object called "StreamReader." This, as its name suggests, reads streams of text. The StreamReader is an object available to System.IO.

14.6.2 File Access Types

A file consists of nothing more than a series of related bytes located on a disk. When your application accesses a file, it must assume what the bytes are supposed to represent (characters, integers, strings, and so on). Depending on what kind of data the file contains, you use the appropriate file access type. There are three types of file access types in VB:

- Sequential—For reading and writing text files in continuous blocks.
- Random—For reading and writing text or binary files structured as fixed-length records.
- Binary—For reading and writing arbitrarily structured files.

We'll focus on how to use sequential access in this discussion. Sequential access is designed for use with plain text files and works best for the files that only consist of texts, such as the files created with a typical text editor. Each character in the file is assumed to represent either a text character or a text formatting sequence.

14.6.3 Opening Files for Sequential Access

When you open a file for sequential access, you open it to perform one of the following operations:

- ReadLine()—Input characters from a file
- WriteLine()—Output characters to a file

For example, you can access a file using the following syntax:

```
Dim FILE_NAME As String = "C:\Users\Owner\Documents\test.
  txt"
Dim objReader As New System.IO.StreamReader(FILE_NAME)
```

The first line just sets up a string variable called FILE_NAME. We store the path and the name of our text file inside the string variable:

```
= "C:\Users\Owner\Documents\test.txt"
```

You set up the StreamReader to be a variable, just like a string or integer variable. But we're setting up this variable differently:

```
Dim objReader As New System.IO.StreamReader(FILE_NAME)
```

We've called the variable *objReader*. Then, after the *As* comes *New*.

This means "Create a New Object." The type of object we want to create is a StreamReader object: System.IO.StreamReader.

System is the main object. *IO* is an object within *System* and *StreamReader* is an object within *IO*. StreamReader needs the name of a file to read. This goes between a pair of round brackets: System.IO.StreamReader(FILE_NAME).

VB.NET will then assign all of this to the variable called *objReader*. So instead of assigning an integer variable, you are assigning a StreamReader to an object type variable. Now that objReader is an object variable, it has its own properties and methods available for use (in the same way that the textbox has a text property). One of the methods available to our new StreamReader variable is the ReadToEnd method. This will read the whole of your text. You can, though, test to see if the file exists. If it does, you can open it; if not, you can display an error message.

```
Dim FILE_NAME As String = "C:\Users\Owner\Documents\test.
  txt"
If  System.IO.File.Exists(FILE_NAME) = True Then
     Dim objReader As New System.IO.StreamReader(FILE_NAME)
     TextBox1.Text = objReader.ReadToEnd
     objReader.Close()
Else
     MsgBox("File Does Not Exist")
End If
```

14.7 Example

Let's use the tree data from our previous course project. First create a text file for input and write your results to another text file.

1. Use notepad to create a file named TreeData.txt and save it in the directory of your VB.NET project. The data must be delimited by commas.

```
1,27,1
2,13,2
3,12,1
4,15,2.5
5,17,2
6,25,2
7,28,0.5
8,10,1
9,29,2.5
10,13,0.5
```

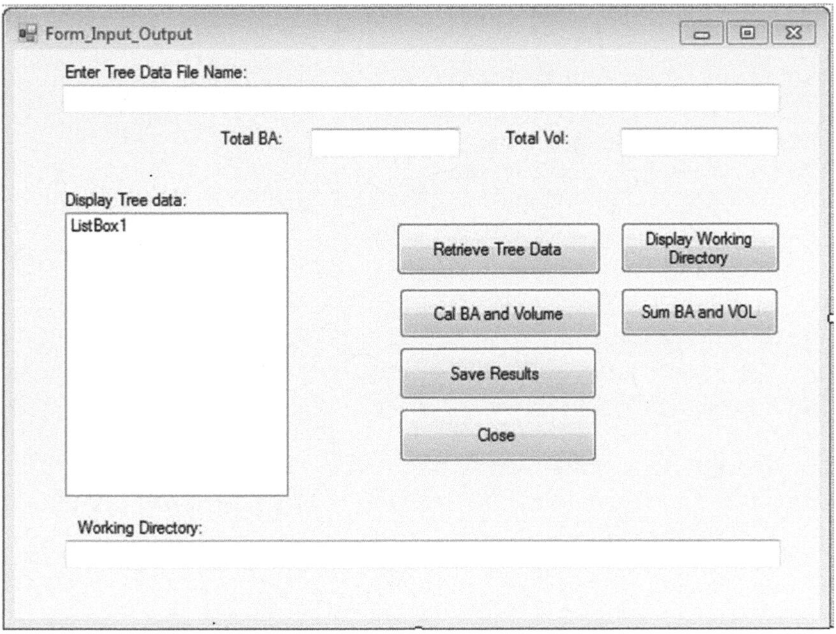

FIGURE 14.2
Interface for calculating BA and volume of trees.

2. Start a new VB.NET project and put the following controls on form1 (Figure 14.2):
   a. Five labels
   b. Three text boxes
   c. A list box
   d. Six command buttons
3. Table 14.3 shows the property setting of the form and other controls.

TABLE 14.3
Property Settings

Control	Property	Setting
Form	Text	Input and output file
Form	Name	Form_input_output
Command1	Text	Retrieve tree data
Command2	Text	Cal BA and volume
Command3	Text	Save result data
Command4	Text	Close

4. Add the following code to the project:

```
Public Class Form_Input_Output

    'In General Declaration
    Dim aryTreeNo() As Integer
    Dim aryDBH(), aryNLogs(), aryBA(), aryVol() As Single
    Dim NofTrees As Integer
    Dim totBA, totVOL As Single

    'Function for calculating BA
    Private Function CalBA(ByVal d As Single) As Single
        CalBA = 0.005454154 * d * d
    End Function

    'Function for calculating volume of a tree
    Private Function CalVol(ByVal d As Single, ByVal l As
      Single) As Single
     CalVol = Math.Round(((0.55743 * l^2 + 41.51275*l -
       29.37337) + (2.78043 - 0.04516*l^2 - 8.77272 * l)*d +
       (0.4177 - 0.01578*l^2 + 0.59042*l)*d^2), 2)
    End Function

    'Procedure for summarizing total basal area and volume
    Private Sub SumBAVol()
        totBA = 0
        totVOL = 0
        Dim i As Integer
        For i = 1 To NofTrees
            totBA = totBA + aryBA(i)
            totVOL = totVOL + aryVol(i)
        Next
    End Sub

    Private Sub Button1_Click(ByVal sender As System.Object,
      ByVal e As System.EventArgs) Handles Button1.Click
        Dim oRead As System.IO.StreamReader
        'oRead = IO.File.OpenText(TextBox1.Text)
        Dim n As Integer
        Dim str As String
        If System.IO.File.Exists(TextBox1.Text) Then
            oRead = IO.File.OpenText(TextBox1.Text)
            str = oRead.ReadLine()
            While Not str Is Nothing
                ListBox1.Items.Add(str)
                n = n + 1
                ReDim Preserve aryTreeNo(n), aryDBH(n),
                  aryNLogs(n), aryBA(n), aryVol(n)
```

```
            Dim start1, start2 As Integer
            start1 = Microsoft.VisualBasic.InStr(1, str,
              ",", CompareMethod.Text)
            start2 = Microsoft.VisualBasic.InStr(start1 +
              1, str, ",", CompareMethod.Text)
            aryTreeNo(n) = Convert.ToSingle(Microsoft.
              VisualBasic.Left(str, start1 - 1))
            aryNLogs(n) = Convert.ToSingle(Microsoft.
              VisualBasic.Right(str, Microsoft.
              VisualBasic.Len(str) - start2))
            aryDBH(n) = Convert.ToSingle(Microsoft.
              VisualBasic.Mid(str, start1 + 1, start2
              - start1 - 1))
            str = oRead.ReadLine()
        End While

        NofTrees = n
        oRead.Close()
    Else
       MsgBox("You either entered a wrong file name or the file
        does not exist!")
    End If

End Sub

Private Sub Button2_Click(ByVal sender As System.Object,
  ByVal e As System.EventArgs) Handles Button2.Click
    Dim n As Integer
    For n = 1 To NofTrees
        aryBA(n) = CalBA(aryDBH(n))
        aryVol(n) = CalVol(aryDBH(n), aryNLogs(n))
    Next
    MsgBox("BA and Volumn were computed!")
End Sub

Private Sub Button4_Click(ByVal sender As System.Object,
  ByVal e As System.EventArgs) Handles Button4.Click
    End
End Sub

Private Sub Button3_Click(ByVal sender As System.Object,
  ByVal e As System.EventArgs) Handles Button3.Click
    'Create a text file
    Dim oWrite As System.IO.StreamWriter
    oWrite = IO.File.CreateText(My.Application.Info.
      DirectoryPath & "\Results.txt")

    'Write to the text file
    Dim i As Integer
    Dim str As String = Nothing
```

```
        For i = 1 To NofTrees
            str = aryTreeNo(i).ToString + "," + aryDBH(i).
              ToString + "," + aryNLogs(i).ToString + "," +
              CalBA(aryDBH(i)).ToString + "," +
              CalVol(aryDBH(i), aryNLogs(i)).ToString
            oWrite.WriteLine(str)
            oWrite.WriteLine()         'Write a blank line to
                                         the file
        Next

        'Close the text file
        oWrite.Close()

        MsgBox("Results were saved!")
    End Sub

    Private Sub Button5_Click(ByVal sender As System.Object,
      ByVal e As System.EventArgs) Handles Button5.Click
        'Displaying current working directory
        TextBox2.Text = Environment.CurrentDirectory()

    End Sub

    Private Sub Button6_Click(ByVal sender As System.Object,
      ByVal e As System.EventArgs) Handles Button6.Click
        'Displaying total BA and Vol
        Call SumBAVol()
        TextBox3.Text = totBA
        TextBox4.Text = totVOL
    End Sub

End Class
```

5. Run your application (Figure 14.3).

Your output file is TreeData.rlt and is located in the working directory. It should look like Figure 14.4.

14.8 Data Access

In this section, you will learn how to use a DataGridView control to retrieve an existing data table in an MS Access database and add new records to the table. You are also allowed to access the report in the MS Access database. Here is an example to demonstrate the whole process.

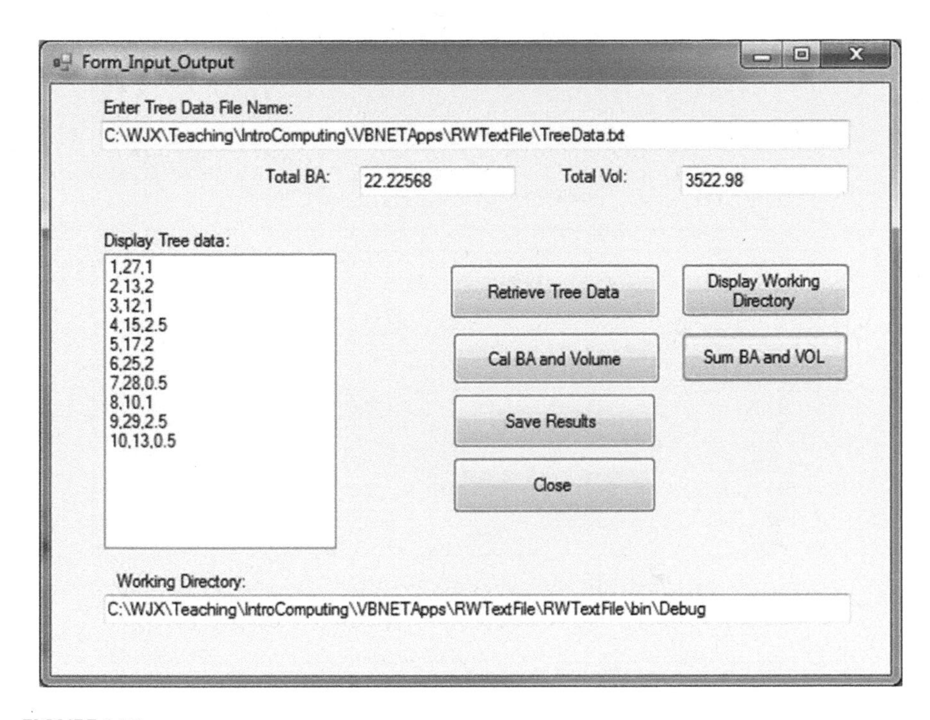

FIGURE 14.3
Display tree data and save results.

test1 - Notepad

File Edit Format View Help

```
1,27,1,3.98,299.05

2,13,2,0.92,57.54

3,12,1,0.79,29.01

4,15,2.5,1.23,105.71

5,17,2,1.58,136.9

6,25,2,3.41,406.92

7,28,0.5,4.28,207.34

8,10,1,0.55,13.96

9,29,2.5,4.59,707.86

10,13,0.5,0.92,26.78
```

FIGURE 14.4
File output of tree data.

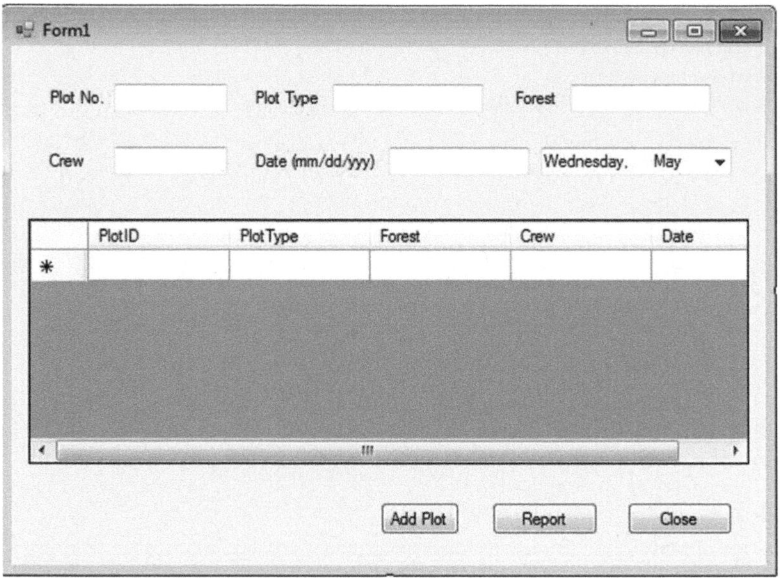

FIGURE 14.5
Interface for retrieving and adding data to database tables.

1. Create a Visual Studio project and name it "VSDataAccess." The interface of Form1 should look like Figure 14.5. Table 14.4 shows all the controls on the form. Another form "frmReport" also needs to be created.
2. Set data source of *DataGridView1.*

   You can follow the instructions described earlier in this chapter to set the data source of DataGridView1. In this example, you need to select the table *tblPlot* as the source table.

TABLE 14.4

Data Controls on Form1

Control	Property	Setting
Label1	Text	Plot no.
Label2	Text	Plot type
Label3	Text	Forest
Label4	Text	Crew
Label5	Text	Date (mm/dd/yy)
Datetimepicker1		
DataGridView1		
Button1	Text	Add plot
Button2	Text	Report
Button3	Text	Close

3. Code in Form1.

   Double click the related controls and add the following code to the corresponding procedures.

```
Public Class Form1

        Private Sub Button3_Click(ByVal sender As System.
          Object, ByVal e As System.EventArgs) Handles
          Button3.Click
                'close the current form
                Me.Close()
        End Sub

 Private Sub Form1_Load(ByVal sender As System.Object, ByVal e
   As System.EventArgs) Handles MyBase.Load
          'This code is automatically generated after you set
            up data connection for the control DataGridView1.
            Your dataset may be different from the code below.
            Make sure that the table "tblplot" is selected to
            fill the datagridview control.
         Me.TblPlotTableAdapter3.Fill(Me._2007DatabaseDataSet.
           tblPlot)

     End Sub

     Private Sub Button2_Click(ByVal sender As System.Object,
       ByVal e As System.EventArgs) Handles Button2.Click

             frmreport.Show()

     End Sub

     Private Sub Button1_Click(ByVal sender As System.
       Object, ByVal e As System.EventArgs) Handles Button1.
       Click
        Dim str1, str2, str3, str4, str5 As String
        str1 = TextBox1.Text
        str2 = TextBox2.Text
        str3 = TextBox3.Text
        str4 = TextBox4.Text
        str5 = TextBox5.Text

        Dim newrow As DataRow =
        _2007DatabaseDataSet.Tables("tblplot").NewRow()
        newrow("plotid") = str1
        newrow("plottype") = str2
        newrow("forest") = str3
        newrow("crew") = str4
        newrow("date") = str5
```

```
            _2007DatabaseDataSet.Tables("tblplot").Rows.
              Add(newrow)
            DataGridView1.DataSource = Me._2007DatabaseDataSet.
              tblPlot
            Me.TblPlotTableAdapter3.
              Update(Me._2007DatabaseDataSet.tblPlot)
            MsgBox("Plot table was successfully updated.")
        End Sub

        Private Sub DateTimePicker1_ValueChanged(ByVal sender As
          System.Object, ByVal e As System.EventArgs) Handles
          DateTimePicker1.ValueChanged
                'Format the date output and give this value to
                  the textbox5.
                TextBox5.Text = Format(DateTimePicker1.Value, "d")
        End Sub
End Class
```

4. Run the application—Form1.

   Click the start button to run this application. You can type new plot data in the four text boxes and choose a date from the date-timepicker control. The running mode is illustrated in Figure 14.6.

5. Report.

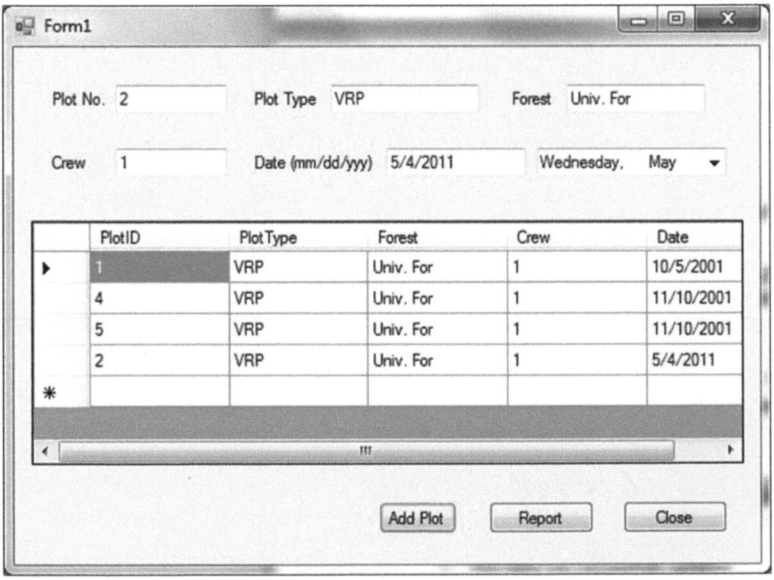

FIGURE 14.6
Running mode of the application for retrieving and adding database data.

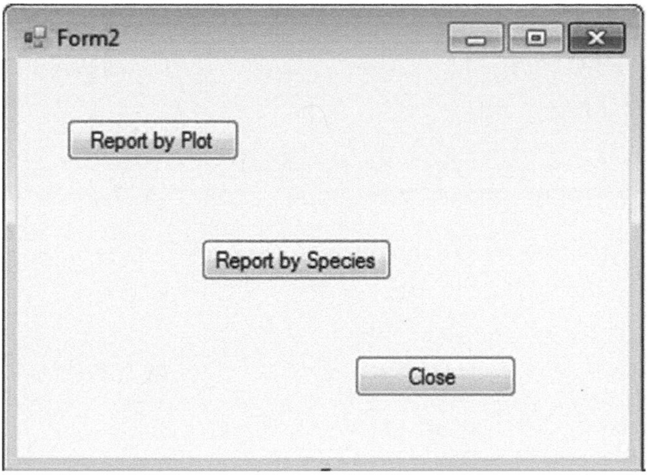

FIGURE 14.7
Design interface of report form.

Click the *Report* button, another form with three buttons will pop up for reporting (Figure 14.7).

6. Code in report form.

```
Public Class frmreport

    Private Sub Button3_Click(ByVal sender As System.Object,
      ByVal e As System.EventArgs) Handles Button3.Click

        Me.Close()

    End Sub

    Private Sub Button1_Click(ByVal sender As System.Object,
      ByVal e As System.EventArgs) Handles Button1.Click

        Call LoadAccessRpt("rptPlot")

    End Sub

    Private Sub Button2_Click(ByVal sender As System.Object,
      ByVal e As System.EventArgs) Handles Button2.Click

        Call LoadAccessRpt("rptSpp")

    End Sub

End Class
```

7. Report procedure in standard module.

```
Public Sub LoadAccessRpt(ByVal rptName As String)
     Dim objAccess As Object

     objAccess = GetObject(My.Application.Info.DirectoryPath &
       "\2007Database.accdb")

     'Open the report in print preview
     objAccess.docmd.openreport(rptName, 2)

     'Make Access visible
     objAccess.Visible = True

     'Maximize the report window
     objAccess.docmd.maximize()

     'End the OLE Automation session
     objAccess = Nothing

End Sub
```

8. Run your application—*frmreport* (Figure 14.7).

14.9 Data Manipulation

ADO.NET is the new database technology of the .NET platform, and it builds on Microsoft ActiveX's Data Objects (ADO). ADO.NET is a language-neutral object model that is the keystone of Microsoft's Universal Data Access strategy.

ADO.NET is an integral part of the .NET Compact Framework, providing access to relational data, XML documents, and application data. ADO.NET supports a variety of development needs. With it you can create database-client applications and middle-tier business objects used by applications, tools, languages, or Internet browsers.

ADO.NET defines DataSet and DataTable objects that are optimized for moving disconnected sets of data across intranets and Internets, including through firewalls. It also includes the traditional Connection and Command objects, as well as an object called a DataReader that resembles a forward-only, read-only ADO record set. If you create a new application, your application requires some form of data access.

ADO.NET provides data access services in the Microsoft .NET platform. You can use ADO.NET to access data by using the new .NET Framework data providers, which are:

- Data provider for Structured Query Language (SQL) server (System.Data.SqlClient).
- Data provider for Object Linking and Embedding Database (OLEDB) (System.Data.OleDb).
- Data provider for Open Database Connectivity (ODBC) (System.Data.Odbc).
- Data provider for Oracle (System.Data.OracleClient).

14.9.1 Example

Let's consider an example that uses a data provider for OLEDB. Suppose we continue to work on our previous project of calculating BA and volume of trees. In the example, instead of using direct file access, we are going to create an Access database to hold the tree data and results, then use OLEDB to access the database and manipulate the data in it.

1. Create a database: create a database containing the following two tables:
   a. tblTreeData
      i. TreeNo, integer, primary key
      ii. DBH, single
      iii. NofLogs, single
   b. tblResult
      i. TreeNo, integer, primary key
      ii. DBH, single
      iii. NofLogs, single
      iv. BA, single
      v. Vol, single
      Remember to input the data of 10 trees in table tblTreeData.
2. Start a new Visual Studio project: add two DataGridView controls, two labels, and three command buttons on form1 (Figure 14.8). The first data grid control is used to display the raw data, including *TreeNo*, *DBH*, and *NofLogs*. The second data grid control is used to display the tblResult table, which contains five fields: *TreeNo*, *DBH*, and *NofLogs*, *BA*, and *Vol*.

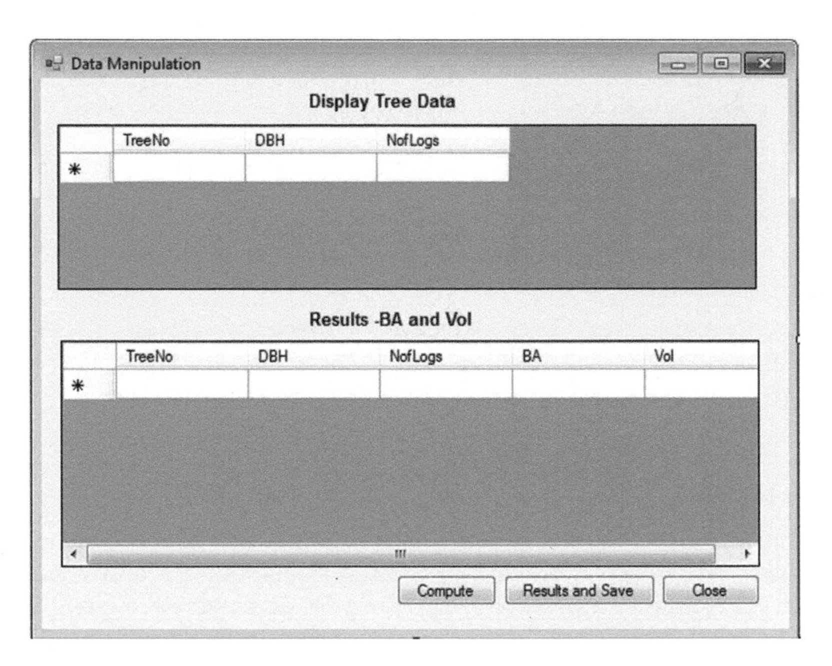

FIGURE 14.8
Design interface of database application.

3. Write code:

```
Imports System.Data.OleDb

Public Class Form1
    'Declare five arrays and one variable at form class level
    Dim aryTreeNo(), aryDBH(), aryNLogs(), aryBA(), aryVol()
    Dim NofTrees As Integer

     'Form loading event procedure
     Private Sub Form1_Load(ByVal sender As System.Object,
       ByVal e As System.EventArgs) Handles MyBase.Load
     'TODO: This line of code loads data into the
       '_2007DatabaseDataSet2.tblResult' table. You can move,
       or remove it, as needed.
             Me.TblResultTableAdapter.
               Fill(Me._2007DatabaseDataSet2.tblResult)
             'TODO: This line of code loads data into the
               '_2007DatabaseDataSet1.tblTreeData' table. You
               can move, or remove it, as needed.
             Me.TblTreeDataTableAdapter2.
               Fill(Me._2007DatabaseDataSet1.tblTreeData)

     Me._2007DatabaseDataSet2.tblResult.Rows.Clear()

     End Sub
```

```
'Function to calculate BA
Private Function CalBA(ByVal dbh As Single) As Single

        CalBA = Math.Round(0.005454154 * dbh * dbh, 2)

End Function

'Function to calculate the volume
Private Function CalVol(ByVal d As Single, ByVal L As
  Single) As Single

        CalVol = Math.Round(((0.55743 * L ^ 2 + 41.51275
          * L - 29.37337) + (2.78043 - 0.04516 * L ^ 2
          - 8.77272 * L) * d + (0.04177 - 0.01578 * L ^ 2
          + 0.59042 * L) * d ^ 2), 2)

End Function

'Reteieving tree data and assigning them to to arrays
Private Sub Button1_Click(ByVal sender As System.Object,
  ByVal e As System.EventArgs) Handles Button1.Click
       Dim query As String = "select * from tblTreeData"
       Dim connstr As String = "Provider=Microsoft.ACE.
         OLEDB.12.0; Data Source=2007Database.accdb"
       Dim conn As OleDbConnection = New
         OleDbConnection(connstr)
       Dim command As OleDbCommand = New
         OleDbCommand(query, conn)
       Dim reader As OleDbDataReader

       conn.Open()
       reader = command.ExecuteReader()

       Dim queryRlt As String = "Delete from tblResult"
       Dim connstrRlt As String = "Provider=Microsoft.
         ACE.OLEDB.12.0; Data Source=2007Database.
         accdb"
        Dim connRlt As OleDbConnection = New
          OleDbConnection(connstrRlt)
        Dim commandRlt As OleDbCommand = New
          OleDbCommand(queryRlt, connRlt)
        Dim readerRlt As OleDbDataReader
        connRlt.Open()
        readerRlt = commandRlt.ExecuteReader()
        readerRlt.Close()
        connRlt.Close()

     Dim n As Integer
        While reader.Read()
            n = n + 1
```

```
                ReDim Preserve aryTreeNo(n), aryDBH(n),
                  aryNLogs(n), aryBA(n), aryVol(n)
                aryTreeNo(n) = reader("TreeNo")
                aryDBH(n) = reader("DBH")
                aryNLogs(n) = reader("NofLogs")
            End While
            NofTrees = n

            reader.Close()
            conn.Close()
            MsgBox("Calculation was done!")
            Button1.Enabled = False
    End Sub

    Private Sub Button3_Click(ByVal sender As System.Object,
      ByVal e As System.EventArgs) Handles Button3.Click

            Me.Close()

    End Sub

    'Computing and saving results
    Private Sub Button2_Click(ByVal sender As System.Object,
      ByVal e As System.EventArgs) Handles Button2.Click

            Dim i As Integer
            For i = 1 To NofTrees
                'Before using the tablerow, you have to
                  create a dataset which contains the
                  datatable (tblResult).
                Dim newrow As _2007DatabaseDataSet2.
                  tblResultRow
                newrow = Me._2007DatabaseDataSet2.tblResult.
                  NewtblResultRow()
                newrow.TreeNo = aryTreeNo(i)
                newrow.DBH = aryDBH(i)
                newrow.NofLogs = aryNLogs(i)
                newrow.BA = CalBA(aryDBH(i))
                newrow.Vol = CalVol(aryDBH(i), aryNLogs(i))
                Me._2007DatabaseDataSet2.tblResult.Rows.
                  Add(newrow)
            Next
                Me.TblResultTableAdaptor.
                  Update(Me._2007DatabaseDataSet2.tblResult)
            Button2.Enabled = False
    End Sub

End Class
```

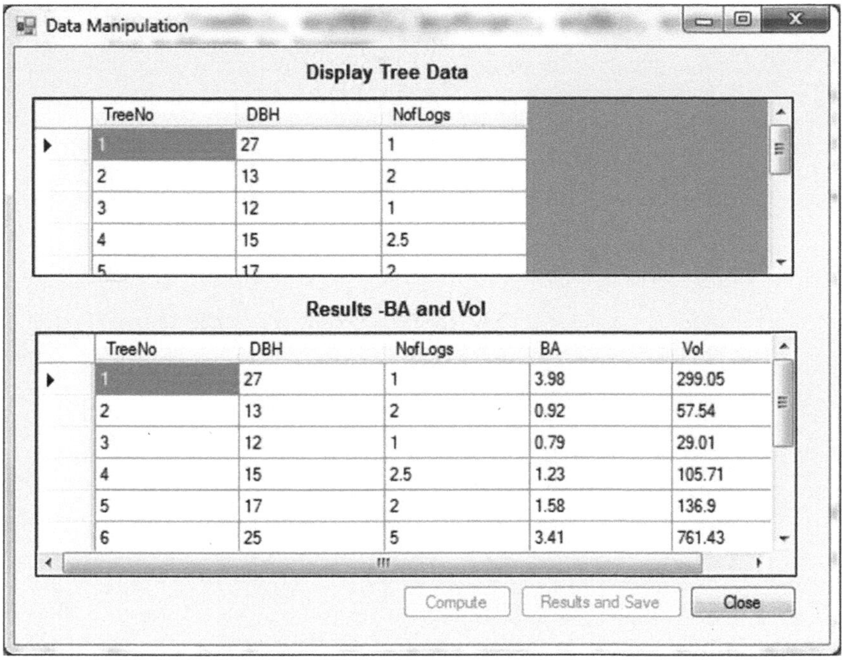

Data Manipulation

Display Tree Data

TreeNo	DBH	NofLogs
1	27	1
2	13	2
3	12	1
4	15	2.5
5	17	2

Results -BA and Vol

TreeNo	DBH	NofLogs	BA	Vol
1	27	1	3.98	299.05
2	13	2	0.92	57.54
3	12	1	0.79	29.01
4	15	2.5	1.23	105.71
5	17	2	1.58	136.9
6	25	5	3.41	761.43

Compute | Results and Save | Close

FIGURE 14.9
Running the application of database.

4. Run your application (Figure 14.9).
   a. Click the *Compute* button.
   b. Click the *Results and Save* button.

Class Exercises

1. We have data for 10 trees in Table 14.5. Create a simple VB.NET project to calculate the total BA and volume of these 10 trees. You should implement the procedures to calculate and sum BA and volume, respectively. Declare three arrays to hold tree data and initialize them in the General Declaration of your project. You are also required to use one command button to invoke calculations of either BA or volume.

   Basal area in ft^2 ($1\ ft^2 = 0.0929\ m^2$):

$$BA = 0.005454154 * (DBH)^2$$

TABLE 14.5
Tree Data

Tree	DBH (in.)	Merchantable Height (Logs)
1	27	1
2	13	2
3	12	1
4	15	2.5
5	17	2
6	25	2
7	28	0.5
8	10	1
9	29	2.5
10	13	0.5

Volume (V) in Doyle board foot (1 Doyle board foot = 2.36 L):

$$V = \left(\left(0.55743 * L \wedge 2 + 41.51275 * L - 29.37337\right) + \left(2.78043 - 0.04516 * L \wedge 2 - 8.77272 * L\right)^* d + \left(0.04177 - 0.01578 * L \wedge 2 + 0.59042 * L\right)^* d \wedge 2\right)$$

where
L = Number of logs
d = DBH in inches (1 in. = 2.54 cm)

References

Barwell, F., R. Blair, R. Case, J. Crossland, and B. Forgey. 2003. *Professional VB.NET* (2nd Edition). Wiley Publishing, Inc., Indianapolis, IN, 985pp.

Microsoft Corporation. 2017. Data type summary (Visual Basic). Available online at https://msdn.microsoft.com/en-us/library/47zceaw7.aspx. Accessed on February 9, 2017.

15

Programming Application Examples in Forest Resource Management

In this chapter, we are going to demonstrate a few forest resource management programming applications and related programming techniques. The application examples, programmed using Visual Basic (VB), VC++, or Java, are:

1. Forest harvesting simulator
2. Timber cruising and inventory
3. Visual Basic for Applications (VBA) for harvesting system production and cost analysis
4. Three-dimensional (3D) log bucking optimization
5. 3D lumber edging and trimming optimization system
6. 3D log sawing optimization system
7. Forest and biomass harvest scheduling and optimization

15.1 Forest Harvesting Simulator

A forest harvesting simulator can help enhance the efficiency and profitability of forest operations by simulating repetitive and complex operational problems under certain working conditions (Wang and LeDoux 2003). Our earliest version of this simulator only modeled one machine type—drive-to-tree feller-buncher under Disk Operating System operating environment. Simulation was conducted by moving a physical machine model in a stand map on a digitizing pad. Since the first version, we have continued to improve the simulator to make it comprehensive, flexible, and easier to use. As new technologies have been introduced into computer platforms and programming tools, we have incorporated these advancements into this simulator.

The current version of the simulator was written with MS VB Version 6.0 for 2D simulations (Wang and Greene 1999) and Visual C++ for 3D operations

(Li 2005) under the Windows environment. Using the newest version of the forest harvesting simulator, you can:

- Edit mapped stands
- Generate stands
- Simulate felling
- Simulate skidding or forwarding
- Monitor the traffic levels of extraction machines
- Analyze simulations
- Retrieve, view, and produce output of simulations

15.1.1 Forest Stand Generation

15.1.1.1 Random Pattern

If a random spatial pattern is requested, a ratio of the stand density to the total number of possible tree locations based on minimum *X* and *Y* spacing is first calculated. Then a random number with a uniform distribution between 0.0 and 1.0 is generated for each possible tree location. If this number is less than or equal to the ratio described, the coordinate location is assigned a tree. If the random number is greater than the ratio, the coordinate location is considered to be unoccupied (Farrar 1981). The minimum spacing of *X* and *Y* is considered in this procedure when we model natural stands. At each location, tree diameter at breast height (DBH) is assigned randomly. The total height and volume of that tree are then calculated based on the assigned DBH (Borders et al. 1990).

15.1.1.2 Uniform Pattern

All possible grids for tree locations are identified based on stand density and *X* and *Y* spacing. If *X* and *Y* both meet the minimum spacing requirements, a tree location is assigned in the center of this *X* by *Y* rectangle. A random variation of a half X_{min} or Y_{min} is allowed in modeling both the *X* and *Y* coordinates for each tree's location.

15.1.1.3 Clustered Pattern

When the clustered pattern is used, the number of cluster centers specified by the user is located randomly within a plot. By generating the *X* and *Y* coordinates randomly, using a pair of random numbers, each tree is given an initial location. The distances from that tree location to each of the cluster centers are determined, and the nearest center is selected. The distance from this center to the tree is then multiplied by a random number between 0.0 and 1.0 to give that tree a new location relative to the cluster center (Farrar 1981).

New coordinates are then calculated for the tree, and the distances between that location and the neighboring trees are checked to assure that the minimum nearest distances are maintained. If a tree location has violated the distance parameter, the procedure is repeated; otherwise, the location is assigned as a tree location.

15.1.2 Felling Operations

The numerical simulation model for chain saw, feller-buncher, and harvester consists of two parts: (1) movement of the logger with a chainsaw from tree to tree or machine movement from tree to tree and (2) tree felling or processing (Figure 15.1).

15.1.2.1 Chainsaw Felling

Walk to tree, acquiring, felling, limbing, and topping are modeled for the chainsaw. The felling direction is first defined within a random variation range and the sawyer is located at one end of a plot. Usually, the logger will move to the nearest tree to be cut and fell the selected trees in a narrow swath. When the logger reaches the far end of the plot, he/she will return felling trees in the next nearest swath.

15.1.2.2 Feller-Buncher Felling

Four functions are modeled for the drive-to-tree feller-buncher: move to a tree, cut the tree, move to dump, and dump. The feller-buncher is first located at one end of the plot and then moves parallel to the rows of trees; the rows are 4.5–6 m wide. Marked trees on either side of the machine are removed. When the machine reaches the end of the row, it turns around and cuts another tree in the nearest swath, continuing until the plot is finished. The system searches for the "cut" tree and adds the tree to the felling head. A solid black circle is drawn at the location of a cut tree to signify the stump. This procedure is repeated until the head is full. The system then moves the machine image to the location of the bunch or pile to be built and drops the trees.

15.1.2.3 Harvester Felling

Six functions were modeled for the cut-to-length harvester: move, boom extend/retreat, cut, swing boom, processing, and dumping. Unlike the feller-buncher, a harvester with a boom can reach several trees at a stop. Trees can be felled and processed with the same harvester. A circle around the harvester is drawn to indicate the reach of the boom. This circle is moved as the harvester moves. The harvester usually runs in a straight trail and works in a 12–15 m wide strip depending on the boom reach. Trees on the trail must be removed for machine travel. Trees on either side of the machine within the

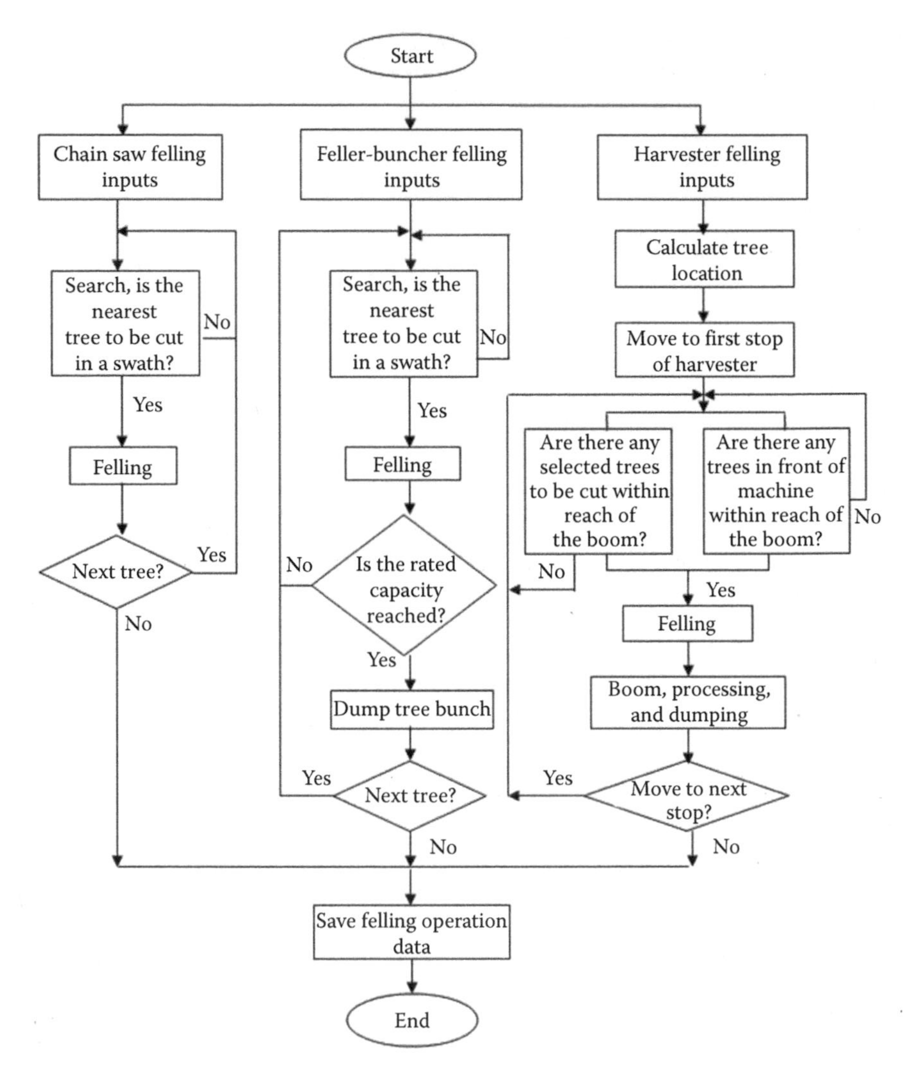

FIGURE 15.1
Flowchart of numerical felling operations.

boom reach can be removed based on the user's choice of harvest method. The processed trees are then dropped on either side of the harvester trail for later forwarding.

15.1.3 Extraction Simulation

15.1.3.1 Extraction Patterns

In an interactive skidding simulation, a landing must first be placed in the logging area that was created by felling a plot a fixed number of times. Tree or log pile data are provided by the felling simulation. The skidder machine will begin at the landing, load the nearest tree pile, and then move to the next closest pile until it is fully loaded. Then the loaded machine will travel back to the landing. While the forwarder is simulated in a similar manner, it follows the harvester's trail and loads logs with a self-mounted boom.

Although interactive simulation allows constant and direct human input to the simulation, it is time-consuming and often repetitive, especially with respect to uniform skidding or forwarding patterns. As a result, a numerical skidding or forwarding simulation was modeled in the system (Figure 15.2). To date, four skidding or forwarding patterns (SP1, SP2, SP3, and FP1) have been modeled (Figure 15.3):

SP1: Freestyle skidding (no designated skid trail)

SP2: Skid trail runs through the center of plot (one trail)

SP3: Skid trails traveling from the landing to the corners of plot (two trails)

FP1: Forwarding along the trails of the harvester (forwarding direct to road)

The program also allows the user to choose the landing location and the machine payload. The landing must be located before performing a simulation. The machine will begin at the landing and move to the nearest tree bunch or log pile, and then move to the next nearest bunch or pile until it is fully loaded. The machine then follows the specified extraction pattern throughout the entire simulation process.

15.1.3.2 Traffic Intensity

The traffic intensity within each smaller grid (e.g., 5 by 5 m) is recorded into a file while the numerical extraction simulation is being performed. Four travel intensity categories for a skidder or forwarder are defined in the system (Carruth and Brown 1996):

TI1: Trees on the plot have been felled.

TI2: Trees that stood on the plot have been removed and no other traffic has passed through the plot.

TI3: Trees that stood on the plot have been removed and trees outside the plot have been skidded through the plot. There have been 3–10 passes with a loaded machine.

TI4: There have been more than 10 passes with a loaded machine through the plot.

After the machine is fully loaded, it will return to the landing by the shortest and easiest route depending on the extraction pattern. The skidding area

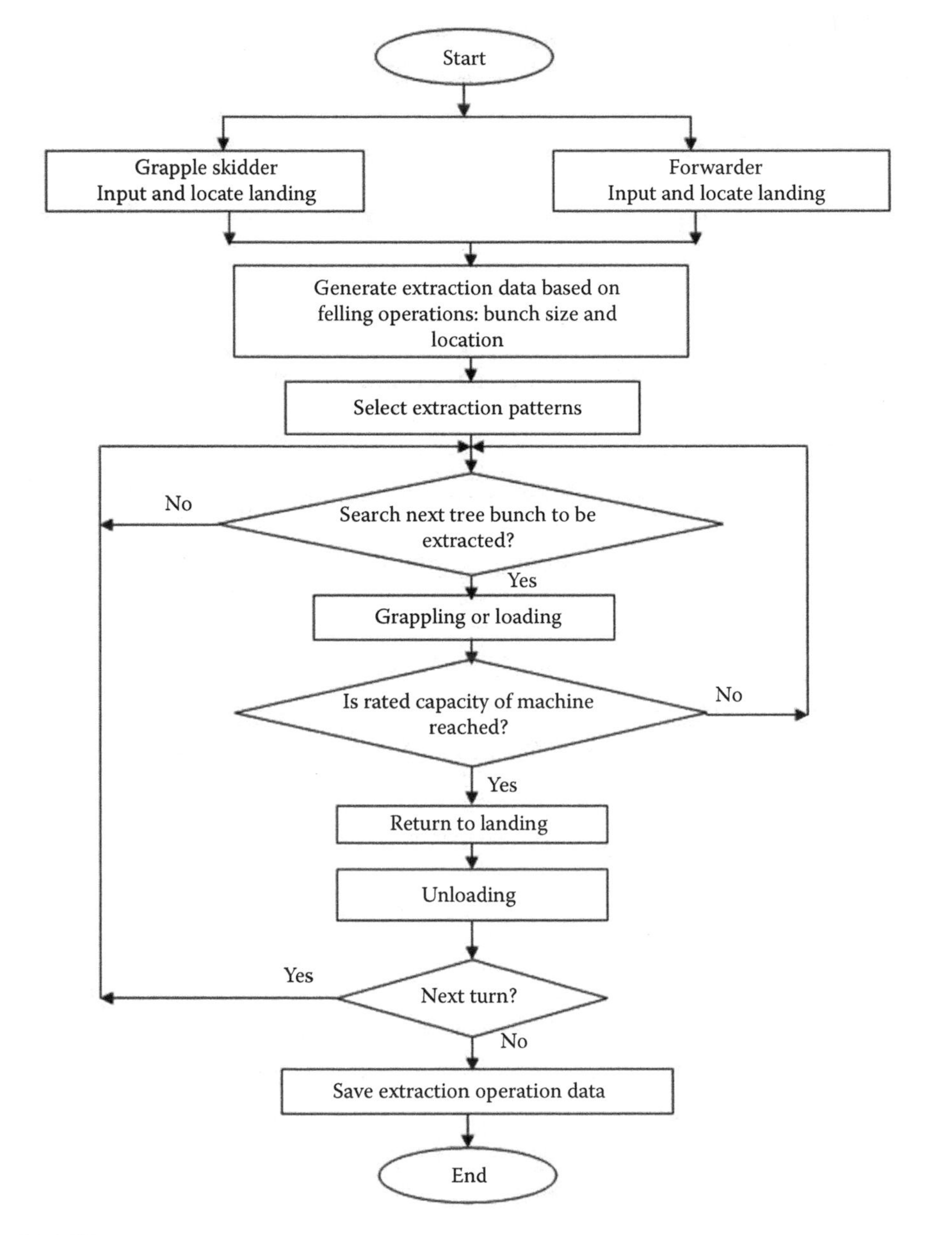

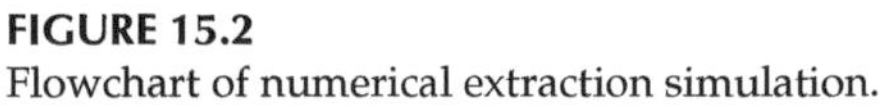

FIGURE 15.2
Flowchart of numerical extraction simulation.

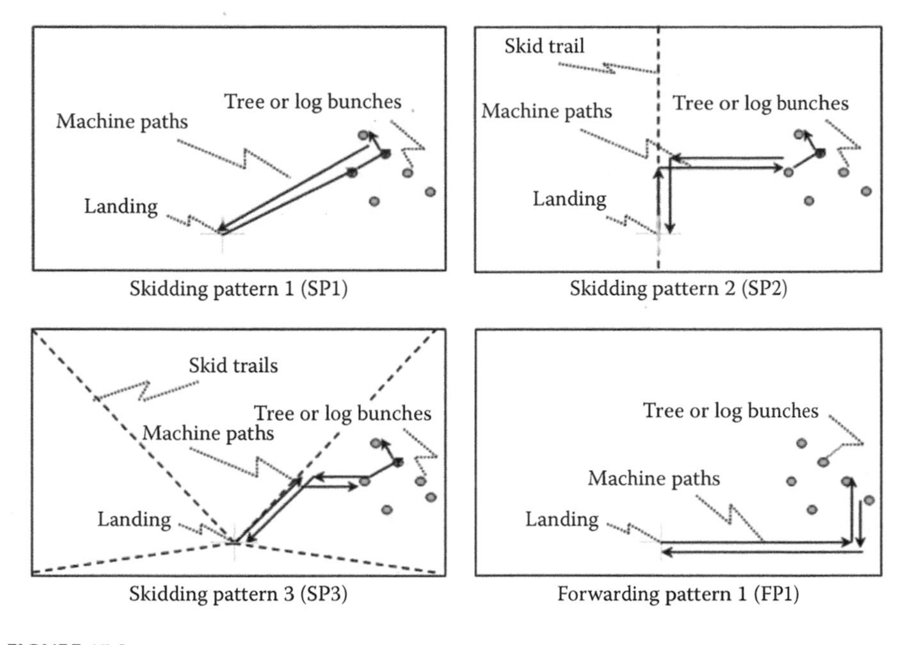

FIGURE 15.3
Diagram of extraction patterns.

is divided into cells (5 by 5 m) for accurate recording of travel intensity. This grid width allows two machines to pass each other on a trail. There are eight possible direction options for a machine to move from its current position to the next position.

This procedure is repeated until the end point (X_i, Y_i) is reached. The number of passes is recorded and accumulated for each grid that the loaded machine passes through as it is traced from the start point to end point. The travel intensity category in each grid is displayed in four colors depending on the intensity level. After the tracking is completed along the entire line segment, the number of passes for the loaded machine in each grid is stored in computer memory, and then saved to a file after the simulation is completed.

15.1.4 Simulation Example

A natural hardwood stand in central Appalachia was generated to illustrate the performance of the stand generation and harvesting simulation. Assume the species composition is yellow poplar 40%, black cherry 18%, red maple 16%, red oak 15%, and other hardwoods 11%. The stand density is 485 trees per hectare with a DBH range from 5 to 110 cm and a plot size of 0.16 ha. Also assume that the stand is about 75 years old and trees are randomly distributed.

Once the above information is entered, a 2D stand map is generated and displayed (Figure 15.4a). Three smaller windows are used to display the stand

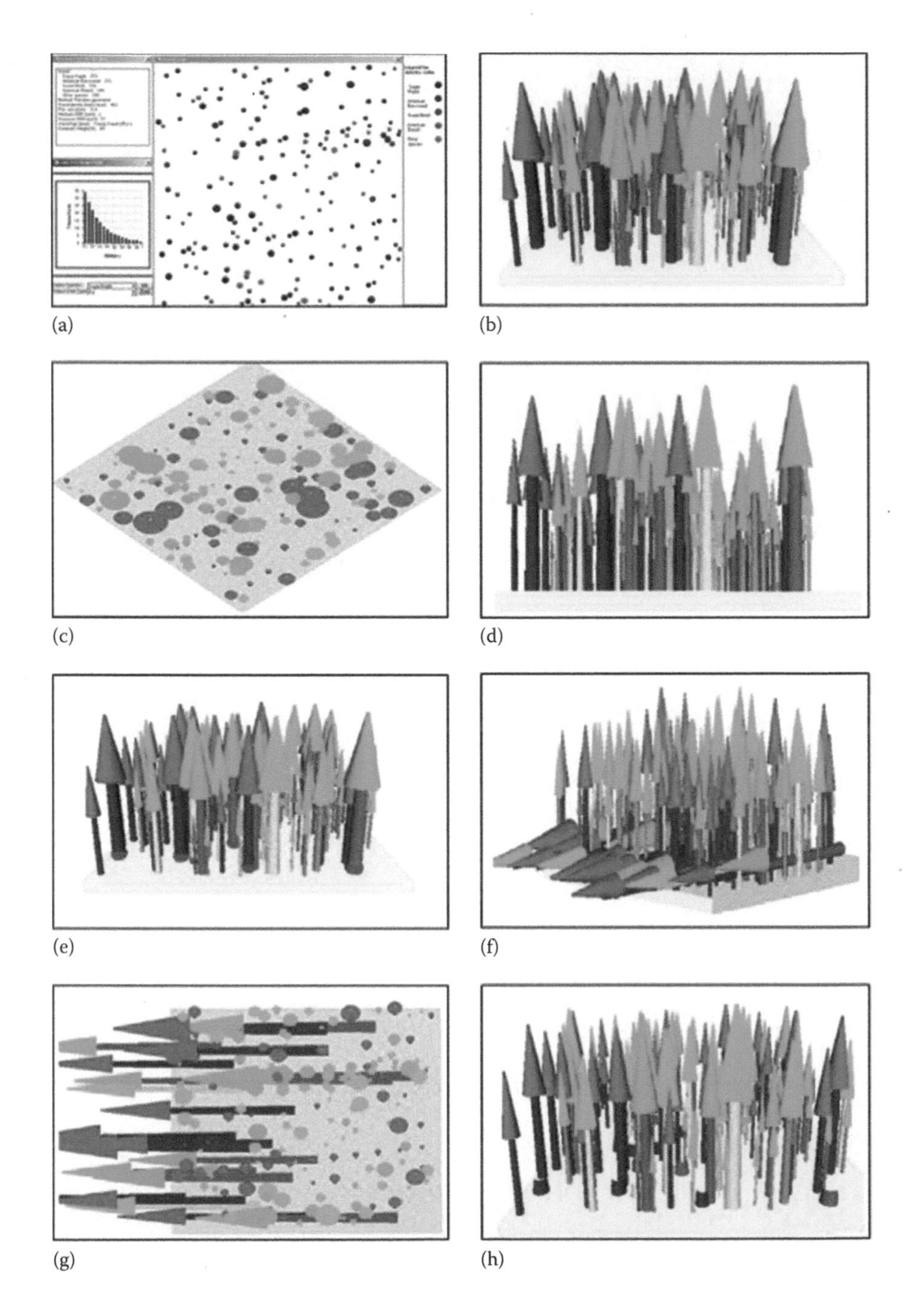

FIGURE 15.4
Stand generation and harvesting simulation. (a) 2-Dimenisonal stand map, (b) perspective projection and projective view, (c) overhead view, (d) profile view, (e) 3D component to mark trees for harvest, (f) felling of marked trees, (g) felling of marked trees, and (h) extraction of felled trees from plot.

information, color legends for species, and the DBH distribution by species or overall. Meanwhile a stand map data file is created and saved in the system. The 3D stand map can be displayed by changing the display mode. Some functionality can be performed on the 3D stand map. For example, the image can be rotated from left or right continuously to allow for examination of the stand structure from different perspectives. The user can also change the crown height and diameter by using the "tree design" module. In order to differentiate species on the map, a unique color is randomly generated and assigned to each species. Additionally, the tree height and DBH are drawn to scale for better visualization.

Two projection modes (perspective projection and parallel projection) and three view modes (projective view, profile view, and overhead view) can be produced by the system. Perspective projection and projective view are the default projection mode and view mode, respectively (Figure 15.4b). Overhead view and profile view can be enabled by changing the view mode (Figure 15.4c and d). The 3D component can also allow the user to mark trees to be harvested (Figure 15.4e). Trees can be marked by species, DBH, or both, or by the user's specified requirements such as marking diseased or poorly formed trees. The marked trees can then be felled (Figure 15.4f and g) and extracted from the plot (Figure 15.4h).

The system saves the operation data into the database for later analyses once a simulation run is complete. The analysis module of the harvesting simulator can analyze the generated stand and compare it with a thinned or partially cut stand. This module provides (1) elemental time summary, (2) machine summary by cycle, (3) summary of harvested stand and extracted site, and (4) production analysis as well as traffic intensity of extraction machines across the site.

15.2 Timber Cruising and Inventory

West Virginia University (WVU) Cruise is especially designed to analyze the field cruising data and provide the results in the format of an MS Access report (Wang 2004a).

15.2.1 Manipulate Field Cruising Data

The field cruising data should be in MS Excel format. In order to import the Excel data to WVU Cruise, you need to do the following things:

1. Open the Excel file that contains the cruising data. The data must contain the following fields in order: Plot#, Species, DBH, merchantable height in the number of logs of a tree, Pulp, Grade, and total

Charlie.xls

	A	B	C	D	E	F	G
1	Plot #	Spp	DBH	MHT	Pulp	Grade	HGT
2	1	RM	34	1.5	0	3	117
3	1	YB	21	0	0	8	0
4	1	YB	13	0	0	10	0
5	1	YB	18	1.5	0	3	97
6	1	SM	12	0.5	0	3	83
7	1	OH	2	0	0	9	0
8	1	SM	19	3	0	3	102
9	1	YB	18	0	0	8	0
10	2	RM	14	1.5	0	3	0
11	2	NRO	23	2.5	0	3	0
12	2	NRO	15	2.5	0	3	0

STATS / SPECIES / PER ACRE SUM

FIGURE 15.5
Format of cruise data.

height in feet (1 ft = 0.305 m). If you have no data for one or more fields, put 0s in the cells (Figure 15.5).

2. Create a Named Range, *DataRange*, in your Excel spreadsheet
   a. Highlight the row(s) and column(s) where your data reside.
   b. On the *Insert* menu, point to *Name* and click *Define*.
   c. Enter the name *DataRange* for the Named Range name.
   d. Click *OK*.
3. Start the WVU Cruise program. The main MDI form is displayed first (Figure 15.6). On the menu bar, there are *File, Report, Tools,* and *Help* menus.

15.2.2 Import Data

1. In the *Tools* menu, click *Import*. A dialog window pops up for importing Excel data into the cruising program (Figure 15.7). Once the file is selected, click *OK*.

15.2.3 Load Data

1. Now you can load the cruising data in the program. In the *File* menu, click *Open*. The cruising data will be loaded for analysis (Figure 15.8).
2. You can browse the loaded data. Notice that there are *View Species and Grade* and *Cruise Design* buttons on the form.

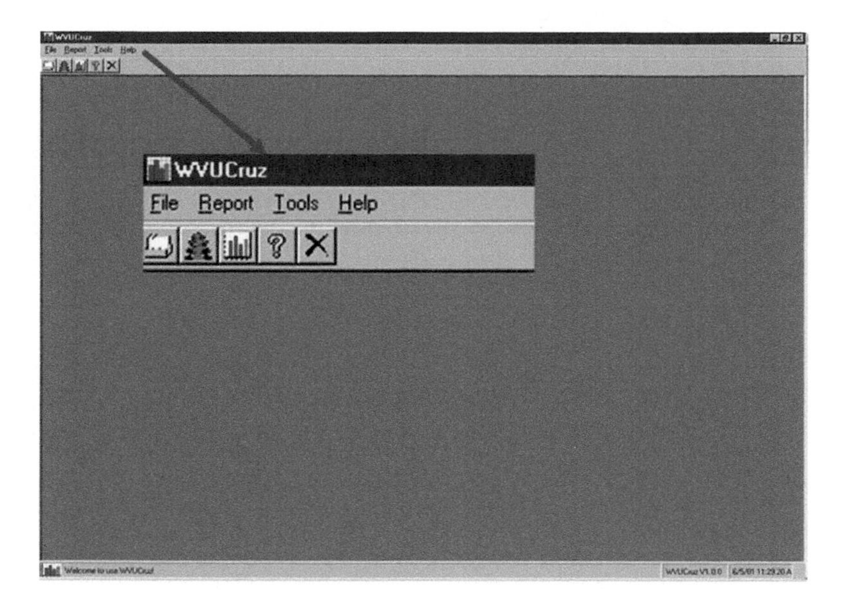

FIGURE 15.6
Menu and toolbars of WVU Cruise.

FIGURE 15.7
Import Excel data into WVU Cruise.

15.2.4 Cruise Design

If the user clicks the *Cruise Design* button, a design form will pop up (Figure 15.9).

In this design window, you can:

1. Design and save cruising information for the report header. Simply enter the required text boxes and then click *Save Cruise Info*.

WVUCruz

File Report Tools Help

Open
Close
Exit

Cruising Data

CRUZ INFORMATION - 204 cruised trees were loaded for analysis!

PointID	Species	DBH	MHT	Grade
	RM	34	1.5	3
1	YB	21	0	8
1	YB	13	0	10
1	YB	18	1.5	3
1	SM	12	0.5	3
1	OH	2	0	9
1	SM	19	3	3
1	YB	18	0	8
2	RM	14	1.5	3
2	NRO	23	2.5	3
2	NRO	15	2.5	3

View Species and Grade | Cruise Design | Close

FIGURE 15.8
Load cruise data for analysis.

Cruise Design

Cruise Name: NONE

Legal Description: NONE

Girard Form Class
Save Cruise Info
Close
OK

Cruise Date: NONE
Cruised Acres: 1
Cruising Allowable Error (%): 10

Plot Configuration
Plot Size (ac): 0.1
Fixed Area
VRP
BAF: 20
100% Tally

Cruise Summary By
Volume
Plot
Cruise Statistics
Stand/Stock Table

Stand/Stock Table By
DBH Class
Species & DBH
Grade, Species, DBH

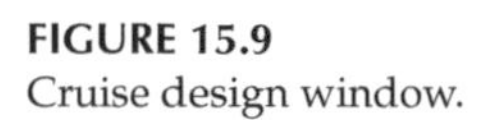

FIGURE 15.9
Cruise design window.

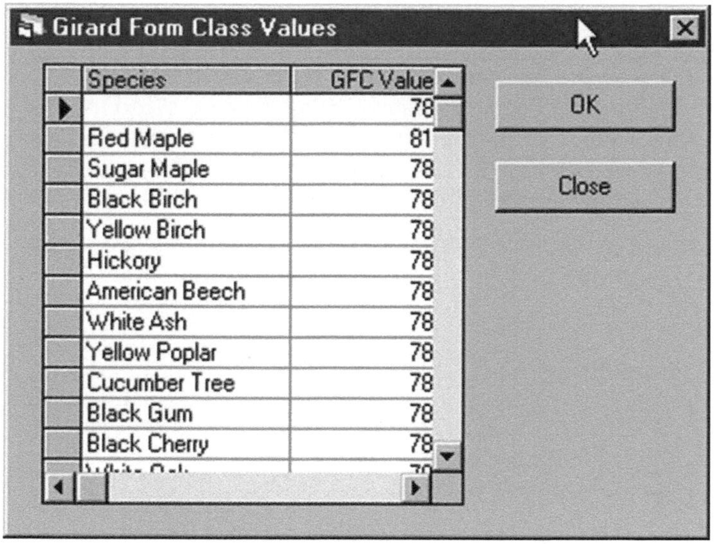

Girard Form Class Values

Species	GFC Value
	78
Red Maple	81
Sugar Maple	78
Black Birch	78
Yellow Birch	78
Hickory	78
American Beech	78
White Ash	78
Yellow Poplar	78
Cucumber Tree	78
Black Gum	78
Black Cherry	78

OK

Close

FIGURE 15.10
Edit Girard form class by species.

2. Select form class for a specific species. Click *Girard Form Class*, and a form is displayed (Figure 15.10).
3. Configure plot. You can use *fixed area, variable radius plot,* or *100% tally.*
4. Summarize cruising data by:
    a. Volume
    b. Plot
    c. Cruise statistics
    d. Stand/stock table by DBH class, Species and DBH, or Grade, Species, and DBH.

15.2.5 Report

You can output the summarized cruising data in the Access format:

1. By volume: Two parts of the volume summary are provided: merchantable volume tract summary and merchantable volume per acre (1 acre = 0.4 ha) summary, which are saved in the database and then reported.
2. The plot-level result is also provided. Once the data are saved, a report by plot will be generated.

3. The cruising statistics are also provided for volume per acre in cubic foot (1 acre = 0.4 ha, 1 ft^3 = 0.0283 m^3), in international board foot (1/4), in Doyle board foot volume, trees per acre (1 acre = 0.4 ha), and basal area per acre. Statistics include mean, standard deviation, standard error, variance, coefficient of variance, confidence interval at 95% level, percent of error, and sample size. The sample size will be especially useful for later cruising design.
4. The stand/stock table report can be reported in three levels by DBH class, Species and DBH, and Grade, Species, and DBH.

15.2.6 Programming

1. Input data: Convert MS Excel data into Access format and then load them into the memory.

```
Dim xlBook As Excel.Workbook
Dim rgSpp As Range

Private Sub Command1_Click()

    On Error GoTo err_Handler

    Dim filePath As String
    Dim i, j, nExcelData As Integer
    Dim DataConn As ADODB.Connection
    Dim rdExcelData As ADODB.Recordset

    If Right(File1.Path, 1) <> "\" Then
        filePath = File1.Path & "\" & File1.fileName
    Else
        filePath = File1.Path & File1.fileName
    End If

    Set DataConn = New ADODB.Connection

    Path = filePath
    DRIVER = "{Microsoft Excel Driver (*.xls)}"
    Db = "DBQ=" & Path & ";"
    Db = Db & "DefaultDir=" & Path & ";"
    Db = Db & "Driver=" & DRIVER & ";"
    DataConn.Open Db
    ChDir App.Path
    Set rdExcelData = DataConn.Execute("SELECT * FROM
      DataRange")

        Set dbCruzData = OpenDatabase("dbCruzData.mdb")
```

```
            Set rdCruzData = dbCruzData.
              OpenRecordset("tblAllTree1")
            Call ClearTreeTable(rdCruzData)

        If rdExcelData.EOF Then
            MsgBox "No data can be loaded from this empty
              file!" & Chr(13) & filePath, vbOKCancel +
              vbInformation, "Excel Conversion"
            Exit Sub
        Else
            j = 0
            nExcelData = 0
            Do While Not rdExcelData.EOF
                nExcelData = nExcelData + 1
                rdExcelData.MoveNext
            Loop
            frmCruzProg.Show
            frmCruzProg.Caption = "Importing cruising
              data ... "
            frmCruzProg.CruzProgBar.Max = nExcelData
            Screen.MousePointer = vbHourglass
            rdExcelData.MoveFirst
            Do While Not rdExcelData.EOF
                i = 0
                j = j + 1
                frmCruzProg.CruzProgBar.Value = j
                rdCruzData.AddNew
                For Each Field In rdExcelData.Fields
                     rdCruzData.Fields(i) = Field.Value
                     i = i + 1
                Next
                rdCruzData.Update
                rdExcelData.MoveNext
            Loop
        End If
        Set rdExcelData = Nothing
        Unload frmCruzProg
        Screen.MousePointer = vbDefault

        Exit Sub

    err_Handler:
        Call Error_Handler

    End Sub
```

2. Process data: Use arrays to hold the data and then compute trees per acre (1 acre = 0.4 ha), basal area, and volume by species, DBH, and others.

Assign cruising data to arrays for processing.

```
Public Sub AssignCruiseData()

    On Error GoTo err_Handler

    Dim i As Integer
    i = 1

    frmCruzProg.CruzProgBar.Max = nRecord
    Screen.MousePointer = vbHourglass
    'nRecord = rdCruiseData.RecordCount
    ReDim Preserve PointNo(nRecord), Species(nRecord),
      DBH1(nRecord), Height(nRecord), PulpW(nRecord),
      Grade(nRecord), THeight(nRecord)
    ReDim SPPCruised(NofSpecies), PointCR(NofPoint),
      DClassCR(NofDClass), GradeCR(NofGrade)
    rdCruzData.MoveFirst
    While Not rdCruzData.EOF
        frmCruzProg.CruzProgBar.Value = i
         PointNo(i) = rdCruzData("PointID")
         If IsNull(rdCruzData("Species")) Then
            Species(i) = "None"
        Else
            Species(i) = rdCruzData("Species")
        End If
         'Species(i) = CSpecies(Species(i))
         If IsNull(rdCruzData("DBH")) Then
            DBH1(i) = 0
         Else
            DBH1(i) = rdCruzData("DBH")
         End If
         If IsNull(rdCruzData("MHT")) Then
            Height(i) = 0
         Else
            Height(i) = rdCruzData("MHT")
         End If
         If IsNull(rdCruzData("Pulp")) Then
            PulpW(i) = 0
         Else
            PulpW(i) = rdCruzData("Pulp")
         End If
         If IsNull(rdCruzData("Grade")) Then
            Grade(i) = "None"
         Else
            Grade(i) = rdCruzData("Grade")
         End If
         If IsNull(rdCruzData("THT")) Then
            THeight(i) = 0
```

```
     Else
        THeight(i) = rdCruzData("THT")
     End If
    'Debug.Print Point(i), Species(i)
     i = i + 1
     rdCruzData.MoveNext
Wend

'Assign species
i = 1
rdSpecies.MoveFirst
While Not rdSpecies.EOF
    If IsNull(rdSpecies("Species")) Then
        SPPCruised(i) = "None"
    Else
        SPPCruised(i) = rdSpecies("Species")
    End If
    'Debug.Print SPPCruised(i)
    i = i + 1
    rdSpecies.MoveNext
Wend
'Assign plot
i = 1
rdNofPlot.MoveFirst
While Not rdNofPlot.EOF
    If IsNull(rdNofPlot("PointID")) Then
        PointCR(i) = 0
    Else
    PointCR(i) = rdNofPlot("PointID")
    End If
    'Debug.Print PointCR(i)
    i = i + 1
    rdNofPlot.MoveNext
Wend
'Assign DBH class
i = 1
rdDBHClass.MoveFirst
While Not rdDBHClass.EOF
    If IsNull(rdDBHClass("NofDClass")) Then
        DClassCR(i) = 0
    Else
        DClassCR(i) = rdDBHClass("NofDClass")
    End If
    i = i + 1
    rdDBHClass.MoveNext
Wend
'Assign grade
i = 1
rdNofGrade.MoveFirst
```

```
        While Not rdNofGrade.EOF
            If IsNull(rdNofGrade("Grade")) Then
                GradeCR(i) = "None"
            Else
                GradeCR(i) = rdNofGrade("Grade")
            End If
            i = i + 1
            rdNofGrade.MoveNext
        Wend

        blCallAssignCruz = True

        Unload frmCruzProg
        Screen.MousePointer = vbDefault

        Exit Sub

    err_Handler:
        Call Error_Handler

    End Sub
```

15.3 VBA for Harvesting System Production and Cost Analysis

Traditionally, the logging system analysis was programmed with either a simple spreadsheet without detailed business functions or with a programming language that was too complicated for many loggers to understand and apply to their operations. The spreadsheet approach often eliminated the detailed business functions, but the calculations could be followed through the worksheets for easy understanding of the process. Approaches using programming language interfaces are usually able to incorporate advanced features and functions; however, many of the calculations are performed outside of the view of the operator, making it difficult for users to follow and understand the entire process. In this section, we introduce a spreadsheet program, the Central Appalachian Harvesting Analyzer (CAHA), that is simple and easy-to-use with more flexible functions for harvesting system production and cost analysis. Specifically, we will discuss (a) development of an integrated MS Excel-based spreadsheet harvesting analysis program with VBA to enhance the logging business management for loggers in central Appalachia and (b) applications of the program to examine the harvesting system configurations and interactions under varied harvest and stand scenarios.

15.3.1 Excel and VBA

VBA follows the syntax of VB and has its own integrated development environment (IDE) running within the process space of a host application, such as Excel. VBA can be used to access and change properties of the object model, handle events intrigued by objects, and call the methods of the objects (Lomax 1998). Both VBA and the stand-alone version of VB use the same language engine, editor, and most supporting tools (Jacobson 1999). A VBA application is always interpreted and cannot run independently of its host application, Excel in our case, whereas VB programs can be compiled into executables and run in their own process space, or compiled into ActiveX components and executed within other applications.

15.3.2 System Design and Implementation

The CAHA was implemented by combining excellence in spreadsheet modeling with VBA. The well-designed worksheet format used in the Auburn Harvesting Analyzer (Tufts et al. 1985, Greene and Lanford 1992) and cost estimates of the harvesting machines based on the machine rate method (Miyata 1980) were used as a foundation for the CAHA. Previously published production/cost equations and stand data from the region were included as built-in functions in the spreadsheet (Rennie 1996, Wang et al. 2002a) (Table 15.1). Other hardwood-specific harvesting productivity/cost data could be used as references, such as information from Huyler and LeDoux (1999), LeDoux and Huyler (2001), LeDoux (1985) for harvester and

TABLE 15.1

Cycle Time Models for Harvesting Machines[a]

Machine	Model (min)	Source
	Felling machines	
Chain Saw	–2.4295 + 0.4222*DBH + 0.002* DistT2	Wang et al. (2004b)
Feller-buncher	0.367 + 0.0008*DBH^2 + 0.00026*MHT^2 + 0.02246*DistT + 0.00679*DistD	Wang et al. (2004c)
Harvester	0.4032 + 0.0022*DBH*2.54* exp(4.85 – 7.82/DBH)*0.3048	Wang et al. (2005a)
	Extraction machines	
Cable skidder	9.918 + 0.0049*Dist – 0.0000006*Dist^2 + 0.0338*TotVol	Wang et al. (2004b)
Grapple skidder	0.844 + 0.00272*Dist + 0.0000007 Dist^2 + 0.022 TotVol	Wang et al. (2004c)
Forwarder	27.5079 + 0.03784*(Dist*0.3048) – 0.00006572*(Dist*0.3048)*(TotVol/35.3145)	Wang et al. (2005a)
Helicopter[b]	1.2648 + 0.01441*TotVol	Wang et al. (2005b)

[a] DBH in inches; DistT—average distance between harvested trees, ft; DistD—distance to dump, ft; Dist—average extraction distance, ft; TotVol—volume per turn, ft^3.

[b] Volume per turn is in MBF.

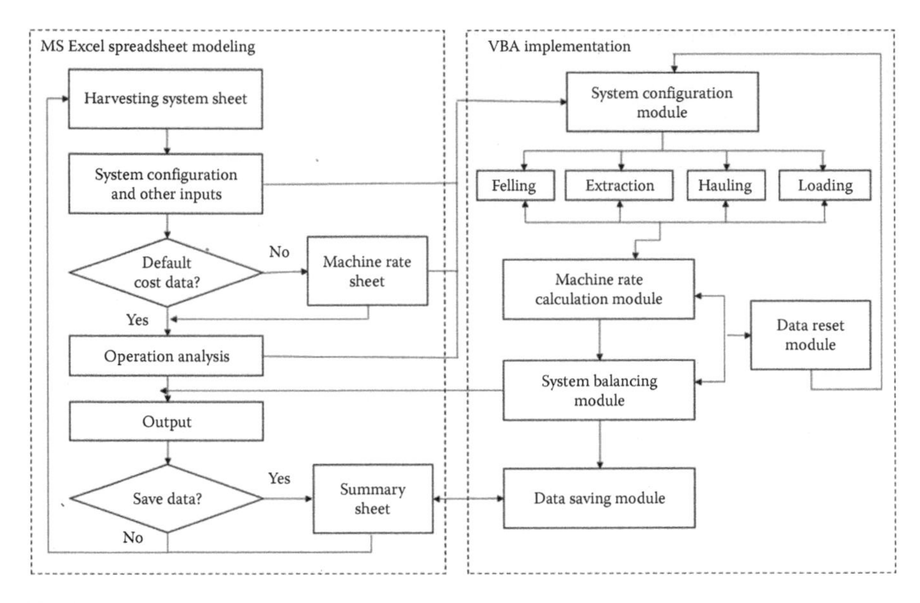

FIGURE 15.11
Flowchart of the Central Appalachian Harvesting Analyzer.

loader and truck, information from Reisinger and Gallagher (2001) for cable skidder and grapple skidder, information from Thompson et al. (1995) for cable skidder and forwarder, Grushecky et al. (2007) for trucking, and information from Sturos et al. (1996) for cable yarding.

The CAHA consists of four Excel worksheets (Figure 15.11). The first two, titled *System_MBF* and *System_CUFT*, contain the main part of the harvesting system analyzer for use with either thousand board feet (MBF) (1 MBF = 2359.73 L) or cubic feet (CUFT) (1 ft^3 = 0.0283 m^3) measurement systems, respectively. The next worksheet is titled *MachineRate* and is used to calculate the machine rate for individual machines. The last sheet is a summary sheet that summarizes the entire harvesting operation. Behind the worksheets are five VBA modules for system configuration: machine rate calculation, system balancing, data reset, and data storage, all of which are implemented through the Excel spreadsheet interface. Inputs to CAHA include stand data, tract size and road condition, logging crew information, harvesting equipment, and cost factors. Outputs are machine hourly production and unit cost, system rate, weekly production, onboard cost, and time required for harvesting a certain tract.

The *System_MBF* sheet is the default sheet of the CAHA, which is organized into three sections similar to Auburn Harvesting Analyzer (Tufts et al. 1985, Greene and Lanford 1992): General Information, Machine, and System (Figure 15.11). The *System_CUFT* sheet uses the same format as the *System_MBF* sheet except for the cubic foot (1ft^3 = 0.0283 m^3) volume unit.

General information: Stand conditions, working crews and schedule, tract size, and other supporting information (equipment moving distance and cost, roads to be built) are included in this section. Stand conditions include DBH classes, merchantable height (number of 4.88 m logs (16 ft logs)), trees per acre (1 acre = 0.4 ha), and volume per tree for each DBH class. Merchantable height and tree volume were calculated using the models developed by Wang et al. (2002b) and Rennie (1996). All other information is directly input by the user.

Machines: Machine productivity and hourly costs can be estimated in this section. Site conditions including average distance between harvested trees and average extraction distance are also specified in this section. Machinery is classified into four categories: felling, extraction, loading, and hauling. Drop-down boxes hold the available machines for each category so that the user may configure a desired system (Table 15.2). Implementation of this harvesting system equipment selection feature is handled by the system configuration module (Figure 15.11). A set of typical data was provided as default values for each machine type used in the region. Once the machine type is selected, it is combined with the other supporting data to automatically calculate and report the productivity based on the built-in functions. The fixed, operating, and labor costs are calculated on an hourly basis for the user to accept as the default cost data before the harvesting operation can then be analyzed. Otherwise, the machine rate sheet will be activated, in which the user is able to calculate the fixed, operating, and labor costs for each individual machine. The machine rate sheet was programmed based on the machine rate method (Miyata 1980). Finally, a data reset module is included to reset the inputs to default values and restart the harvesting system configuration (Figure 15.11).

System: Harvesting systems can be balanced (a) manually by directly inputting the number of machines employed in each function or (b) automatically through the system balancing module (Figure 15.11). This section displays the summarized utilization rate, production, and costs for each harvesting function as well as weekly production and onboard cost for the balanced

TABLE 15.2

Harvesting System Configurations[a]

	Harvesting Functions			
System	**Felling**	**Extraction**	**Loading**	**Hauling**
1	Chain saw (2)	Cable skidder (1)	Larger loader (1)	Long-log truck (3)
2	Feller-buncher (1)	Grapple skidder (2)	Larger loader (1)	Long-log truck (3)
3	Harvester (1)	Forwarder (1)	Larger loader (1)	Long-log truck (3)
4	Chain saw (2)	Helicopter (1)	Larger loader (1)	Long-log truck (3)

[a] Value in parenthesis indicates the number of machines used for that function of the balanced harvesting system.

harvesting system. The weekly production rate is provided both in ft^3 per productive machine hour (PMH) (or MBF/PMH for the MBF version) and in the number of truckloads. Additionally, total time required for harvesting the given tract is estimated and reported based on the tract size and system rate.

A hidden summary sheet, which saves the results from scenario runs to make analysis and comparison easier, can be invoked via the data storage module when clicking the *Save Sheet* command button. Stand conditions, machine rate/costs, and system rate/costs are saved on this summary sheet. Three bar charts for hourly felling production and unit cost, hourly extraction production and unit cost, and weekly system production and onboard cost are automatically generated and displayed in the summary sheet. The user can graphically compare harvesting systems under different stand and harvest conditions.

15.3.3 Application Example

To demonstrate the application of the CAHA, four harvesting systems were examined for the harvesting operations on a 20 ha (50 acre) tract by using the Excel spreadsheet program. A knuckleboom loader was used for loading and a long log truck with a loading capacity of 26 tons (1148 ft^3) per truck was assumed for hauling operation. The same stand conditions of a second-growth hardwood forest about 75 years old were examined for all four harvesting systems. The stand density was 390 trees/ha (156 trees/acre) with the average DBH of 32.84 cm (12.93 in.) and volume per ha 222.5 m^3 (3155 ft^3/acre).

A typical diameter-limit cut of removing trees greater than 30.48 cm (12 in.) of DBH was applied, in which we assumed that average distance between harvested trees was 9.88 m (32.44 ft) for chainsaw and feller-buncher felling, and 6.96 m (22.86 ft) for harvester. Turn payload was assumed at 2.95, 2.41, and 11.58 m^3 (104, 85, and 409 ft^3) for cable skidder, grapple skidder, and forwarder, respectively, with an average extraction distance of 304.8, 304.8, and 457.2 m (1000, 1000, and 1500 ft). It was assumed that turn payload of the helicopter was 2,993.71 kg (1186.2 m^3) (6,600 lb (41,892 ft^3)). One harvesting scenario could be examined with one program run, with the results saved to the summary sheet after each run of the program. Automatic harvesting system balancing was specified by invoking the system balancing module. System rate, weekly production, and onboard cost together with the days required to cut that tract were calculated for the balanced system. The production/cost information for the felling and extraction machines as well as for the balanced harvesting systems was summarized on the summary sheet. Additionally, three bar chart figures related to felling productivity and cost, extraction productivity and cost, and harvesting system weekly production and onboard cost for each harvesting scenario were automatically generated on the summary sheet through implementation of the data storage module (Table 15.3).

Harvesting systems were balanced for Systems 1–3 based on the machine productivity and utilization rate. Two chainsaws and one cable skidder, one

TABLE 15.3

Harvesting Operation Summaries by Harvesting System

System	System Rate (ft³/SMH)	Weekly Production		Onboard Cost ($/cunit)	Days Required to Cut Tract
		ft³	Truck Loads		
1	229.08	9,163	8	67.6	61
2	662.22	26,489	23	42.0	21
3	398.94	15,958	14	60.8	35
4	484.53	19,381	17	213.3	29

feller-buncher and two grapple skidders, and one harvester and one forwarder were utilized for harvesting Systems 1, 2, and 3, respectively. For harvesting System 4, it was assumed that two chainsaws and one helicopter were used and that the helicopter started the extraction after felling was complete. One loader and three trucks were assumed for all four harvesting systems. Miscellaneous costs including equipment relocation, harvesting operations support, and road building were calculated as $25.21, $14.38, $18.01, and $18.88 per cunit for Systems 1–4, respectively. System onboard cost was the sum of felling, extraction, loading, and miscellaneous costs. System rate was estimated at 6.48, 18.75, 11.30, and 13.73 m³ (229, 662, 399, and 485 ft³) per SMH with an onboard cost of $68, $42, $61, and $213 per cunit for harvesting Systems 1–4, respectively. System 2 was the most productive combination in the region with a weekly production of 750.08 m³ (26,489 ft³) or 23 truckloads (Table 15.4). It only took 21 business days to cut the tract using Harvesting System 2, while the cutting time would be nearly tripled if System 1 were chosen. It took 6 more business days to cut the tract using System 3 than using System 4 (Table 15.3).

TABLE 15.4

Descriptive Statistics of the Inventoried Stands Used in the Case Study

	N	Mean	StdDev	Maximum	Minimum	Median
Tree height (m)	14,008	18	11	42	2	22
DBH (cm)	14,008	36	15	132	3	36
Quadratic mean diameter (cm)	14,008	28	3	36	21	28
Trees per ha	92	497	210	1505	232	439
Basal area (m²/ha)	92	30	11	72	11	28
Merchantable volume (m³/ha)	92	1784	625	4802	557	1668
Forest carbon stock (Mg/ha)	92	147	49	363	74	136
Merchantable carbon stock (Mg/ha)	92	69	24	170	21	64

15.4 3D Log Bucking Optimization

15.4.1 System Design

15.4.1.1 System Structure

The optimal bucking system consists of three major components: (1) data manipulation/storage, (2) 3D modeling, and (3) bucking optimization, including the following functional requirements: data acquisition, data standardization, value calculation, bucking optimization, 3D environmental normalization, 3D image display, 3D image manipulation, and data storage and analysis (Wang et al. 2009). Component object model was employed to integrate the system that was designed using the principle of object-oriented programming. The system was programmed with Microsoft Foundation Class (MFC) and Open Graphics Library (OpenGL). Users can easily build a Windows-compatible graphical user interface and link the system to data objects through MFC and ActiveX controls. The 3D objects created can be rotated, scaled, and translated by performing OpenGL transformation. MFC's IDE facilitates the management of the bucking system during the development process.

15.4.1.2 Data Manipulation and Storage

ActiveX Data Object (ADO) was employed to retrieve data from and save the bucking results to an Access database. ADO consists of seven basic objects: Connection, Recordset, Command, Error, Field, Property, and Parameter, and they are related interactively.

The entity–relationship (ER) model for the optimal bucking system was implemented via Microsoft Access, including five entity types: *Stems* for storing stem number, and basic stem information; *Shapes* for storing stem sweeps and diameters' data at each 1.22 m (4 ft) intersection; *Grades and Prices* for storing grading rules and price matrix; *Defects* for storing defects data associated with each stem; and *Logs* for bucking results (Figure 15.12). Five relationships among these entity types were defined, which reflect the interrelationships among these entities.

15.4.1.3 3D Stem Modeling

In order to provide the user with a realistic tree-stem, 3D modeling techniques were used together with OpenGL primitive drawing functions to generate a 3D tree-stem visualization, which is composed of simple triangle strips filled with stem images, such as bark or the butt-end image (Figure 15.13). The user can perform rotate, translate, and scale functions to get a better understanding of the stem's superficial characteristics.

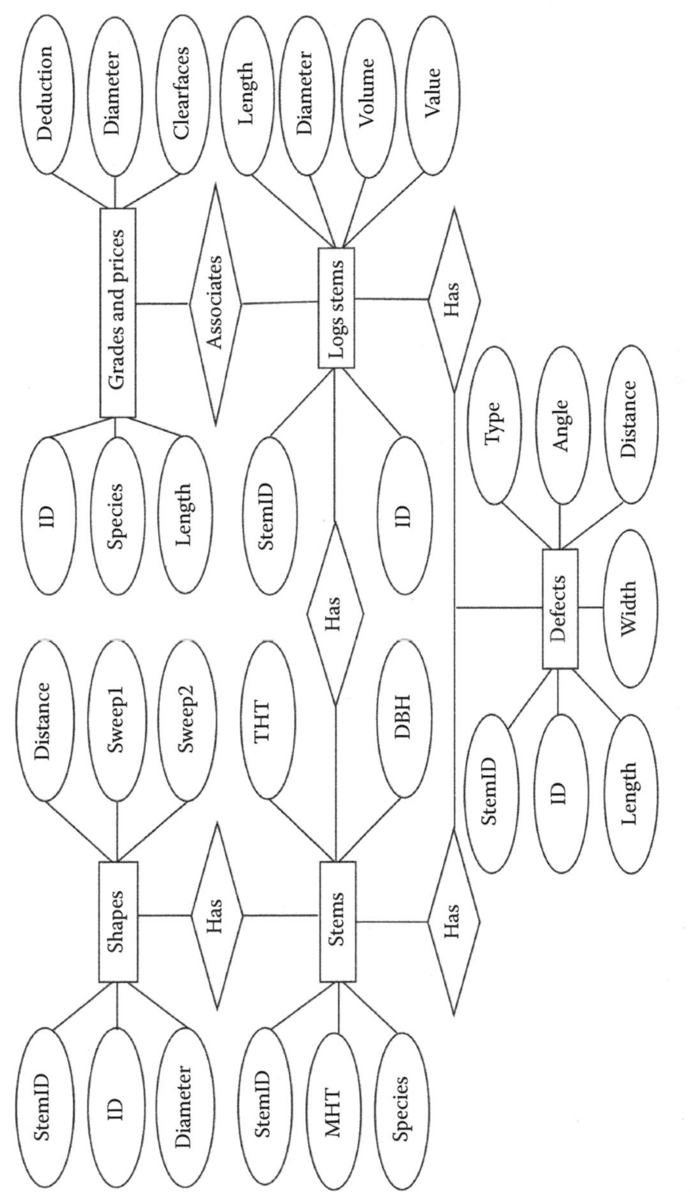

FIGURE 15.12
Diagram of the entity–relationship model for the 3D optimal bucking system.

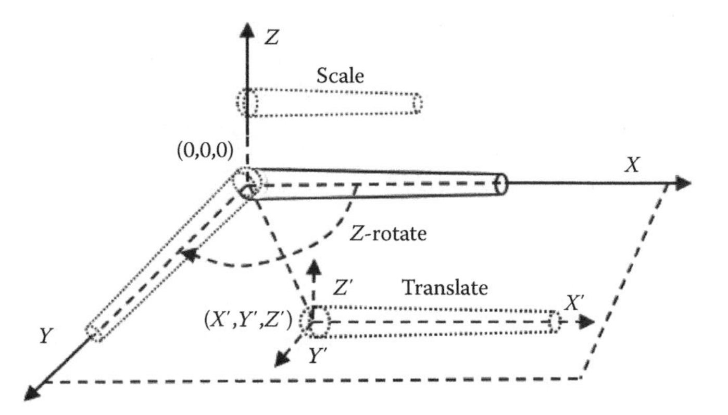

FIGURE 15.13
Diagram of the modeling transformation for a 3D tree-stem.

15.4.1.4 Optimal Bucking Algorithm

A network analysis technique was implemented to generate the optimal bucking patterns. The principle of Dijkstra's algorithm for the single-source shortest path problem was adopted. It was used to find the longest path in the weighted, directed graph of tree-stem bucking, which maintains a set of Y cutting points or nodes whose final longest-path weights from the origin X_1 already have been determined. The algorithm repeatedly selects the potential cutting point $X_i \in X{-}Y$ with the maximum longest path estimate, adds X_i to Y, and relaxes all edges or arcs leaving X_i.

The efficiency or running time of Dijkstra's algorithm depends on how the maximum-priority queue of potential cutting points or nodes is implemented. If we maintain the maximum-priority queue by taking advantage of the cutting points or nodes being numbered 1 to n, we can simply store these nodes associated with weights into a 1D array with n elements. The running time of this algorithm consists of three parts: (1) searching the cutting point with maximum weight, (2) adding this point or node to the point set Y, and (3) removing this node from the set $X{-}Y$. Since adding or removing a node from a point set takes constant time, $O(1)$, and searching a point takes $O(n)$, the efficiency or running time of the algorithm for n potential cutting points along a tree-stem can be expressed as:

$$
\begin{aligned}
T(n) &= \left[2O(1) + O(n)\right] \times n \\
&= O\left(n^2\right)
\end{aligned}
$$

where

$O(n^2)$ represents the asymptotic upper bound of n^2

$O(1)$ is the constant time

15.4.2 Bucking System Implementation

Once a tree-stem is selected, a 3D image can be generated (Figure 15.14). The 3D display dialog consists of three major sections: display area, information area, and command area. Display area is for displaying a 3D stem image and viewing the bucking results of a selected stem. Text in the upper left corner of the display area is used to update inside bark diameter and length from

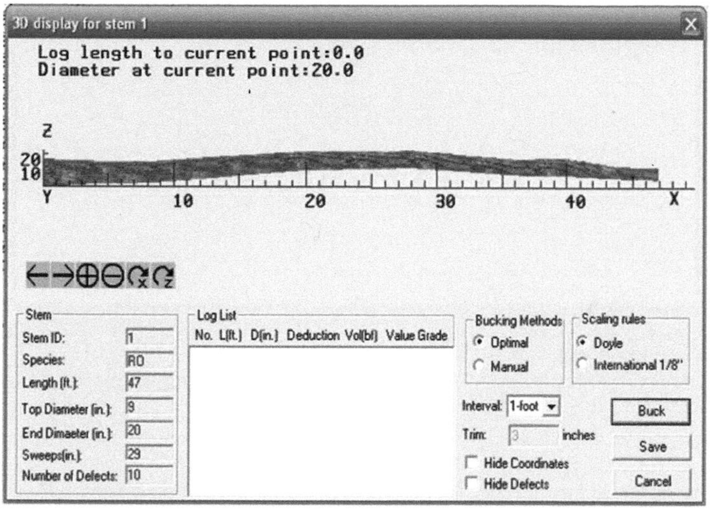

(a)

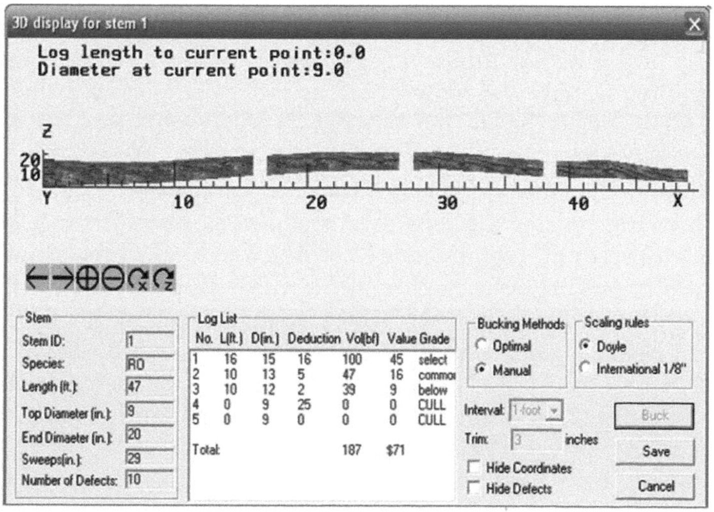

(b)

FIGURE 15.14
(a) 3D bucking system for a tree-stem with (b) manual bucking solution and (c) optimal solution.

(Continued)

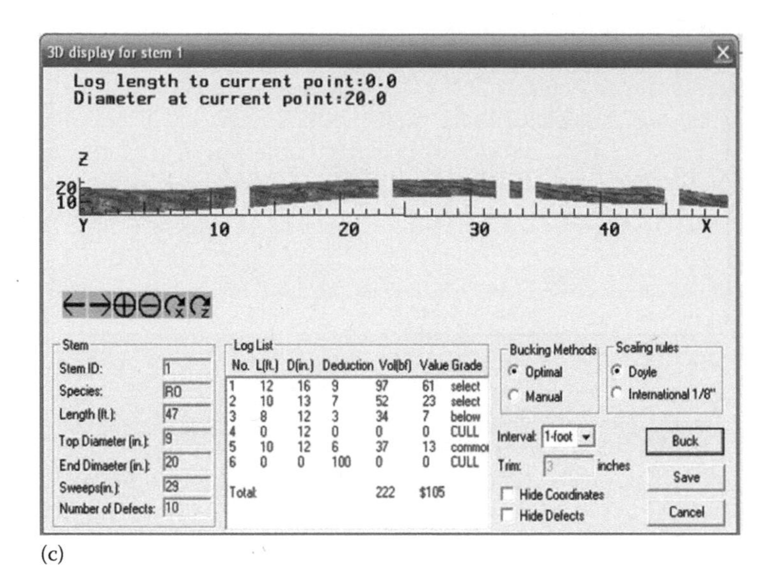

(c)

FIGURE 15.14 (*Continued*)
(a) 3D bucking system for a tree-stem with (b) manual bucking solution and (c) optimal solution.

a previous cut to the current cut position. In the currently displayed image case, the saw is at the butt-end of the stem and the associated inside bark diameter of this stem is 50.8 cm (20 in.). Six command buttons in the display area (*Left move, Right move, Zoom in, Zoom out, Rotate along x-axis,* and *Rotate along z-axis*) are used for implementing zoom, projection, and perspective view functions. Defects are represented as red rectangles with their actual sizes and locations on the stem, which can be either measured in the field or entered by the user.

Information area shows the basic information for a selected stem and the detailed bucking results. Options are provided to users for log rules and bucking methods in the command area. When the user selects optimal bucking, a stage interval should be chosen from the drop-down list. If manual bucking is selected, the drop-down list for stage interval selection is disabled. All the log volume is calculated based on the selected log rule: Doyle or international 1/4 in. (1 in. = 2.54 cm). At the bottom left of the command area, there are two checkboxes for displaying or hiding coordinates and defect data. Three command buttons, *Buck, Save,* and *Cancel,* are used respectively for bucking, saving bucking results, and closing the display dialog.

In Figure 15.14b, stem 1 was recorded as red oak in the field and was manually bucked into 4 logs (one 4.88 m (16 ft), and three 3.05 m (10 ft) in length). The total log value for stem 1 is $71 and the total volume of logs bucked from stem 1 is 187 bf. The log length for the fourth log in the log list is 0 instead because this log is not a grade log. Accordingly, stem 1 was optimally bucked using a stage interval of 1 ft (1 ft = 0.305 m). The optimal bucking yielded a total log value of $105 and a total log volume of 222 bf for this stem, which included four logs with lengths of 3.66 m (12 ft), 3.05 m (10 ft), 2.44 m (8 ft), and 3.05 m (10 ft). The user has the option to save or perform alternative bucking processes using different methods or stage intervals.

15.5 3D Lumber Edging and Trimming System

The optimal lumber edging and trimming system consists of four major components: (1) data manipulation/storage, (2) 3D modeling, (3) lumber grading, and (4) edging and trimming optimization (Figure 15.15) (Lin et al. 2011a). A component object model using the principles of object-oriented programming was used to integrate the system. The system was programmed with MFCs and OpenGL. MFC provides a user-friendly interface and can be easily connected to the database and transplanted to any other Microsoft Windows application, whereas OpenGL provides color images of 3D objects and offers the 3D virtual simulation environment (Wang et al. 2009). The software system can be implemented either on a desktop or a laptop and run on a Windows platform.

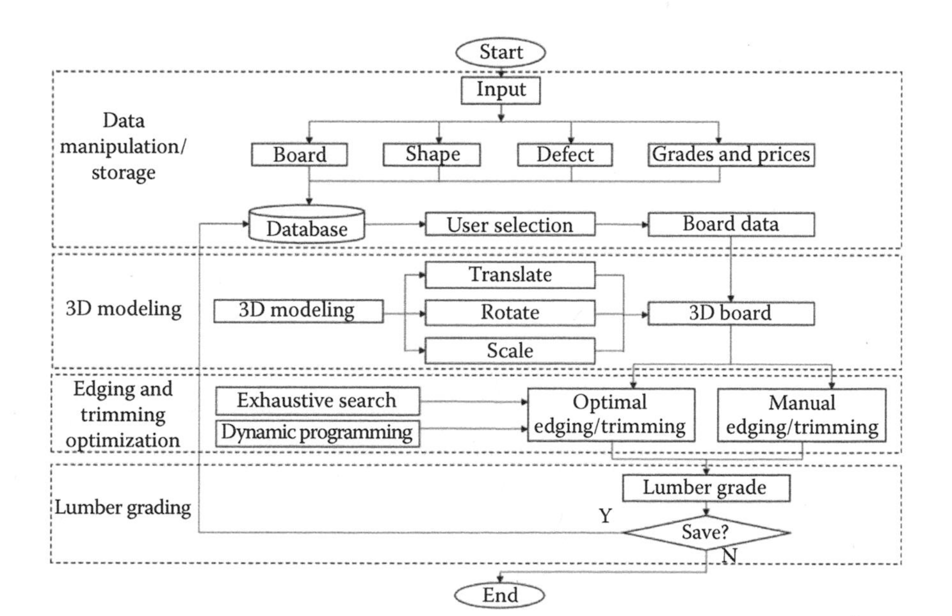

FIGURE 15.15
Architecture of the optimal lumber edging and trimming system.

15.5.1 Data Manipulation and Storage

Microsoft ADOs enable client applications to access and manipulate data from a variety of sources through an Object Linking and Embedding Database provider (Microsoft Developer Network Platforms 2010). The primary benefits of ADO are ease of use, high speed, low memory overhead, and a small disk footprint. In this study, ADO was applied to retrieve data from, and to save edging and trimming results to, a Microsoft Access database. The simple way to incorporate ADO into programming is through the use of ActiveX controls, and it is very convenient to link the system database with MFC and ActiveX controls. The ER model for the optimal edging and trimming system was implemented via Microsoft Access and included four entity types: (1) board, (2) shape, (3) defect, and (4) defect type. Once a board has been edged and trimmed, the results, including surface measure (SM), lumber grade, and lumber value, can be stored in a summary table within the database.

15.5.2 3D Lumber Modeling

Three-dimensional modeling techniques together with OpenGL primitive drawing functions were used to generate 3D lumber visualizations. OpenGL is a powerful yet flexible and standard tool to create high-quality, multidimensional graphics (Woo et al. 1999). Two OpenGL libraries, OpenGL Utility Library and OpenGL Utility Toolkit, were used to make a visual representation of lumber and of the edging and trimming process. A board is visualized using simple triangular strips filled with a digital image of an actual board. The user can rotate, zoom in/out, and/or move the board around to facilitate visualization of the board and to better understand the superficial characteristics at different scales. Three basic transformations, rotate, scale, and translate, were modeled by using the functions glRotatef(), glScalef(), and glTranslatef(), respectively. For example, rotation is performed by calling glRotatef(a, x, y, z), which generates the rotation matrix by defining the degrees to be rotated (a) and the axis to be rotated about (x-axis, y-axis, or z-axis). The generic matrix of rotation of an angle around x-axis can be derived and expressed as (Woo et al., 1999:

$$R_x(\alpha) = \begin{bmatrix} 1 & 0 & 0 & 0 \\ 0 & \cos\alpha & -\sin\alpha & 0 \\ 0 & \sin\alpha & \cos\alpha & 0 \\ 0 & 0 & 0 & 1 \end{bmatrix}$$

Let the coordinates of a board originally drawn on screen be (x_1, y_1, z_1), (x_2, y_2, z_2), …, (x_n, y_n, z_n), respectively. If that piece of lumber is rotated by α around the x-axis and coordinates are transformed to (x_1', y_1', z_1'), (x_2', y_2', z_2'), …,

(x'_n, y'_n, z'_n), then the coordinate matrix after rotating by α degrees around the *x*-axis can be expressed as (Wang et al. 2009):

$$\begin{bmatrix} x'_1 & x'_2 & \cdots & x'_{n-1} & x'_n \\ y'_1 & y'_2 & \cdots & y'_{n-1} & y'_n \\ z'_1 & z'_2 & \cdots & z'_{n-1} & z'_n \\ 1 & 1 & \cdots & 1 & 1 \end{bmatrix} = R_x(\alpha) \times \begin{bmatrix} x_1 & x_2 & \cdots & x_{n-1} & x_n \\ y_1 & y_2 & \cdots & y_{n-1} & y_n \\ z_1 & z_2 & \cdots & z_{n-1} & z_n \\ 1 & 1 & \cdots & 1 & 1 \end{bmatrix}$$

$$TS' = R_x(\alpha) \times TS$$

where *TS* is the matrix containing locations of different coordinates for shape, defects, and other visual controls before transformation and *TS′* is the matrix of coordinates after transformation. Similarly, the coordinate matrices for the triangle strip can be rotated around the *y*- and *x*-axes.

The scale and translation are performed by calling glScalef (*Sx, Sy, Sz*) and glTranslatef (*dx, dy, dz*) functions that generate the scale and translation matrices. *Sx, Sy,* and *Sz* are the scales to the *x, y,* and *z* coordinates of each point of measurement for a board while *dx, dy,* and *dz* are the values needed to be translated along the *x*-axis, *y*-axis, and *z*-axis, respectively.

15.5.3 Lumber Grading

The lumber grading component is based on Klinkhachorn's hardwood lumber grading routine (Klinkhachorn et al. 1988) and the National Hardwood Lumber Association (NHLA) lumber grading rules. To determine a possible grade for a lumber, the width, length, and SM of the lumber are computed. A potential grade from the highest to the lowest is assigned to the poor face, then the potential number of clear cuttings and cutting units (CUs) can be calculated (Lin et al. 2011a). By comparing the number of cuttings and CUs obtained from a piece of lumber, a final grade can be determined based on the requirements of the NHLA grading rules (NHLA 2008). Potential grades used in the current version include First and Seconds, SELECT, 1Common (1COM), 2Common (2COM), and 3Common (3COM). After a board is edged and trimmed, the processed board data including dimension, shape, and defect are recalled by the lumber grading routine, and a lumber grade is assigned to this board. Using stored lumber price data by grade and species, the lumber value can be determined.

15.5.4 Optimal Edging and Trimming Algorithm

A cutting pattern that yields the maximum value is the optimum edging and trimming solution. The exhaustive searching process to find this pattern can be very time-consuming. Since there are numerous ways of edging and trimming a flitch, an optimal computer procedure was developed to aid in this searching process including exhaustive search and dynamic programming (DP). The exhaustive search algorithm explores all possible combinations of edging and trimming lines within the original size of the board and is guaranteed to find the optimal solution. The shape of the board is determined by different combinations of edging and trimming lines. Information regarding board length, width, SM, and defects is then recalled by the lumber grading component, and a lumber grade for that board can be assigned. The board's value is determined based on the grade, SM, species, and the lumber price.

DP is a more efficient search procedure that can be used to achieve the optimum edging and trimming solution. All potential edging and trimming line positions are predefined by dividing a board into equidistant levels in both horizontal and vertical directions. This allows the lumber edging and trimming problem to be formulated as a set packing problem with the objective being to maximize the total lumber value. The key to solving the edging and trimming problem by DP is to recognize the recursive relationship (Bhandarkar et al. 2008). An original board can be divided into $N_e = ER/c_1$ horizontal edging lines and $N_t = TR/c_2$ vertical trimming lines, where ER and TR are edging range and trimming range, respectively, and c_1 and c_2 are the edging and trimming intervals, respectively. Let $s^*(i,j)$ be the optimal edging and trimming patterns for the horizontal edging lines from 1 to i and vertical trimming lines from 1 to j, and $v^*(i,j)$ be the corresponding lumber value. If $v^*(k,l)$ and $s^*(k,l)$ for all $k \leq i$ are known, then the combined edging and trimming flitch problem can be formulated as a recursive function:

$$v^*(i+1,j+1) = \max_{k\in[1,m]}\left(\max_{l\in[1,n]}\left(\left(v^*\left(i+1-\left\lceil\frac{W_k}{c_1}\right\rceil-\left\lceil\frac{K}{c_1}\right\rceil, j+1-\left\lceil\frac{L_l}{c_2}\right\rceil-\left\lceil\frac{K}{c_2}\right\rceil\right)\right.\right.\right.$$
$$\left.\left.\left.+g\left(i+1-\left\lceil\frac{W_k}{c_1}\right\rceil, i+1, j+1-\frac{L_l}{c_2}, j+1\right)\right)\right)\right)$$

where

$W_k = \{W_1, W_2, \ldots, W_m\}$ is the allowed set of lumber width
$L_l = \{L_1, L_2, \ldots, L_n\}$ is the allowed set of lumber length
K is the saw kerf
$g(i,j,k,l)$ is the lumber value between edging lines i and j, and trimming lines k and l

The requirements for the lumber are that its width be ≥7.6 cm (3 in.) and its length be ≥1.22 m (4 ft).

15.5.5 Optimal Edging and Trimming System Implementation

All the computer simulations were performed on a regular desktop PC equipped with a 3.16 GHz CPU, 3.25 GB of RAM, and a 300 GB hard drive under the Microsoft Windows platform. The edging and trimming process was implemented by a 3D-based Windows dialog box with four tab controls labeled *Board, Shape, Defect,* and *Defect Type*. The *Board* tab is used to display all the board data saved in the database. To view the shapes and defects information associated with a selected board, the user can click the corresponding tab controls. A defect on a board is measured by two lengths (left and right) and two widths (low and up). Each board can be divided into nine possible sections named from 1 to 9 from the top left corner all the way through the bottom right corner. The section determination for each cutting board is illustrated in Figure 15.16 and the measurements of shape and defect information are illustrated in Figure 15.17.

Once a board is selected, its 3D image can then be generated (Figure 15.18). The interface consists of three major sections: (1) display area (right top area), (2) results area (right bottom area), and (3) control and command area (left area). The display area displays the 3D board image and the edging and trimming results of a selected board. Information provided by an NHLA grader is displayed in the upper portion of the display area and includes lumber length, width, thickness, grade, SM, and value. This information is used to compare the edging and trimming results produced by the optimal system. On top of the control and command area are two control checkboxes

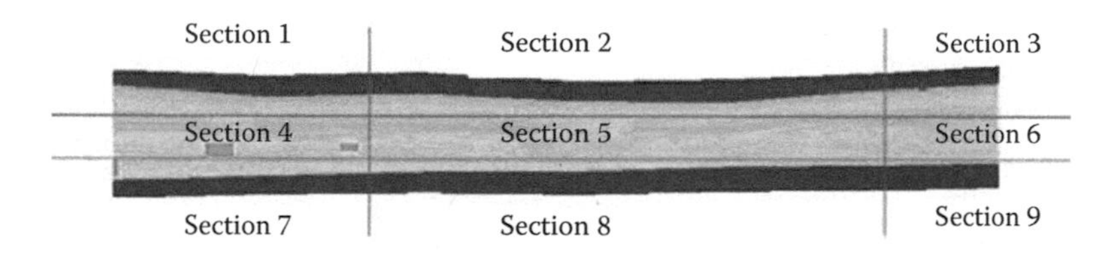

FIGURE 15.16
Section determination for a cutting board.

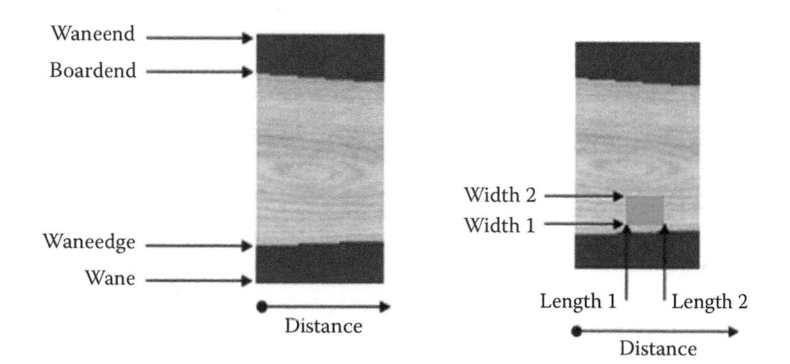

FIGURE 15.17
Illustration of measuring shape and defect information.

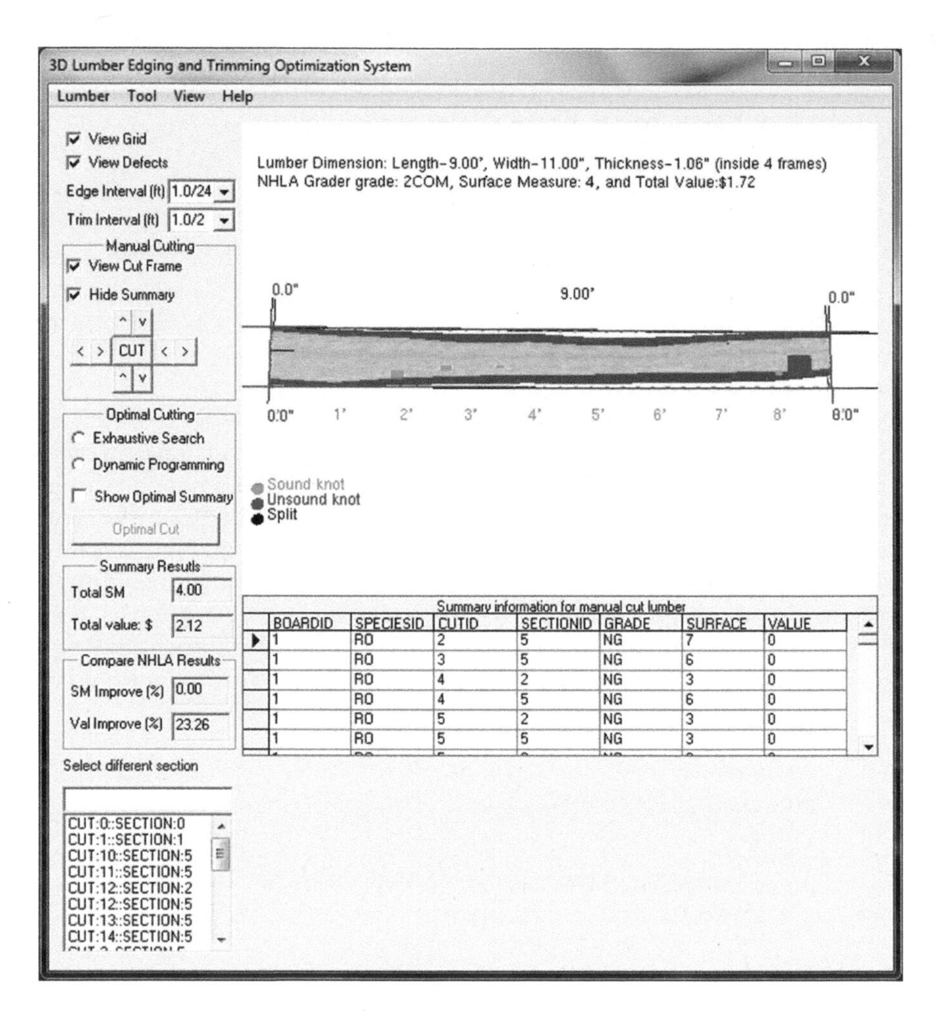

FIGURE 15.18
Lumber edging and trimming system with options of exhaustive search algorithm solution, DP algorithm solution, and manual solution.

(*View Grid* and *View Defect*). By default, both checkboxes appear unchecked. The first is used to display the grid along *x*, *y*, and *z* axes to show the length, width, and thickness, respectively, of the lumber in inches (1 in. = 2.54 cm), and the second is used to display the defect with legend in different colors. Two control combo boxes are used to change the intervals for edging lines and trimming lines. By default, the interval is 1.27 cm (0.5″) for edging lines and 15.24 cm (6″) for trimming lines. The user can also manually change the interval values. For the edging line interval, 0.63, 1.27, and 2.54 cm (0.25, 0.5, and 1 in.) are available for use, whereas 5.08, 15.54, and 30.48 cm (2, 6, and 12 in.) are available for trimming intervals.

Edging and trimming simulations can be performed by two approaches: (1) optimal cutting and (2) manual cutting. For optimal cutting, an exhaustive search or DP algorithm is available to optimize the edging and trimming process for the selected board. During the optimal simulation, the program shows the searching progress and, finally, the total running time. For example, for Board 1 (red oak [*Quercus rubra*] lumber), the lumber grade, SM, and total lumber value were 2COM, 5, and US$2.15, respectively, when using exhaustive search; but the grade, SM, and lumber value were 1COM, 4, and US$2.12, respectively, using the DP algorithm.

The controls and commands in the manual cutting group can be used to train edger and trimmer operators. When the user clicks the *View Cut Frame* checkbox, the edging and trimming function will be activated and the *CUT* button enabled. At this stage, the board is bounded by four red frames, the edging and trimming lines, with the horizontal lines representing the edging lines and the vertical lines representing the trimming lines. These frames can be moved by clicking the up- and down-arrow buttons. The two left buttons can be used to move the left trimming lines, and the two right buttons can be used to move the right trimming lines. Similarly, the upper and lower buttons can be used to control the moving directions of the edging lines. Every time a frame is moved, the board is regenerated, and the updated lumber length, width, and SM are displayed. Once the frames are set up for desired sections, users can press the *CUT* button to cut the board. If unsatisfied with the current operation, the user can delete the generated lumber and process the board again.

15.6 3D Log Processing Optimization System

This example is to show you the development of a computer-aided edging and trimming, sawing, and grading system. Our goal was to develop a cost-effective computer-aided log processing simulation system for lumber manufacturing optimization to maximize lumber value recovery based on log defect, dimension, and other superficial characteristics (Lin et al. 2011b, Lin and Wang 2012). This system can simultaneously optimize the primary and secondary processing to achieve a global optimal solution. In addition, the system can be used as a decision-making system for lumber production and a training tool for novice sawyers.

15.6.1 System Components and Data Management

15.6.1.1 System Components

The system consists of six major components: 3D log generation, opening face determination, headrig log sawing, flitch edging and trimming, cant

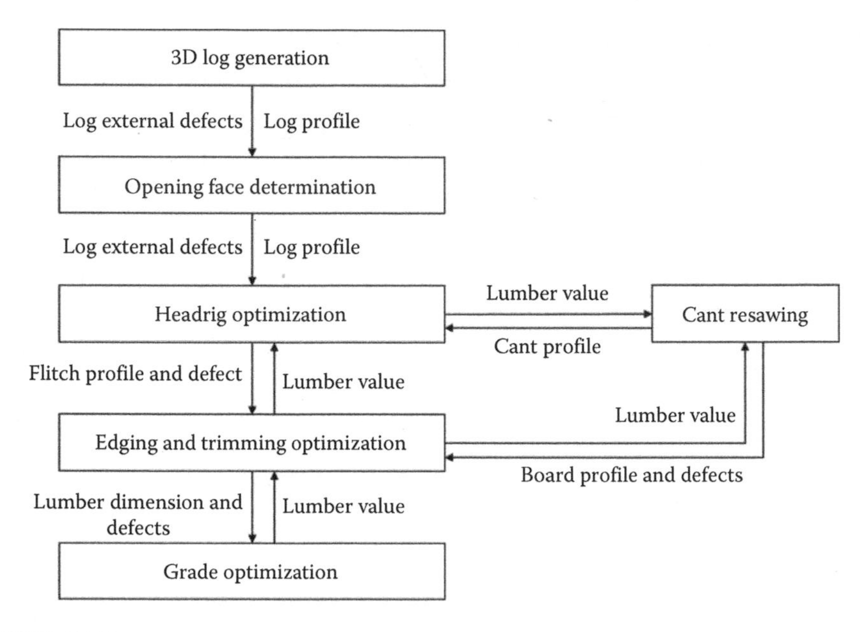

FIGURE 15.19
Components' log processing optimization system.

resawing, and lumber grading (Figure 15.19). Each component accomplishes its own task and is linked to related components by transferring arguments and/or global variables, which will make modifications and maintenance easier. The 3D log generation component generates a real 3D-shaped log that can be rotated, scaled, and translated based on log data and performance requirements. The opening face component simulates the log position, opening face position, and opening face size. The headrig optimization component simulates sawing the log into slabs, flitches, and/or cants and simulates the optimal sawing patterns with maximum log value by applying either heuristic or DP algorithms. The optimum value of each flitch or cant cut from the log can be determined as well. If cant resawing is simulated, the boards generated from the cant also need to be edged and/or trimmed. The edging and/or trimming optimization component calls the headrig optimization or cant resaw component for flitch/board information and defect profiles exposed on the board faces. The optimal edging and/or trimming patterns are then simulated by either an exhaustive search or the DP algorithm. All the generated lumber will be simulated and processed by the lumber grading component. Based on lumber dimensions, defects, lumber price, and species, the optimum lumber value is obtained. Finally, the total lumber value along with the corresponding optimum simulated sawing and edging and/or trimming pattern is recorded in the system.

15.6.1.2 System Data Management

Microsoft ADOs are used to retrieve data from and save simulated sawing results to an MS Access database. ADO enables client applications to access and manipulate data from a variety of sources through an Object Linking and Embedding Database provider (MSDN 2010). The simple way to incorporate ADO into programming is through the use of ActiveX controls, so the user can link the system database conveniently by MFC and ActiveX controls. An MS Access database (which includes four entity types: logs, shapes, defects, and grades) is created to hold the log and lumber information in the system. The log entity type stores log number and basic log information, such as species, log position, log length, small-end and large-end diameters. The shapes entity type stores log sweep and diameter data at one-foot (1 ft = 0.305 m) intervals. The defects entity type contains defects data associated with each log. The grades entity type stores lumber grading rules and lumber price. An ER model is implemented via the MS database.

15.6.2 System Modeling and Algorithms

15.6.2.1 3D Log and Internal Defect Modeling

Log shape modeling is very important in determining the optimum log breakdown. A circular cross section model is adopted to represent a log, which uses a series of cross sections at designated intervals along the log length (Figure 15.20a). This model is much closer to real log shape because the data at each cross section are collected as well as log sweep and log crook. 3D modeling techniques together with OpenGL primitive drawing functions are used to generate 3D log visualizations. The OpenGL functions such as translation, rotation, and scaling are used to facilitate log visualization, and the related mathematical modeling was described by Woo et al. (1999). Studies have

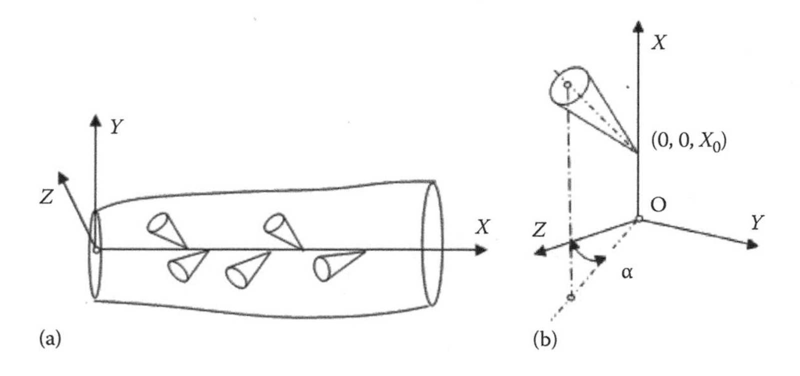

FIGURE 15.20
3D log and defect model. (a) A 3D log and knots, (b) Knot represented as a cone arbitrarily positioned in the *XYZ* space.

shown that strong correlations exist among surface defect indicators such as overgrown knots, overgrown knot clusters, sound knots, unsound knots, and internal knot defects (Thomas 2008). We only consider knots as internal log defects in this model, because they are the most commonly found on board surfaces and can have significant impacts on log quality and lumber value. A cone model is used to represent an internal log knot with apex assumed at the central axis of the log (Thomas 2008) (Figure 15.20b). The vertex of the cone lies on the x-axis at a distance X0 from the origin of the coordinates, and α is the knot angle between the z-axis and the projection of the knot axis on the y–z plane.

15.6.2.2 Determining Opening Face

During lumber production, the first cut determines the remaining cuts that must be either parallel or perpendicular to the first cut. Therefore, the initial saw cut has direct impact on the lumber grade and volume yield (Denig 1993). In this model, the opening face is simulated with consideration of log surface defects and log profile. Since no logs are absolutely straight, log sweep is considered to describe the curvature of a log. If a log's sweep is less than 76 mm, the log is treated as if it had no sweep; otherwise, it is deemed a "sweepy" log and sweep is considered in the modeling process.

15.6.3 Primary Log Sawing Simulation

The integrated primary and secondary log breakdown optimization is solved by linking two recursive relationships. The primary log breakdown simulates producing a flitch that is sent to the secondary breakdown to determine the value. An optimal edging and/or trimming simulation solution is then generated for the produced flitch. Specifically, once the log opening face is determined, the system uses either a heuristic or DP algorithm to achieve the optimum simulated sawing pattern at the headrig. The generated flitches are then edged and trimmed using the optimal edging and trimming algorithms. The optimum value of a flitch is then returned to the headrig log-sawing simulation, and the log simulated sawing pattern is finalized.

15.6.3.1 Heuristic Algorithm

Heuristic refers to experience-based techniques for problem solving. It is more easily adaptable to a complex restriction problem, such as a log grade sawing process. In this study, the heuristic algorithm for log-sawing simulation is developed based on a modified Malcolm's (1965) simplified procedure for developing grade lumber from hardwood logs. The basic principle is that the log is not rotated unless another log face can yield a higher grade of lumber than the current face or the current face reaches the central cant. Then the log is rotated to the next face with potential for the highest lumber grade. This simulated sawing process is repeated until a specified size cant is produced (Lin 2011).

```
Algorithm to determine log grade sawing pattern:
begin
     cutting from the opening face
     repeat
          if (the lumber grade from current face <  the
            remaining face(s) )
               let the log rotate to the face that
                 generates the highest lumber grade
          else if (the current face reaches the central
            cant)
               assign a flag to current face to prohibit
                 cutting current face and rotate the log
                 to the face that generates the highest
                 lumber grade
     until-  all faces are cut and a central cant left
end
```

15.6.3.2 Dynamic Programming Algorithm

The primary log breakdown problem can be easily solved using DP, which separates a large problem into a series of tractable smaller problems. The key to DP is to find the recursive relationship. In a log grade sawing simulation, a log is divided into four log sawing faces. Then an optimal simulated sawing pattern can be found for each face by solving the recursive function:

$$f_n(i) = \max\{f_{n+1}(j) + v_{ij}\}$$

where n is the current stage, $n = 1, 2, 3$. Each stage is corresponding to one log sawing face, i is the current state at stage n, j is the state at stage $n+1$, v_{ij} is the lumber value contributed to the objective function, $f_{n+1}(j)$ is the contribution value at stages $n+1, \ldots, 4$ to the objective function if log sawing simulation is in state j at stage $n+1$.

For each stage in the above equation, the optimal lumber value from each sawing face can be obtained using this equation:

$$v(n) = \max\{v(m) + g(m,n)\}$$

where m and n are possible sawing line positions at current sawing face ($1 \le m \le n \le N$, m and n are discrete values), N is the total potential sawing line positions between the opening face and central cant. The term $g(m,n)$ is the lumber value generated between sawing lines m and n, which is determined through the edging and/or trimming optimization. $v(m)$, $v(n)$ are portions of the optimal lumber value at the current face from the opening face to sawing line m and n, respectively.

If lumber thicknesses, sawing kerf width, and simulated sawing resolution are known, $v(n)$ can also be expressed as follows (a modified recursive function based on Bhandarkar et al. (2008)):

$$v^*(i+1) = \max_{i\in[1,m]}\left(v^*\left(i+1-\frac{T_j}{c}-\frac{K}{c}\right)+g\left(i+1-\frac{T_j}{c},i+1\right)\right)$$

where

$T_j=(T_1,T_2,\ldots,T_m)$ is a set of lumber thicknesses
m is the total number of lumber thickness
c is the *sawing* plane resolution (mm)
K is the kerf thickness (mm)
$v^*(i)$ represents the optimal lumber value between cutting planes 1 and i
$g(i,j)$ is the lumber value from the sawing line i through j, depending on flitch edging and trimming optimization

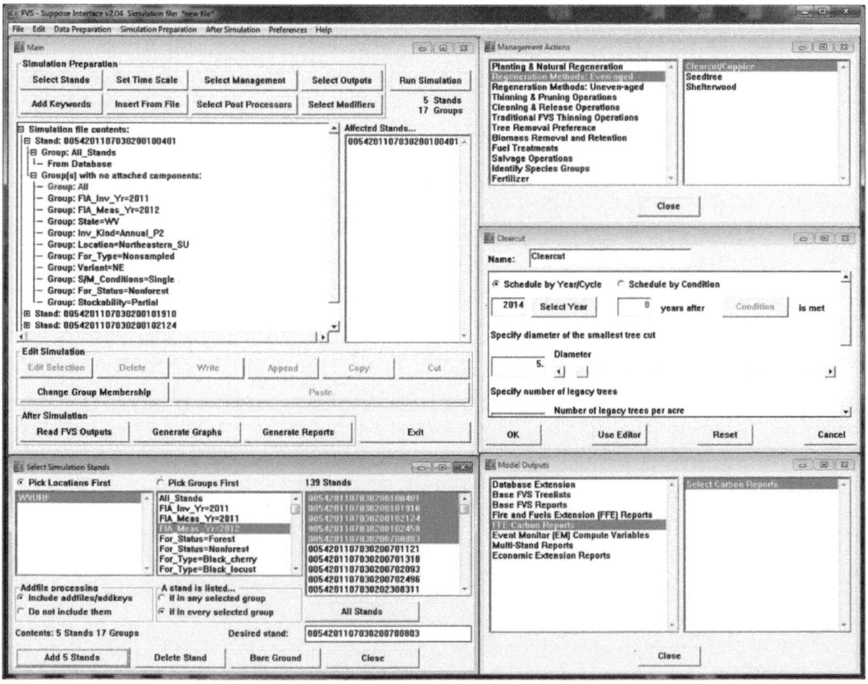

FIGURE 15.21
Prototype of the 3D optimal log sawing system.

15.6.3.3 Example

The system was implemented using MS Visual C++. It can be used to perform 3D log generation, opening face determination, headrig log sawing simulation, flitch edging and trimming simulation, and lumber grading (Lin et al. 2011a,b, Lin and Wang 2012) (Figure 15.21).

15.7 Forest and Biomass Harvest Scheduling and Optimization

Forest harvest scheduling has been extensively studied worldwide. Traditionally, it was applied to optimize timber production from the forest considering some potential operational factors. Recent scheduling examples typically consider the multiple objectives of forest management. The following example used the mixed linear programming approach to model forest harvest scheduling and carbon sequestration in terms of maximizing the total revenue of forests from timber, biomass, and carbon (Liu 2015). This model was also applied to a mixed hardwood forest in the central Appalachian region to analyze the effects of carbon-to-timber price ratio, biomass-to-timber price ratio, and harvest area on carbon sequestration.

15.7.1 Forest Inventory Data

Data for a case study of the model application were collected from WVU Research Forest, a mixed hardwood forest of 3042 ha, located approximately at 39.66°N, 79.78° near Morgantown, West Virginia, USA. The forest has 92 cutting units (i.e., equivalent to stands) with area varying from 7 to 41 ha. Recent forest inventory data were acquired from WVU Division of Forestry and Natural Resources. Each stand had at least 5 cruise points and all together 14,008 tree records were available. A description of these stand parameters is given in Table 15.4.

15.7.2 Forest Stand Growth Simulation

Forest Vegetation Simulator (FVS) Northeast Variant (NE) with Fire and Fuels Extension (FFE) program was used to generate the inventoried stand data to simulate the growth of each stand (USDA 2015). The FVS is a whole-stand growth and yield model that was developed by USDA Forest Service. The original prognosis model introduced by Stage (1973) was developed for northern Idaho and western Montana. The software system is available at http://www.fs.fed.us/fmsc/fvs/software/complete.php. Once the software is downloaded and installed on a computer, we can open it and import inventory data using the *Select Stands* button. Then we set time-scale, management, and post processors. After we select the output, we can press *Run Simulation* to obtain the final report (Figure 15.22). The growth of every stand

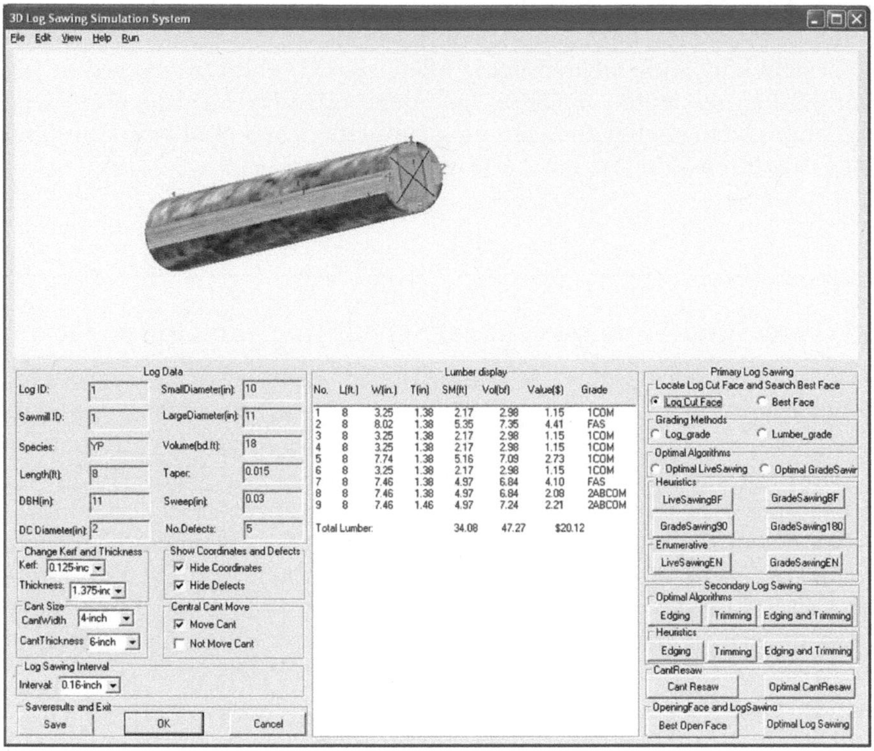

FIGURE 15.22
Interface of the Forest Vegetation Simulator.

can be viewed graphically by clicking *Generate Graphs* and then *Pick Variable* (Figure 15.23). Further detailed guides for FVS are available at http://www.fs.fed.us/fmsc/fvs/documents/index.shtml.

Outputs from the simulations are tree records, basal area, volume, and carbon (C) in different forest components. A planning horizon of 50 years for clear-cutting treatment was examined for harvest simulations to simulate short-term management length. Harvest simulation was conducted for each of 92 stands for the planning period beginning in 2014 on a 5-year interval. Clear-cutting simulation included the removal of all the trees from the stand. Volumes were estimated using the National Cruise System available in FVS-NE-FFE. Natural regeneration was assumed to take place beginning in the same planning period after harvest.

15.7.3 Harvest Scheduling Model Development

Forest harvesting typically causes some inherent risks of land erosion and disruption of wildlife habitats (Barahona et al. 1992). However, these risks

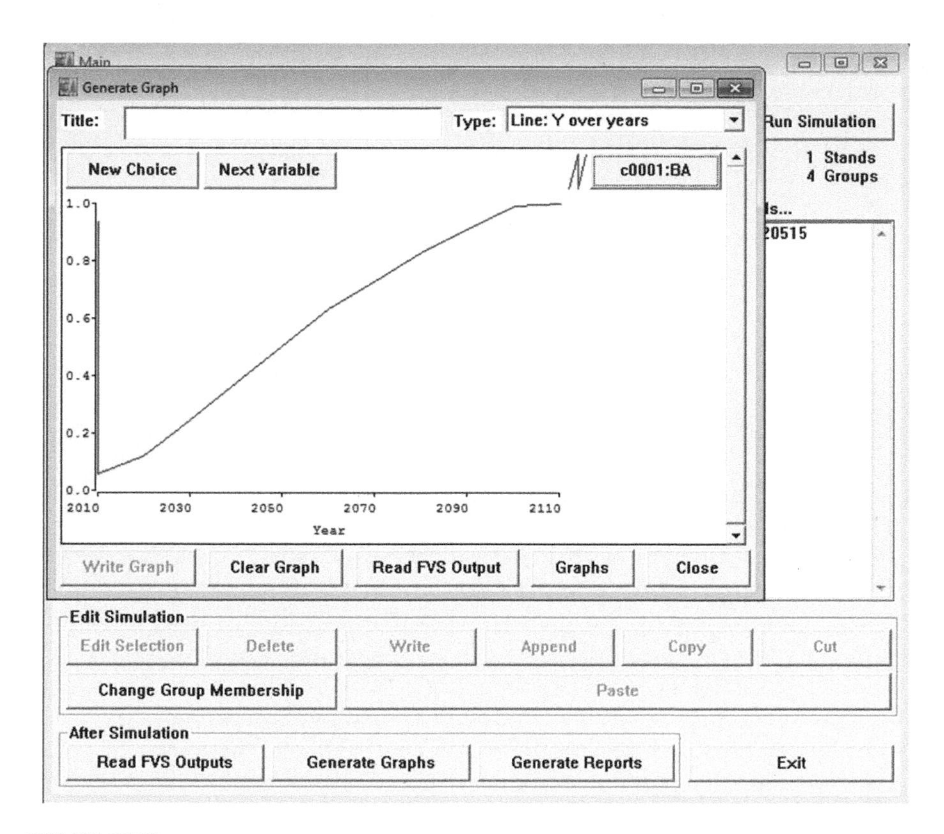

FIGURE 15.23
Increment of stand basal area.

can be effectively mitigated through better scheduling, planning, and implementation of forests' best management practices, such as harvest area limits and buffer sizes of streamside management zones. To achieve this, an optimization model was developed to maximize the total revenue of forests in terms of carbon, timber, and biomass values (Liu 2015). The objective function of the model is formulated as:

$$\max Z = C + W + B$$

where C is the monetary value of carbon sequestered and is calculated by the following equation:

$$C = r_{CO_2} p^{CO_2} \sum_{i=1}^{S} \sum_{t=1}^{T} \left\{ f_{ci}\left(a_{it}\right) - r_{dry} \delta x_{it} \left[G_{i,t-1} + f_{bi}\left(a_{i,t-1}\right) \right] \right\}$$

Similarly, W is the value of timber and B is the value of biomass. The values for timber and biomass are calculated by the following two equations, respectively.

$$W = p^W \sum_{i=1}^{S} \sum_{t=1}^{T} \eta_T x_{it} \left[G_{i,t-1} + f_{bi}\left(a_{i,t-1}\right) \right].$$

$$B = \rho \cdot p^B \sum_{i=1}^{S} \sum_{t=1}^{T} \eta_B x_{it} \left[G_{i,t-1} + f_{bi}\left(a_{i,t-1}\right) \right]$$

where

$f_{bi}(a_{it})$ is the growth function of the aboveground dry biomass of stand i at period t
$f_{ci}(a_{it})$ is the stand carbon storage function of stand i at period t
p^B is the present price of biomass ($/tonne)
p^{CO_2} is the present carbon price in terms of carbon dioxide ($·$CO_2$/tonne)
p^W is the average present price of wood product ($·dry/tonne)
r_{CO_2} is the coefficient used to convert carbon into CO_2 equivalent
r_{dry} is the coefficient used to convert dry biomass into carbon
δ is the percentage of wood products other than long-lived wood products
η_B is the percentage of woody residue in total aboveground biomass
η_T is the percentage of raw timber in total aboveground biomass
ρ is the percentage of biomass that is economically available

A harvest decision for a stand at a given time is denoted by a binary variable:

$$x_{it} = \begin{cases} 1, \text{if stand } i \text{ is harvested at period } t; \\ 0, \text{otherwise.} \end{cases}$$

where $t = 1 \ldots T$ and $i = 1 \ldots S$. T is the total management periods. S is the total number of stands. An integer variable a_{it} represents the stand age of stand i at time period t. A continuous variable G_{it} is the aboveground dry biomass in metric tons (Mg) of stand i at period t. The constraints include maximum permissible continuous harvest area, even flow of timber supply, and average stand age requirement.

15.7.4 Case Study

The base case scenario of this model assumed the timber product price at $200/dry Mg according to timber market report (Appalachian Hardwood Center 2014), carbon price at $10/Mg CO_2 eq according to historical data of Chicago Climate Exchange (2011), and average woody residue price at $1/dry Mg (Wu et al. 2011a,b). A clear-cutting with an area limit of 40 ha was used in the base case management scenario. The configuration of all the other parameters was listed in Table 15.5. The model in the case study was solved using ILOG CPLEX

TABLE 15.5

Parameter Configuration for the Case Study

Name	Value	Reference
G_{io}		Inventory
r_{CO_2}	3.667	
r_{dry}	0.5	Wit et al. (2006)
ρ	0.65	Wu et al. (2010c)
δ	82%	US DOE (2007)
η_B	60%	US DOE (2007)
η_T	60%	US DOE (2007)

12.5 on a computer with 8 GB of memory and a 2.93 GHz CPU. Necessary programs were written in JAVA to implement the model (Figures 15.24 and 15.25), and 5000 s were set as the time limit to obtain a gap less than 1%.

The optimized carbon sequestration rate of the base case scenario over the planning horizon of 50 years was 0.408 Mg/ha/year. Among different carbon components of the forest, aboveground living stands were the major contributor (59.6%) to the total carbon storage, followed by belowground living

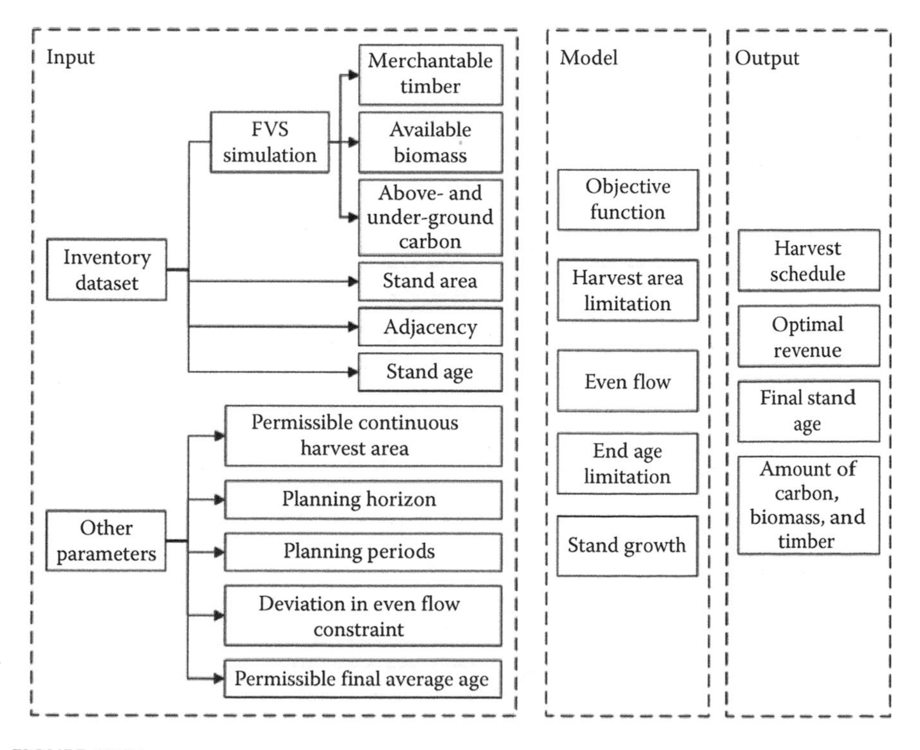

FIGURE 15.24
Flowchart of the model implementation.

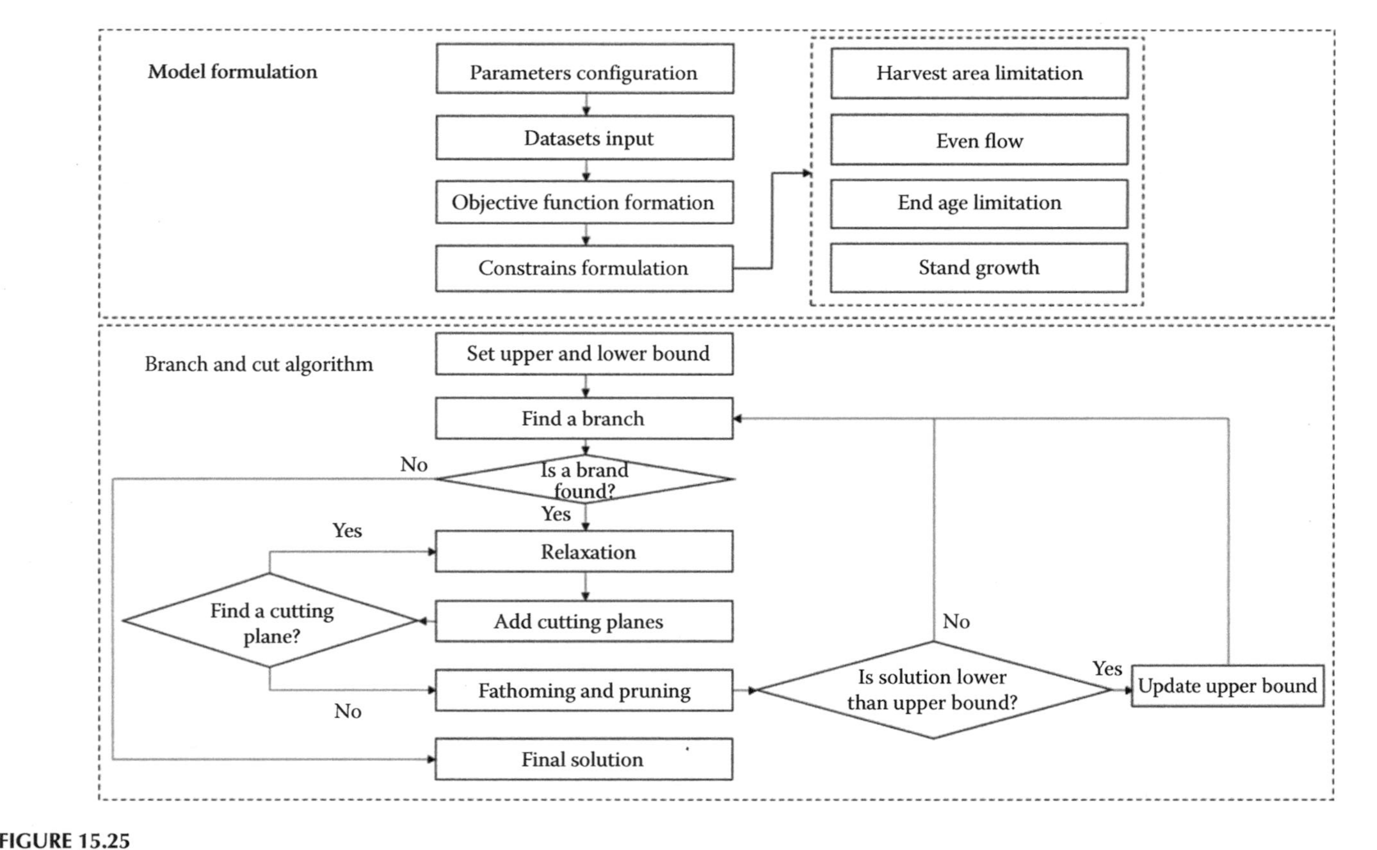

FIGURE 15.25
Branch and cut algorithm to solve the model.

component (15.6%). The forest carbon sequestration rate decreased right after each harvest. However, it would return to the previous rate after sufficient time of growth. The revenue was $42.4 ha/year where carbon sequestration accounted for 40%, and timber and biomass for 59% and 1%, respectively.

The increment of carbon sequestration rate (marginal rate) generally increased until the carbon-to-timber price ratio rose to 0.45, and decreased from 0.322 to 0.0 Mg/ha/year if the ratio was greater than 0.45 (Figure 15.26a).

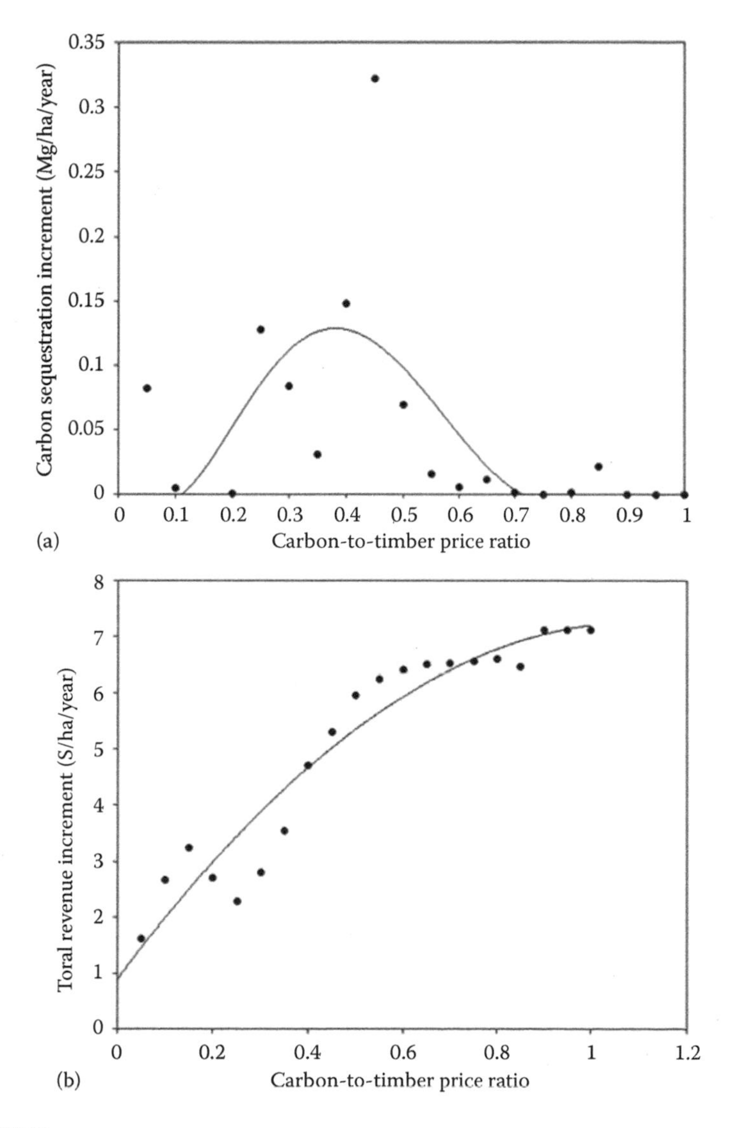

FIGURE 15.26
Variations of (a) carbon sequestration rate and (b) total forest revenue by carbon-to-timber price ratio ($\Delta = 0.05$).

The rate increment reached 0 when the carbon-to-timber price ratio was greater than 0.8. Accordingly, the revenue steadily increased from $1.6 to $7.1 ha/year as the carbon-to-timber price ratio increased from 0.0 to 1.0 (Figure 15.26b). When the price ratio was greater than 0.8, the increment of forest revenue attained a flat plateau.

The carbon sequestration rate slightly varied from 0.325 to 0.323 Mg/ha/year when the biomass-to-timber price ratio increased from 0.0 to 0.7 with a carbon-to-timber price ratio of 0.0 (Figure 15.27a). As woody biomass

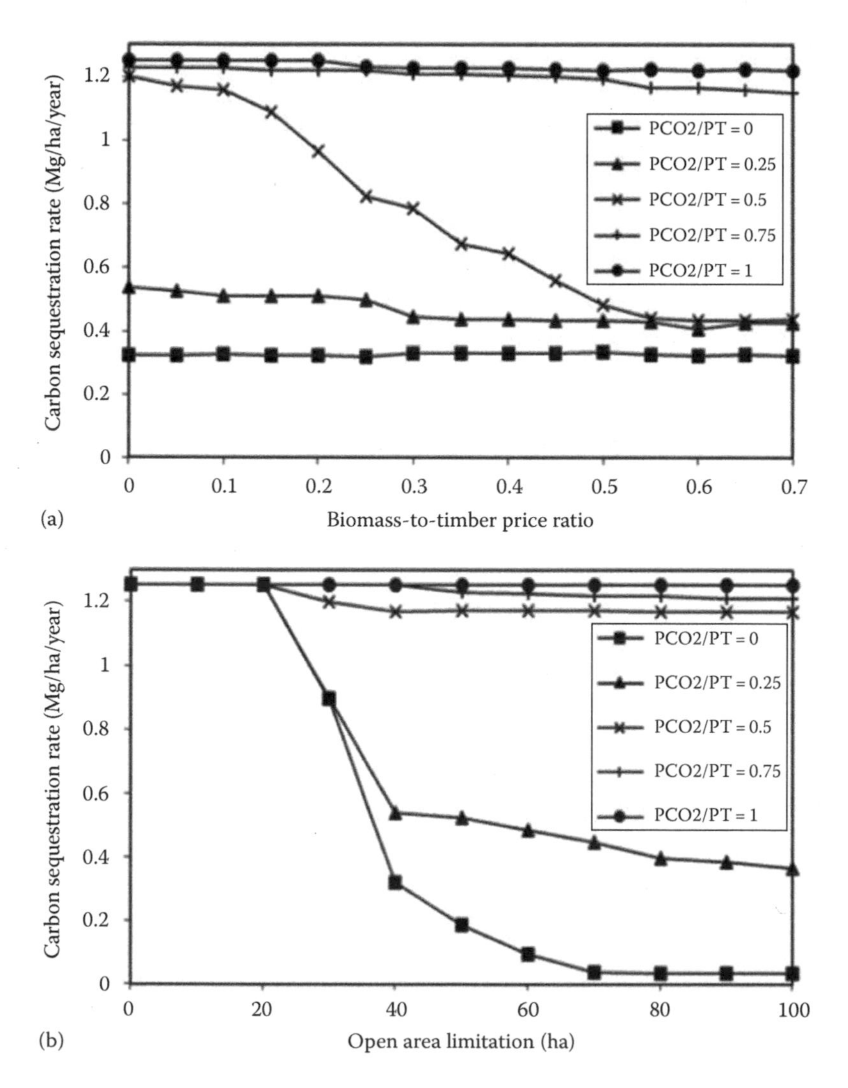

FIGURE 15.27
Carbon sequestration rate (Mg/ha/year) by (a) biomass-to-timber price ratio; (b) harvest area size (ha).

price increased, the carbon sequestration rate declined. When the carbon-to-timber price ratio was 0.0 or 1.0, the carbon sequestration rate would decline approximately 2% considering the biomass-to-timber price ratio changed from 0.0 to 0.7. However, an obvious change of the carbon sequestration rate was noticed when the carbon-to-timber price ratio was 0.5.

Harvest area limitation is important to preserve wildlife habitat in the forest and prevent soil erosion. For a given carbon-to-timber price ratio, the size restriction of continuous harvest areas becomes a primary factor affecting the amount of carbon sequestrated in a forest stand. The maximum potential carbon sequestration rate of 1.253 Mg/ha/year was achieved when the harvest area limit was less than 20 ha for lower carbon-to-timber price ratio (Figure 15.27). Assuming the carbon-to-timber price ratio was 0.0, the carbon sequestration rate steadily declined from 1.25 to 0.03 when the harvest area had changed from 0 to 100 ha. When the carbon-to-timber price ratio was higher, the carbon sequestration rate became stable though the harvest area varied.

Class Exercises

1. Describe the structure of a VB project in forest and natural resource applications.
2. Explain sequential file access and give three operations of its *Open* statement.
3. How do you implement a VB function or general procedure?
4. Compare and contrast data control vs. data access object.

References

Appalachian Hardwood Center. 2014. WV timber market report. Available online at http://ahc.caf.wvu.edu/ahc-resources-mainmenu-45/timber-market-report-mainmenu-62. Accessed on October 7, 2014.

Barahona, F., A. Weintraub, and R. Epstein. 1992. Habitat dispersion in forest planning and stable set problem. *Operations Research* 40(1): 14–21.

Bhandarkar, S.M., X. Luo, R. Daniels, and E.W. Tollner. 2008. Automated planning and optimization of lumber production using machine vision and computer tomography. *IEEE Transactions on Automation Science and Engineering* 5(4): 677–695.

Borders, B.E., W.M. Harrison, D.E. Adams, R.L. Bailey, and L.V. Pienaar. 1990. Yield prediction and growth projection for site-prepared loblolly pine plantations in the Carolinas, Georgia, Florida, and Alabama. School of Forest Resources, University of Georgia, Athens, GA. PMRC Technical Report 1990-2.

Carruth, J.S. and J.C. Brown. 1996. Predicting the operability of South Carolina coastal plain soils for alternative harvesting systems. USDA Forest Service, St. Paul, MN. General Technical Report NC-186, pp. 47–53.

Chicago Climate Exchange. 2011. CCX historical price and volume [online]. Available from: https://www.theice.com/ccx.jhtml. Accessed on September 2, 2013.

Denig, J. 1993. *Small Sawmill Handbook*. Miller Freeman, San Francisco, CA.

Farrar, K.D. 1981. In situ stand generator for use in harvesting machine simulations. MS Thesis, Virginia Tech, Blacksburg, VA, 211pp.

Greene, W.D. and B. Lanford. 1992. Logging cost analysis. Center for Continuing Education and Warnell School of Forest Resources, The University of Georgia, Athens, GA.

Grushecky, S.T., J. Wang, and D.W. McGill. 2007. Influence of site characteristics and costs of trucking and extraction on logging residue accumulations in southern West Virginia. *Forest Products Journal* 57(7/8): 63–67.

Huyler, N.K. and C.B. LeDoux. 1999. Performance of a cut-to-length harvester in a single-tree and group-selection cut. USDA Forest Service, Radnor, PA. Research Paper NE-711.

Jacobson, R. 1999. *Microsoft Excel 2000 Visual Basic for Applications: Fundamentals*. Microsoft Press, Redmond, WA.

Klinkhachorn, P., J.P. Franklin, C.W. McMillin, R.W. Conners, and H.A. Huber. 1988. Automated computer grading of hardwood lumber. *Forest Products Journal* 38(3): 67–69.

LeDoux, C.B. 1985. Stump-to-mill timber production cost equations for cable logging eastern hardwoods. USDA Forest Service, Northeastern Station, Morgantown, WV. Research Paper NE-566.

LeDoux, C.B. and N.K. Huyler. 2001. Comparison of two cut-to-length harvesting systems operating in eastern hardwoods. *Journal of Forest Engineering* 12(1): 53–59.

Li, Y. 2005. Modeling operational forestry problems in central Appalachian hardwood forests. PhD Dissertation, Division of Forestry, West Virginia University, Morgantown, WV.

Lin, W. 2011. Development of a 3D log processing optimization system for small-scale sawmills to maximize profits and yields from central Appalachian hardwoods. PhD Dissertation. West Virginia University, Morgantown, WV.

Lin, W. and J. Wang. 2012. An integrated 3D log processing optimization system for hardwood sawmills in central Appalachia, USA. *Computers and Electronics in Agriculture* 82(2012): 61–74.

Lin, W., J. Wang, and B. Sharma. 2011a. Development of an optimal three-dimensional visualization system for rough lumber edging and trimming in central Appalachia. *Forest Products Journal* 61(5): 401–410.

Lin, W., J. Wang, and E. Thomas. 2011b. Development of a 3D log sawing optimization system for small sawmills in central Appalachian, US. *Wood and Fiber Science* 43(4): 379–393.

Liu, W. 2015. Economic and environmental analyses of biomass utilization for bioenergy products in the Northeastern United States. Dissertation, West Virginia University, Morgantown, WV.

Lomax, P. 1998. *VB & VBA in a Nutshell: The Language* (1st Edition). O'Reilly & Associates, Inc., Sebastopol, CA.

Miyata, E.S. 1980. Determining fixed and operating costs of logging equipment. USDA Forest Service, St. Paul, MN. General Technical Report NC-55.

MSDN. 2010. Microsoft ActiveX Data Objects (ADO). Available from: http://msdn.microsoft.com/en-us/library/ms675532(v=vs.85).aspx. Accessed on August 16, 2010.

National Hardwood Lumber Association (NHLA). 2008. *Rules for the Measurement and Inspection of Hardwood and Cypress*. NHLA, Memphis, TN. Available from: http://www.nhla.com/pdf/2008_Rules_all.pdf. Accessed on April 8, 2011.

Reisinger, T.W. and T.V. Gallagher. 2001. Evaluation and comparison of two tree-length harvesting systems operating on steep slopes in West Virginia. In *Proceedings of the 24th Annual COFE Meeting*, July 15–19, 2001, Snowshoe, WV, pp. 56–64.

Rennie, J.C. 1996. Formulas for Mesavage's cubic-foot volume table. *NJAF* 13(3): 147.

Stage, A.R. 1973. Prognosis model for stand development. U.S. Dept. of Agriculture, Forest Service, Intermountain Forest and Range Experiment Station, Ogden, UT. Research Paper INT-137, 32pp.

Sturos, J.A., M.A. Thompson, C.R. Blinn, and R.A. Dahlman. 1996. Cable yarding as a low-impact alternative on sensitive sites in the lake states. In *Proceedings of Papers Presented at the Joint Meeting of the Council on Forest Engineering and International Union of Forest Research Organizations*, July 29–August 1, 1996, Marquette, MI, pp. 109–116.

Thomas, E. 2008. Predicting internal Yellow-Poplar log defect features using surface indicators. *Wood and Fiber Science* 40(1): 14–22.

Thompson, M.A., J.A. Sturos, N.S. Christopherson, and J.B. Sturos. 1995. Performance and impacts of ground skidding and forwarding from designated trails in an all-age northern hardwood stand. In *Proceedings of the 18th Annual Meeting of the Council on Forest Engineering: Sustainability, Forest Health & Meeting the Nation's Needs for Wood Products*, June 5–8, 1995, Cashiers, NC.

Tufts, R.A., B.L. Lanford, W.D. Greene, and J.O. Burrows. 1985. Auburn harvesting analyzer. *Compiler* 3(2): 14–15.

U.S. DOE. 2007. Technical guidelines voluntary reporting of greenhouse gases (1605(b)) program. United States Department of Energy, Washington, DC.

USDA FS. 2015. Forest Vegetation Simulator (FVS). Available online at http://www.fs.fed.us/fmsc/fvs/. Accessed on November 10, 2015.

Wang, J. and W.D. Greene. 1999. An interactive simulation system for modeling stands, harvests, and machines. *Journal of Forest Engineering* 10(1): 81–99.

Wang, J., S. Grushecky, and J. Brooks. 2004a. An integrated computer-based cruising system for central Appalachian hardwoods. *Computers and Electronics in Agriculture* 45(2004): 133–138.

Wang, J., S.T. Grushecky, and J. McNeel. 2005a. Production analysis of helicopter logging in West Virginia: A preliminary case study. *Forest Products Journal* 55(12): 71–76.

Wang, J. and C. LeDoux. 2003. Estimating and validating ground-based timber harvesting production through computer simulation. *Forest Science* 49(1): 64–76.

Wang, J., C.B. LeDoux, and Y. Li. 2005b. Simulating cut-to-length harvesting operation in Appalachian hardwoods. *International Journal of Forest Engineering* 16(2): 11–27.

Wang, J., Y. Li, and G. Miller. 2002a. A 3D stand generator for Appalachian hardwood forests. In *IUFRO S 4.11 Symposium on Statistics and Information Technology in Forestry*, September 8–12, 2002, Virginia Tech, Blacksburg, VA.

Wang, J., Y. Li, and G. Miller. 2002b. Development of a 3D stand generator for central Appalachian hardwood forest. In *Proceedings of the IUFRO S4.11 Symposium on Statistics and Information Technology in Forestry*, IUFRO Secretariat, Vienna, Austria, 10pp.

Wang, J., J. Liu, and C.B. LeDoux. 2009. A 3D bucking system for optimal bucking of central Appalachian hardwoods. *International Journal of Forest Engineering*. 20(2): 26–35.

Wang, J., C. Long, and J. McNeel. 2004c. Production and cost analysis of a feller-buncher and grapple skidder in central Appalachian hardwood forests. *Forest Products Journal* 54(12): 159–167.

Wang, J., C. Long, J. McNeel, and J. Baumgras. 2004b. Productivity and cost of manual felling and cable skidding in central Appalachian hardwood forests. *Forest Products Journal* 54(12): 45–51.

Wit, H.A., T. Palosuo, G. Hylen, and J. Liski. 2006. A carbon budget of forest biomass and soils in southeast Norway calculated using a widely applicable method. *Forest Ecology and Management* 225: 15–26.

Woo, M., J. Neider, T. Davis, and D. Shreiner. 1999. *OpenGL Programming Guide: The Official Guide to Learning OpenGL, Version 1.2*. Addison-Wesley Longman Publishing Co., Inc., Boston, MA.

Wu, J., M. Sperow, and J. Wang. 2010c. Economic feasibility of a woody biomass-based ethanol plant in central Appalachia, USA. *Journal of Agricultural and Resource Economics* 35(3): 522–544.

Wu, J., J. Wang, Q. Cheng, and D. DeVallance. 2011a. Assessment of coal and biomass to liquid fuels in central Appalachia, USA. *International Journal of Energy Research* 36(7): 856–870. DOI:10.1002/er.1838.

Wu, J., J. Wang, and J. McNeel. 2011b. Economic modeling of woody biomass utilization for bioenergy and its application in central Appalachia. *Canadian Journal of Forest Research* 41(2011): 165–179.

16

Programming for Mobile Devices and Applications in Time Study of Timber Harvesting Machines

16.1 Programming for Mobile Devices

As discussed in Chapter 9, mobile devices can be smartphones or any other PDAs or handheld PCs, which run under iOS, Windows Mobile, or Android operating systems.

16.1.1 iPhone/iPad App Programming

For iPhone/iPad programming, you will need a newer Mac computer with the latest version of OSX (the operating system for Mac computers). Once you have your Mac in place you may sign up for the Apple Developer Program at http://developer.apple.com. From there, you may download the related tools for programming. The primary tools that you should download include XCode and Interface Builder. XCode is the integrated development environment (IDE) for iPhone/iPad development and is where you will edit your iPhone/iPad code and keep your app software projects organized. Interface Builder is a program that you can use to build a graphical user interface (GUI) for your iPhone/iPad without using code at all.

16.1.1.1 Programming Languages

iPhone/iPad programming requires you use one or more of the following programming languages: C, C++, Objective-C, or Swift. C is the basic programming language used in many software systems and is primarily used to work with the low-level operating system functions in iPhone OS (iOS). C++ is used in games and to leverage code from other platforms in iOS. Objective-C is the object-oriented programming language used to work with the majority of iPhone components. Swift is a new language Apple created to develop apps for iPhone/iPad, which is the result of the latest research on programming languages, combined with decades of experience building Apple platforms (https://developer.apple.com/swift/).

16.1.1.2 iPhone Development Frameworks

The primary framework that you will use with iPhone programming is the Cocoa-Touch framework (http://developer.apple.com). This is used to create the buttons, labels, and text boxes that appear on the iPhone. Programmers will generally refer to other foundational frameworks such as NSFoundation as part of Cocoa-Touch.

iOS (formerly iPhone OS) is a mobile operating system developed and distributed by Apple Inc. It was originally released in 2007 for the iPhone and iPod Touch and has since been extended to support other Apple devices such as the iPad and Apple TV (www.apple.com/ios). Unlike Windows Mobile and Android, Apple does not license iOS for installation on non-Apple hardware. There are four increasingly complex abstraction layers in iOS: the fundamental lower-level Core OS layer, the Core Services layer, the Media layer, and the higher-level Cocoa-Touch layer (Figure 16.1a). The latest version of the operating system is iOS 10.x.

16.1.2 Android Programming

Android is a Linux-based operating system for mobile devices such as smartphones and tablet computers that was developed by Google (http://developer.android.com/guide/basics/what-is-android.html). It allows developers to write code on Java-based language that utilizes Java libraries. Android has a large community of developers writing applications ("apps") that extend the

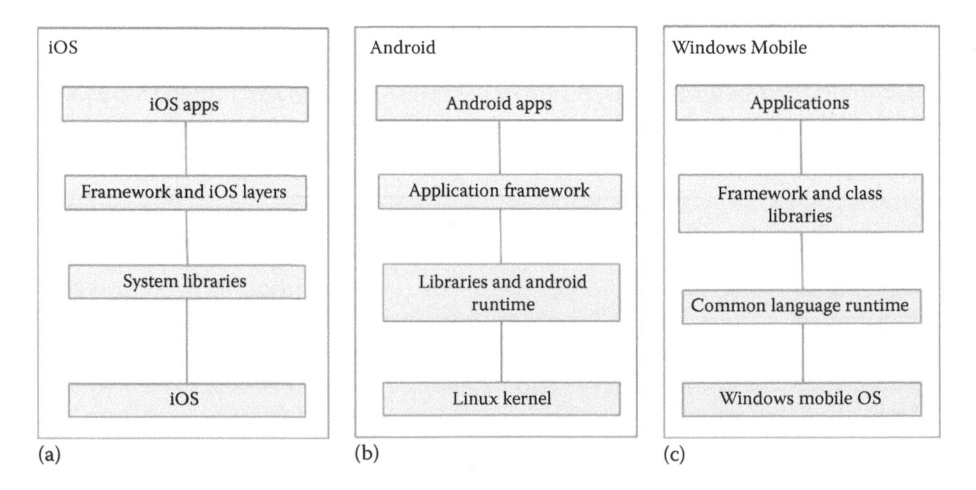

FIGURE 16.1
Development framework of (a) iOS, (b) Android, and (c) Windows Mobile .NET Compact.

functionality of the devices. Apps can be downloaded from third-party sites or through online stores such as Google Play.

16.1.2.1 Android Application Development Framework

Android was created on top of the open-source Linux 2.6 kernel (Felker and Dobbs 2011) and provides immense flexibility and the opportunity to develop diverse smart mobile applications. Developers have full access to the same framework application programming interfaces (APIs) used for the core applications. The application architecture is designed to simplify the reuse of components. The Android framework has been developed with many different features consisting of four major components: Linux kernel at the very bottom, Libraries and Android Runtime, Application framework, and Applications (Figure 16.1b). The Application component sits at the very top of the Android development framework. We can develop apps here using interactive tools with other components.

16.1.3 Windows Mobile

Windows Mobile is a mobile operating system developed by Microsoft for smartphones, pocket, tablet, and handheld PCs (Evers 2005). It is based on the Windows CE kernel and first appeared as the Pocket PC 2000 operating system (Boling 2003). Most versions of Windows Mobile have a set of standard features such as multitasking and the ability to navigate a file system with support for many of the same file types. Windows Mobile can perform basic tasks, including Internet Explorer, MS Mobile Office, and others.

16.1.3.1 .NET Compact Framework

The Microsoft .NET Compact Framework is a hardware-independent environment that supports building and running managed applications on resource-constrained computing devices (Wigley and Sutton 2003). The .NET Compact Framework inherits the full .NET Framework architecture of the common language runtime for running managed code. It provides interoperability with the Windows CE or Mobile operating system of a device. The .NET Compact Framework is a subset of the .NET Framework and also contains features exclusively designed for the .NET Compact Framework. It provides the features and ease of use that facilitate bringing native device application developers to the .NET Framework and bringing desktop application developers to devices.

If you have experience with Microsoft Visual Studio like VB.NET for desktop applications, it should be easier for you to adapt to the Windows Mobile applications. Handheld device development in Visual Studio includes a set of project types and emulators that target development for Pocket PC,

Smartphone, and handheld PCs based on the .NET Compact Framework platform architecture (Figure 16.1c) (http://msdn.microsoft.com/en-us/netframework/aa497273.aspx).

16.2 Visual Basic .NET for Windows Mobile

The VB.NET for Windows Mobile is Microsoft's answer to simplifying the process of creating applications for handheld computers. By augmenting their popular VB.NET product with the Windows Mobile Software Development Kits (SDKs), Microsoft has provided a tool that enables anyone experienced with VB.NET to immediately begin to develop Windows Mobile applications.

At present, to program Windows Mobile applications, we need to install Microsoft Visual Studio first (a later version preferred), then install Windows Mobile SDK 8, 10, or a later version. The Windows Mobile SDKs add documentation, sample code, header and library files, emulator images, and tools to Visual Studio that let you build applications for Windows Mobile (Microsoft Corporation 2012, 2016). The SDKs will allow you to build Universal Windows apps for tablet, phone, PC, or Xbox on the Universal Windows Platform or on Classic Windows applications.

16.2.1 Windows Mobile SDK

Windows Mobile SDK is an add-in for the VB.NET design and development environment. Once the Windows Mobile SDK is installed on your computer, the VB.NET integrated development environment, or IDE, is modified to provide a set of tools for creating Windows Mobile applications. Windows Mobile SDK only adds functionality to the VB.NET IDE and does not remove any existing functionality that is not applicable for use when creating Mobile applications.

16.2.2 New Project Types

When you first start VB.NET in the Windows Mobile SDK, you will notice a few new project types. Depending on the version of Visual Studio and SDK, the new project types could include Windows Phone App, Windows Phone Databound App, Windows Phone Class Library, and a few others.

Of all of the modifications that the Windows Mobile SDK makes to the VB.NET IDE, the most notable are those that are made to the menu structure. Several of the menus (the *File*, *Project*, *Debug*, *Run*, and *Tools* menus) contain new items. The modified menu structure only appears when you are working with a Windows Mobile project.

16.2.3 Features in VB.NET for Windows Mobile

It is best to know some of the different features that VB.NET and Windows Mobile SDK can provide, especially their potential impacts on how we create Windows Mobile applications. Here are some of them (Roof 1998, https://msdn.microsoft.com/en-us/library/bb847935.aspx, Microsoft Corporation 2016):

1. VB.NET for mobile devices uses different ActiveX controls. There are a number of controls included in VB.NET that might not be applicable for Windows Mobile applications. The controls themselves are physically different, and in many cases have different properties, methods, and events. The point is not to assume that the control for mobile devices will work just like the equivalent VB.NET control does.
2. You will use a different method to exit your application—both *End* and *Close()* in VB.NET for Windows Mobile.
3. There is no multiple document interface form in VB.NET for Windows Mobile.

16.2.4 Programming Examples of VB.NET for Windows Mobile

Let's consider a simple example that demonstrates how to program a Windows Mobile App for the basal area calculation of a tree in forest management. We start a VB.NET project and select *Windows Phone App* project type. A Windows phone interface will be displayed together with the *Project* and *Properties* windows as with a regular VB.NET project (Figure 16.2). In this project, we enter a diameter at breast height (DBH) value in a textbox and then click the *Calculate* button. The basal area of that tree will be calculated and displayed in another textbox. The code listing is as follows:

```
Imports System
Imports System.Threading
Imports System.Windows.Controls
Imports Microsoft.Phone.Controls
Imports Microsoft.Phone.Shell

Partial Public Class MainPage
    Inherits PhoneApplicationPage

    ' Constructor
    Public Sub New()
        InitializeComponent()

        SupportedOrientations = SupportedPageOrientation.
          Portrait Or SupportedPageOrientation.Landscape
```

```
        ' Sample code to localize the ApplicationBar
        'BuildLocalizedApplicationBar()

    End Sub

    Private Sub Button_Click(sender As Object, e As
      RoutedEventArgs)
        Dim dbh, ba As Single
        dbh = txtBoxDBH.Text
        ba = 0.005454154 * dbh * dbh
        txtBoxBA.Text = ba
    End Sub

End Class
```

Let's try another example to develop a timber cruising program for Windows Mobile devices. The program interface should consist of three tabs: (1) cruise setup and plot information, (2) tree data collection, and (3) about the program (Figure 16.3).

The program can be executed through an emulator (Figure 16.4) or on a Windows Mobile device.

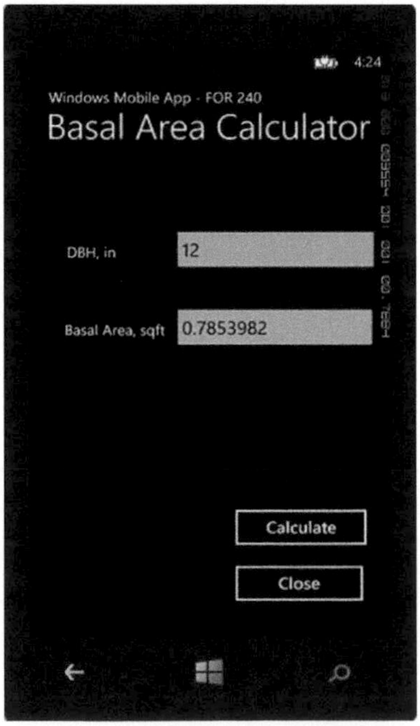

FIGURE 16.2
Windows Phone App for tree basal area calculation.

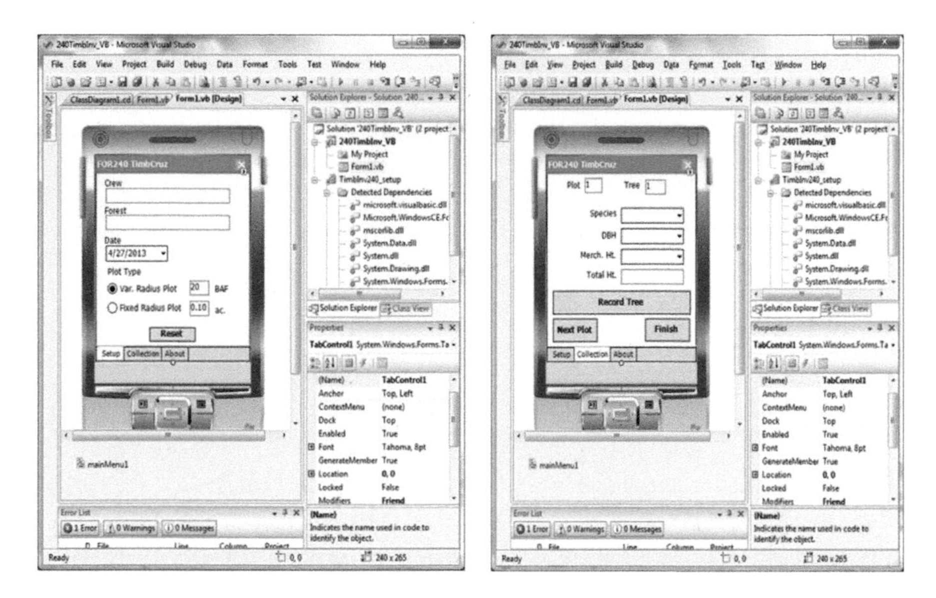

FIGURE 16.3
Interface design of a Windows Mobile application.

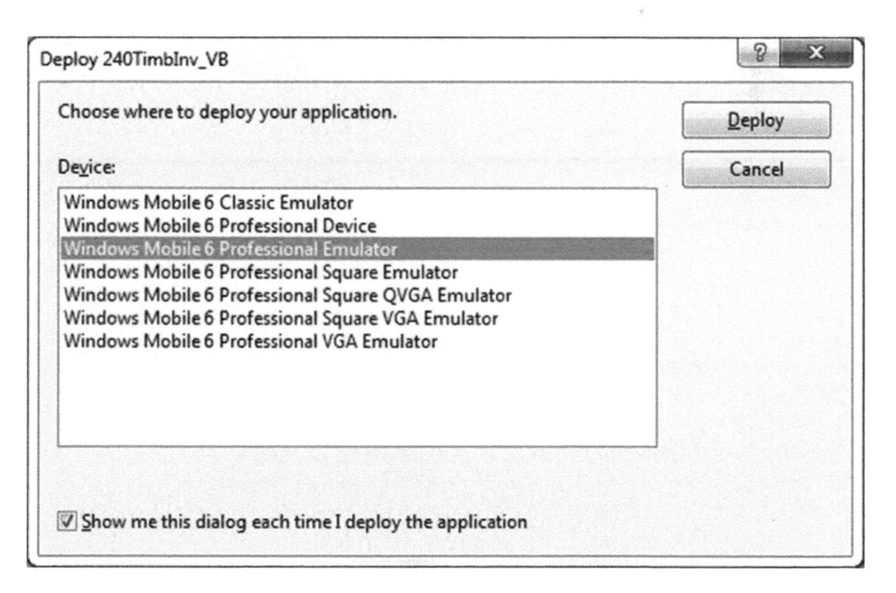

FIGURE 16.4
Emulators for Windows Mobile applications.

The code listing is as follows:

```
Public Class Form1
    Shared Trees As New List(Of String)()
    Shared Plots As New List(Of String)()
    Shared pupdate As Integer
    Shared finish As Integer = 0

    Private Sub cboMHT_SelectedIndexChanged(ByVal sender As
      System.Object, ByVal e As System.EventArgs) Handles
      cboMHT.SelectedIndexChanged
        If cboSpecies.Text = "" AndAlso cboDBH.Text = "" AndAlso
          cboMHT.Text = "" AndAlso txtTHT.Text = "" Then
            btnNextPlot.Enabled = True
            btnEnd.Enabled = True
        ElseIf cboSpecies.Text = "" OrElse cboDBH.Text = "" Then
            btnNextTree.Enabled = False
            btnNextPlot.Enabled = False
            btnEnd.Enabled = False
        ElseIf cboMHT.Text = "" AndAlso txtTHT.Text = "" Then
            btnNextTree.Enabled = False
            btnNextPlot.Enabled = False
            btnEnd.Enabled = False
        Else
            btnNextTree.Enabled = True
            btnNextPlot.Enabled = False
            btnEnd.Enabled = False
        End If
    End Sub

    Private Sub btnReset_Click_1(ByVal sender As System.
      Object, ByVal e As System.EventArgs) Handles btnReset.
      Click
        txtCrew.Text = ""
        txtForest.Text = ""
        dtDate.Value = Date.Now
        txtBAF.Text = "20"
        txtFRP.Text = "0.10"
        rdoVRP.Checked = True
    End Sub

    Private Sub btnNextTree_Click(ByVal sender As System.
      Object, ByVal e As System.EventArgs) Handles
      btnNextTree.Click
        Dim plot As String = txtPlot.Text
        Dim tree As String = txtTree.Text
        Dim species As String = cboSpecies.Text
        Dim DBH As String = cboDBH.Text
        Dim MHT As String = cboMHT.Text
```

```
        Dim THT As String = txtTHT.Text
        Dim obs As String() = {plot, tree, species, DBH,
          MHT, THT}
        Dim record As String = String.Join(",", obs)
        Trees.Add(record)
        Dim tid As Int16 = Convert.ToInt16(tree)
        tid += 1
        txtTree.Text = Convert.ToString(tid)
        cboSpecies.Text = ""
        cboDBH.Text = ""
        cboMHT.Text = ""
        txtTHT.Text = ""
        btnNextTree.Enabled = False
        btnNextPlot.Enabled = True
        pupdate = 1
        finish = 1
    End Sub

    Private Sub btnEnd_Click_1(ByVal sender As System.Object,
      ByVal e As System.EventArgs) Handles btnEnd.Click
        Dim type As String = Nothing
        Dim plot As String = txtPlot.Text
        Dim crew As String = txtCrew.Text
        Dim forest As String = txtForest.Text
        Dim [date] As String = Convert.ToString(dtDate.Value.
          [Date])

        If pupdate = 1 Then

            If rdoVRP.Checked = True Then
                type = "VRP"
                Dim factor As String = txtBAF.Text
            End If
            If rdoFRP.Checked = True Then
                type = "FRP"
                Dim factor As String = txtFRP.Text
            End If
            Dim info As String() = {plot, type, forest, crew,
              [date]}
            Dim record As String = String.Join(",", info)
            Plots.Add(record)
        End If
        If finish = 1 Then
            Dim folder As String = Environment.
              GetFolderPath(Environment.SpecialFolder.Personal)
            Dim plot_path As String = "\Plot_" & forest & "_"
              & Convert.ToString(dtDate.Value.Month) &
              Convert.ToString(dtDate.Value.Day) & Convert.
              ToString(dtDate.Value.Year) & ".txt"
```

```
        Dim tree_path As String = "\Trees_" & forest & "_"
          & Convert.ToString(dtDate.Value.Month) &
          Convert.ToString(dtDate.Value.Day) & Convert.
          ToString(dtDate.Value.Year) & ".txt"
        Dim plot_save As String = folder & plot_path
        Dim tree_save As String = folder & tree_path
        If Not System.IO.File.Exists(plot_save) Then
            Using plotstream As System.IO.StreamWriter =
              System.IO.File.CreateText(plot_save)
                For Each value As String In Plots
                    plotstream.WriteLine(value)
                Next
            End Using
        Else
            System.IO.File.Delete(plot_save)
            Using plotstream As System.IO.StreamWriter =
              System.IO.File.CreateText(plot_save)
                For Each value As String In Plots
                    plotstream.WriteLine(value)
                Next
            End Using
        End If
        If Not System.IO.File.Exists(tree_save) Then
            Using treestream As System.IO.StreamWriter =
              System.IO.File.CreateText(tree_save)
                For Each value2 As String In Trees
                    treestream.WriteLine(value2)
                Next
            End Using
        Else
            Using treestream As System.IO.StreamWriter =
              System.IO.File.CreateText(tree_save)
                For Each value2 As String In Trees
                    treestream.WriteLine(value2)
                Next
            End Using
        End If
    End If
    Close()
End Sub

Private Sub btnNextPlot_Click(ByVal sender As System.
  Object, ByVal e As System.EventArgs) Handles
  btnNextPlot.Click
    Dim type As String = Nothing
    Dim plot As String = txtPlot.Text
    Dim crew As String = txtCrew.Text
    Dim forest As String = txtForest.Text
    Dim [date] As String = Convert.ToString(dtDate.Value.
      [Date])
```

```
        If rdoVRP.Checked = True Then
            type = "VRP"
            Dim factor As String = txtBAF.Text
        End If
        If rdoFRP.Checked = True Then
            type = "FRP"
            Dim factor As String = txtFRP.Text
        End If
        Dim info As String() = {plot, type, forest, crew, [date]}
        Dim record As String = String.Join(",", info)
        Plots.Add(record)
        Dim pid As Int16 = Convert.ToInt16(plot)
        pid += 1
        txtPlot.Text = Convert.ToString(pid)
        txtTree.Text = "1"
        btnNextTree.Enabled = False
        btnNextPlot.Enabled = False
        pupdate = 0
    End Sub

    Private Sub txtTHT_TextChanged(ByVal sender As System.
      Object, ByVal e As System.EventArgs) Handles txtTHT.
      TextChanged
        If cboSpecies.Text = "" AndAlso cboDBH.Text = ""
          AndAlso cboMHT.Text = "" AndAlso txtTHT.Text = ""
          Then
            btnNextPlot.Enabled = True
            btnEnd.Enabled = True
        ElseIf cboSpecies.Text = "" OrElse cboDBH.Text = ""
          Then
            btnNextTree.Enabled = False
            btnNextPlot.Enabled = False
            btnEnd.Enabled = False
        ElseIf cboMHT.Text = "" AndAlso txtTHT.Text = "" Then
            btnNextTree.Enabled = False
            btnNextPlot.Enabled = False
            btnEnd.Enabled = False
        Else
            btnNextTree.Enabled = True
            btnNextPlot.Enabled = False
            btnEnd.Enabled = False
        End If
    End Sub

    Private Sub cboDBH_SelectedIndexChanged(ByVal sender As
      System.Object, ByVal e As System.EventArgs) Handles
      cboDBH.SelectedIndexChanged
        If cboSpecies.Text = "" AndAlso cboDBH.Text = ""
          AndAlso cboMHT.Text = "" AndAlso txtTHT.Text = ""
```

```
        Then
          btnNextPlot.Enabled = True
          btnEnd.Enabled = True
    ElseIf cboSpecies.Text = "" OrElse cboDBH.Text = "" Then
          btnNextTree.Enabled = False
          btnNextPlot.Enabled = False
          btnEnd.Enabled = False
    ElseIf cboMHT.Text = "" AndAlso txtTHT.Text = "" Then
          btnNextTree.Enabled = False
          btnNextPlot.Enabled = False
          btnEnd.Enabled = False
    Else
          btnNextTree.Enabled = True
          btnNextPlot.Enabled = False
          btnEnd.Enabled = False
    End If
End Sub

Private Sub cboSpecies_SelectedIndexChanged(ByVal sender
  As System.Object, ByVal e As System.EventArgs) Handles
  cboSpecies.SelectedIndexChanged
    If cboSpecies.Text = "" AndAlso cboDBH.Text = ""
      AndAlso cboMHT.Text = "" AndAlso txtTHT.Text = ""
      Then
        btnNextPlot.Enabled = True
        btnEnd.Enabled = True
    ElseIf cboSpecies.Text = "" OrElse cboDBH.Text = ""
      Then
        btnNextTree.Enabled = False
        btnNextPlot.Enabled = False
        btnEnd.Enabled = False
    ElseIf cboMHT.Text = "" AndAlso txtTHT.Text = "" Then
        btnNextTree.Enabled = False
        btnNextPlot.Enabled = False
        btnEnd.Enabled = False
    Else
        btnNextTree.Enabled = True
        btnNextPlot.Enabled = False
        btnEnd.Enabled = False
    End If
End Sub

Private Sub rdoFRP_CheckedChanged(ByVal sender As System.
  Object, ByVal e As System.EventArgs) Handles rdoFRP.
  CheckedChanged
    If rdoFRP.Checked = True Then
        txtFRP.Enabled = True
        txtBAF.Enabled = False
    End If
End Sub
```

```
    Private Sub rdoVRP_CheckedChanged_1(ByVal sender As
      System.Object, ByVal e As System.EventArgs) Handles
      rdoVRP.CheckedChanged
        If rdoVRP.Checked = True Then
            txtFRP.Enabled = False
            txtBAF.Enabled = True
        End If
    End Sub
End Class
```

16.3 VB.NET for Mobile Device Application in Time Study of Timber Harvesting

Time study is "a set of procedures for determining the amount of time required, under certain standard conditions of measurement, for tasks involving some human, machine, or combined activity" (Mundel and Danner 1994). Through the years, time studies have been conducted in different ways: (1) stopwatches and paper, (2) video cameras, (3) disk operating system–based handheld devices, and (4) global positioning systems.

This example is to demonstrate how to (1) develop a handheld time study system with an MS Windows-based GUI for timber harvesting operations, (2) adapt a relational database as the back end for time and factor data storage in the system, and (3) build an interface module for time and factor data communication between the desktop PC and the handheld PC using ActiveX Data Object (ADO) (Wang et al. 2003).

16.3.1 System Structure

This time study system consists of two major components: the handheld system, and a GUI and data storage component on the desktop PC (Figure 16.5). The handheld system is used to edit species, design harvesting functions and variables, and collect site, elemental time, and variable data. The GUI component on the desktop provides the interfaces and functions needed to transfer data between the handheld and the desktop PC and to manipulate and export the data for later analyses. The data storage component is a typical relational database containing tables of the time study data.

The handheld system was written with Microsoft VB CE, which runs under the Microsoft Windows CE environment. It contains two modules—design and collect (Figure 16.6). The design module in this system includes functions that allow the design work to be done on either the PC or the handheld. It provides the user with the option to enter or edit tree species to be used in the study site.

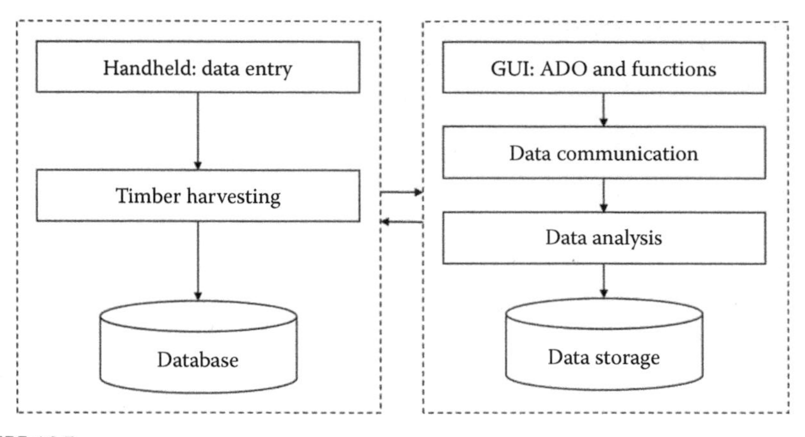

FIGURE 16.5
Architecture of the time study system.

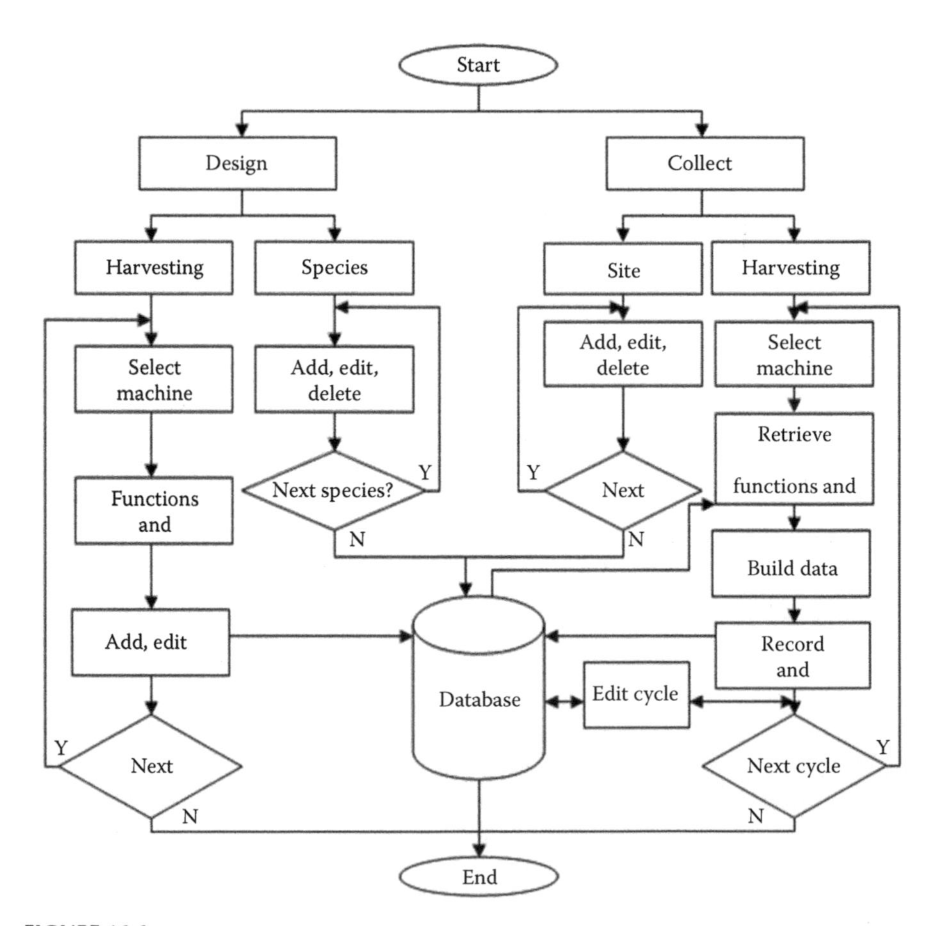

FIGURE 16.6
Flowchart of the handheld-based time study system.

Harvesting functions refer to the procedures or steps involved in the work cycle of a harvesting machine. For example, chainsaw felling may require four steps: walk to tree, acquire, fell, and delimb/top. The system allows the users to define functions for a specific machine. Harvesting variables are the factors that affect harvesting operations and elemental times. For example, DBH, height of the tree, and distance between harvested trees are the variables for chainsaw felling (in addition to site effects). Once the time study design is completed, the collect module can be invoked and will retrieve the information entered in the design module. Supporting help files using html-based architecture are also provided for this handheld time study system.

The data transfer interface module was written with MS VB V6.0 under the MS Windows 2000, XP, or Windows 7 environment. ADO CE API was employed to conduct data transfer via a dynamic link library (DLL)—adofiltr.dll. This DLL contains two functions, `DesktopToDevice()` and `DeviceToDesktop()`, which are used to transfer data or copy tables. It runs on the desktop PC, not the handheld. The desktop initiates and controls the transfer process. The key requirement for this transfer process is to have the same table schemas on both desktop and handheld. The ADO CE data transfer feature has a solid set of tools for transferring data. While the manual method for copying tables does not offer the control needed by most applications, the programmatic method does. This feature allows the transfer of complete tables between devices rather than requiring the synchronization of individual records.

A relational database model was used for holding harvesting functions, variables, and time study data in the handheld system, which was implemented based on the entity-relationship model. The relational database model presents the data as a collection of tables. Instead of modeling the relationships in the data according to the way that the data are physically stored, the structure is defined by establishing relationships between data entities. Basically, there are six data entity types in the model—*harvesting functions, variables, site, species, time track,* and *felling/skidding/forwarding/yarding*. Each entity type has its own attributes. For example, the *harvesting functions* entity has *function ID* and *name*, and *machine type* attributes. Entity types are related using relationships such as *has, contains,* and *associates* in the model. A derived attribute *Elapsed_time* was used in *time track* entity, which is derived from *Start_time* and *End_time* attributes. Attributes belonging to a key are underlined for an entity set. *Cycle number, machine type, function start time, end time, elapsed time,* and associated harvesting variables are automatically recorded. *Harvesting functions, variable data,* and *species* are stored in separate data tables in the design module and are identified by their primary keys and harvesting machine types. In the collect module, *harvesting functions* and *variables* can be queried and retrieved for a specific machine type on which another data table is created for storing functions, variables, and elemental times. *Species* information is also retrieved for data entry. The site data table contains general information such as *site number, name, location, slope,* and *weather conditions* about the logging site. *Site number* is used as a foreign key to associate site information with other data tables created in the collect module.

The *time track* entity type is used to track the start and end times of each element in a work cycle. It can also be used as a backup data table for felling, skidding/forwarding, or yarding data entities. The *felling/skidding/forwarding/yarding* entity type is designed to store time study data of felling, skidding, forwarding, or yarding depending on the type of logging operation being studied. Data schemas of the main data storage on the desktop PC are the same as those used in the handheld system in order to facilitate the data transfer.

16.3.2 System Implementation

16.3.2.1 Design Module

Three functions were implemented in this module for designing and editing species, and harvesting functions and variables (Figure 16.7). Data were edited and saved into the database in the design phase and can then be retrieved for use in the data collection phase. The system also allows the user to navigate the database for a species, function, and variable. While navigating to a specific data record in the list, the user can edit or delete the current record.

Harvesting functions and variables were implemented to allow the user to design and edit their names and associate them with machine types. Data fields, including function/variable ID, function/variable name, and harvesting type, are used in such a data table.

16.3.2.2 Collect Module

To collect data for a site, three fields are required: *site name, site slope,* and *study date* (Figure 16.7). The site number will automatically be increased and recorded when a new site is added. It will be retrieved later when the user starts to collect time study data and will be saved together with these data as a foreign key in the database.

Collecting elemental times and variable data is the ultimate objective of time studies. When invoking the collect module to collect harvesting elemental times and variable data, all the useful information entered under the design module can be retrieved and employed (Figure 16.7).

Elemental times and variables are saved in a database table whose data structure is created based on the parameters entered in the design module. The handheld system can check the data table status to determine whether the table was created or if the existing table's data structure is consistent with

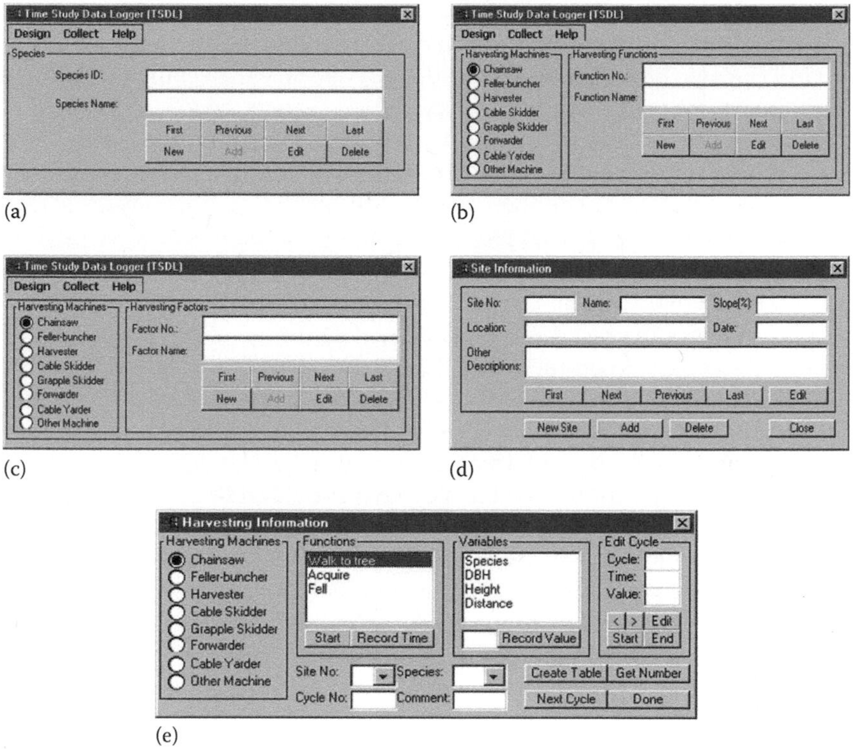

(a)

(b)

(c)

(d)

(e)

FIGURE 16.7
Main interface forms in the handheld-based time study system. (a) Design tree species, (b) design harvesting functions for associated machines, (c) design harvesting variables for associated machine, (d) design site related variables, and (e) interface for time study data collection.

the current data. If the table does not exist or its structure is not consistent, a new table can be created.

In order to associate the site information with the time study data, a site number in the *Site No.* combo box must be selected. To record elemental time for a function, the user needs to select the function from the *Function* list box. Click *Start* when this function begins and click *Record Time* once the function ends. Repeat the above procedures for any other functions in the list. If the user is not sure what function the logger is going to perform next, the program allows the user to click the *Start* button first, then select the correct function when it is identified, and finally click *Record Time* when the function is completed. The system also allows the user to record a function multiple times in sequence, and the elapsed time for this function is then calculated accordingly. To record a value for a variable, the user needs to select the variable in the *Variable* list box using the same procedures outlined for selecting a function. Then the user can simply type a value in the text box beside the

Record Value button and click this button. The value for the selected variable is recorded. Repeat the procedures for other harvesting variables. A brief comment can also be recorded in the system, which is especially useful when a comment for a cycle (reason for delay, etc.) is warranted.

Another option is also provided for entering the variable's value. If the handheld does not have a keyboard or the user does not like to use the keyboard on the handheld, she/he can use the *Species* combo box to select a species by clicking the species required if the variable is tree species. If the variable is numeric, the user can click the *Get Number* button and a data input form will be invoked. Then, the user simply clicks number buttons and the *Enter* button to get the required number.

Once the recording is completed for the work cycle, the user can click the *Next Cycle* button to save the current work cycle data to the database. Elemental time is recorded in seconds and can be converted to minutes when the user exports the data for analysis. Units for harvesting variables can be defined by the user.

An editing function for the data in a cycle is provided in the collect module. The system is implemented to allow the user to go back to any previously recorded work cycle and edit the elemental times and variables or fill out the missed data in the cycle. Since it requires connecting to the same data object from two different processes at the same time, the system was implemented not to allow performing two or more data manipulation events concurrently, but rather sequentially in order to avoid mutating table errors.

This functionality is especially useful for the time study of chainsaw felling. For example, the sawyer may harvest one tree and complete all the functions before walking to the next tree to be harvested. In some cases, however, the sawyer might acquire and fell a tree, then walk to another tree and acquire and fell it, and then go back to the first tree to delimb and top it. To edit the cycle data, the user first needs to navigate to a cycle using the "<" or ">" buttons and then use the following procedures: (a) select the function or variable from the list boxes, (b) modify the values in the *Time* and *Value* boxes, and (c) click *Edit* to update the cycle data (Figure 16.7e). The user can also use the *Start* and *End* buttons in the Edit Cycle frame box to record the elapsed time for the selected function instead of entering a value in the *Time* box.

16.3.3 Transfer Data

A data transfer module was implemented on the desktop PC, which provides three basic functions: (a) transfer data from the HPC to the PC, (b) update main data storage, and (c) empty data tables copied on the HPC. For the sake of data security, this module was designed to run the above events sequentially. First the data tables on the HPC are copied to a temporary database on the PC. Then the tables in the temporary database are appended to the related tables in the main database. Finally, the system empties the data tables on the handheld in preparation for the next time study. Meanwhile, the user can decide whether or not the event should be activated in each step.

Class Exercises

1. List the ways of programming mobile devices.
2. What are the features of VB.NET for Windows Mobile devices that differ from regular VB.NET?

References

Boling, D. 2003. *Programming Microsoft Windows CE .NET* (3rd Edition). Microsoft Press, Redmond, WA, 1224pp.

Evers, J. 2005. Microsoft to phase out Pocket PC, Smartphone brands. InfoWorld, IDG. November 8, 2015. http://www.infoworld.com/article/2668041/computer-hardware/microsoft-to-phase-out-pocket-pc--smartphone-brands.html

Felker, D. and J. Dobbs. 2011. *Android Application Development for Dummies*. Wiley Publishing, Inc., Indianapolis, IN, 388pp.

Microsoft Corporation. 2012. Windows Mobile 6 Professional and Standard Software Development Kits Refresh. http://www.microsoft.com/en-us/download/details.aspx?id=6135. Accessed on April 26, 2012.

Microsoft Corporation. 2016. Downloads and tools for Windows 10. https://developer.microsoft.com/en-us/windows/downloads. Accessed on April 15, 2016.

Mundel, M. and D. Danner. 1994. *Motion and Time Study—Improving Productivity*. Prentice Hall, Upper Saddle River, NJ, 770pp.

Roof, L. 1998. *Professional Visual Basic Windows CE Programming*. Wrox Press Ltd., Birmingham, U.K.

Wang, J., J. McNeel, and J. Baumgras. 2003. A computer-based time study system for timber harvesting operations. *Forest Products Journal* 53(3): 47–53.

Wigley, A. and M. Sutton. 2003. *Microsoft .NET Compact Framework*. Microsoft Press, Redmond, WA, 860pp.

Section VI

Web-Based Applications

17

Introduction to HTML

HTML stands for hypertext markup language. It was invented for formatting online documents and to be used by web browsers to display documents.

HTML was invented by Tim Berners-Lee in the late of 1980s at the European Laboratory for Particle Physics in Geneva, Switzerland. It has been in use by the World Wide Web (WWW) global information initiative since 1990.

HTML is a very simple markup language used to create hypertext documents. You can simply use Notepad or a professional editor such as Microsoft Visual InterDev or Expression Web to edit HTML files. While you can create a web page on your own PC, if you would like to broadcast your pages to the world, you have to save them on a server.

17.1 Terms and HTML Files

To better understand how HTML is used to create web pages, we need to know the following terms:

Hypertext—text stored in electronic form with cross-referenced links between pages.

SGML—standard generalized markup language, a standard for describing markup languages.

Java—a programming language developed by MicroSun System.

VB—stands for Visual Basic developed by Microsoft.

VB.NET—stands for Visual Basic .NET, an object-oriented programming language by Microsoft.

JavaScript—scripting language that is run on the client.

VBScript—a scripting language developed by Microsoft.

ASP—active server page, server-based Microsoft technology.

ASP.NET—active server page .NET, an open-source server-side web application framework for development of dynamic web pages by Microsoft.

XML—extensible markup language, a meta-language like SGML but without the complexity. It is a much easier way to describe online information.

Typically, web documents use a naming convention to identify the type of file being used or transferred on the Internet. Usually, HTML files have a name followed by the extension `.html` or `.htm`. Image files will have names ending with extensions such as `.gif`, `jpg`, `mpg`, etc. It is recommended that you follow these naming conventions in order to minimize confusion and problems while creating your web documents. It is also recommended that the actual file name be as descriptive of its contents as possible in order to facilitate later work on these files. Examples of recommended names for html documents include:

```
ProjectModule1.html
Presentation.htm
FinalReport.html
```

Examples of less desirable names for html documents would be:

```
a.html
myFile.htm
aProject.html
```

File location depends on the server systems. On Windows systems, all files should be put under `c:\inetpub\wwwroot\user_defined_directory`. Basically, the starting page can be anything. However, on most systems, it is assumed that a file named `index.html` or `index.htm` or `default.html` is the point of entry in a directory. That is, if only a directory name is provided by a person browsing the web, the web server will search for a file called `index.html` or `default.html` and return that file to be displayed if it exists. For example, the following two web links will direct to the same starting page of WVU Wood Science and Technology:

`http://www.wdsc.caf.wvu.edu/index.html` or

`http://www.wdsc.caf.wvu.edu`

17.2 HTML Structure

The following is the basic structure of HTML codes:

```
<HTML>
<HEAD>
<TITLE>HTML example</TITLE>
</HEAD>
<BODY>
<H1>HTML example</H1>
```

```
<P>The following is an example of HTML markup code:</P>
...
</BODY>
</HTML>
```

HTML codes usually come in pairs, enclosing the text which they format. The entire file is bracketed by HTML tags. Within this there are two main sections, the <TITLE> and <BODY> blocks. Paragraphs in the body text are enclosed within <P> tags, and so on. Linear white space is ignored; any number of blank lines may be left between codes to make it easier to read the raw HTML file. Links are easy to code, though they have been omitted from the above example for simplicity.

Let's create a simple web page that links to another page. You should name the first page "index.html" and include in it the following code:

```
<html>
<title>First Web Page</title>
<h1>Web Site for First Page</h1>
Welcome to my web site. Click <a href="jxPage2.html"> here
  </a> to access another page.
</html>
```

Name the second file "jxPage2.html" and its code should read:

```
<html>
<title>Second Web Page</title>
<h1>Another Page</h1>
Welcome to this page. From here, you can browse more pages.
</html>
```

The above code, when viewed on a web browser, will generate an output as shown in Figure 17.1.

HTML uses keywords to identify specific formatting and coding operations. For example, the keyword <html> signifies that the text that follows is an html document. Note, however, there is a keyword </html> at the end of the file. This indicates the end of the html document. So the <html> keyword begins an html document, and the keyword </html> concludes it.

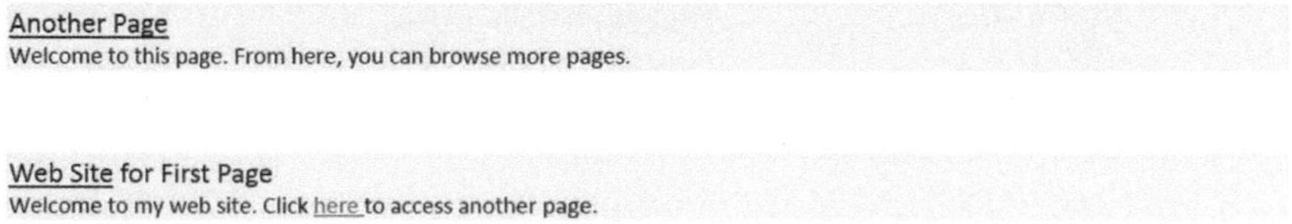

Another Page
Welcome to this page. From here, you can browse more pages.

Web Site for First Page
Welcome to my web site. Click here to access another page.

FIGURE 17.1
An output example.

In general, in HTML language, keywords are paired in this fashion: one begins a formatting action, the other terminates it. In the above example, `<title>` precedes the title given to the page and `</title>` follows it. Likewise, the keyword `<h1>` is used to begin a phrase to be displayed as a header (size 1) and `</h1>` ends the phrase.

Keywords such as `<html>`, `<title>`, and `<h1>` only have formatting implications, while some keywords such as `<a href=" ">` bring about actions. The `<a>` keyword is referred to as an anchor. It is used to identify active text, that is text which, when clicked, will make the browser load a new page (i.e., the page identified by the hypertext reference, e.g., `href="jingxinwang.html"`). Here it is assumed that the file `jingxinwang.html` exists in the same directory as the `index.html` file. If so, it will be shown on the user's browser when the user clicks on the word "`here`" imbedded between `<a>` and `</a>`.

17.3 Applications of HTML Tags

The example above introduced HTML keywords most frequently used. Table 17.1 lists other common HTML keywords.

An HTML web page always starts with `<HTML>` and then follows with the `<Head>` section and the `<Body>` section. The following can further help you understand these tags.

TABLE 17.1

Commonly Used HTML Keywords or Tags

Tag	Description
`<html> </html>`	Begin, end of html document
`<title></title>`	Document title
`<body></body>`	Web page contents
`<h1></h1>`	Header size 1
`<h2></h2>`	Header size 2 (smaller than size 1)
`<h3></h3>`	Header size 3 (smaller than size 2)
`<p></p>`	Paragraph delimiter
`<a href=""></a>`	Link identifier
`<img> </img>`	Image identifier/loader
`<table></table>`	Table formatting
`<tr></tr>`	Table row
`<td></td>`	Table column
`<form></form>`	Interactive date entries or retrievals

17.3.1 HTML

This tag tells your browser that the file contains HTML-coded information. The file extension .html also indicates this is an html document.

17.3.2 Head

The head element identifies the first part of your HTML-coded document that contains the title. The title is shown as part of your browser's window.

17.3.3 Title

The title tag contains your document title and identifies its content in a global context. The title is typically displayed in the title bar at the top of the browser window, but not inside the window itself. It is what the system will add to your favorites when browsing a site.

17.3.4 Body

The second and largest part of your html document is the body. It contains the content of your document and is displayed within the text area of the browser window.

These four tags can be seen at work in the following example:

```
<html>
<head>
<title>My Web Page</title>
</head>
<body>
This is the body of my web page.
</body>
</html>
```

If the above file is saved as "`tagtest1.html`" and you browse it, you will have output like in Figure 17.2.

The following tags are used within the body of your html document.

17.3.5 Headings

The syntax of the heading tag is:

```
<hi> text string </hi>
```

where **`i`** is between 1 and 6 from largest to smallest. For example,

```
<h2>a page title</h2>
```

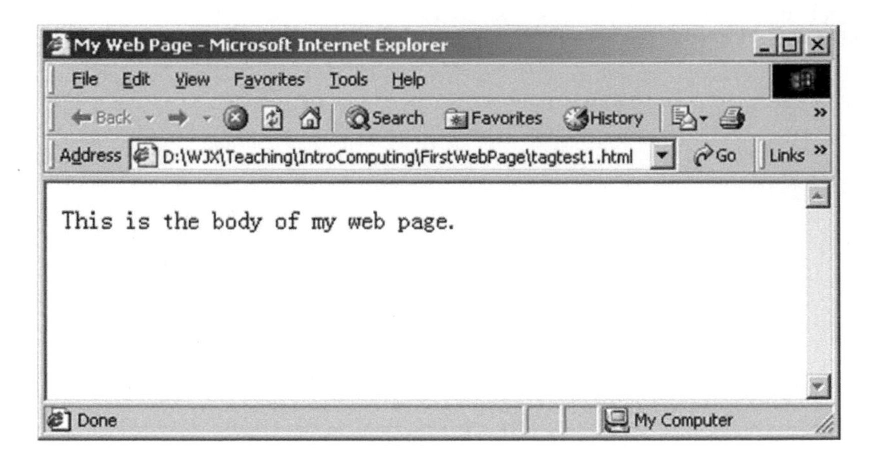

FIGURE 17.2
Output of tagtest1.html.

17.3.6 Paragraphs

Web browsers ignore line breaks in documents. The text wraps through all white spaces unless the `<p></p>` tag is used, in which a new paragraph is formed.

For example, if you typed the following into the body of your html document:

```
This is
a test
of the
p.
```

the output would look like this in a browser window:

```
This is a test of the p.
```

However, if the above file was rewritten like this:

```
<p>This is</p>
a test
of the
p.
```

you would see:

```
this is
a test of the p.
```

you can also justify the text. For example, if you typed:

```
<p align = center>
This is a centered paragraph.
</p>
```

you would see:

```
                This is a centered paragraph.
```

Let's try an exercise to show you how to use the heading and paragraph tags. Open Notepad, type the following code, and save it as "`HPTags.html`." The browser window will look like Figure 17.3.

```
<html>
<title>Headings and Paragraphs</title>
<h1>FOR 470V</h1>
<h2>Introduction To Computing</h2>
<h3>In Natural Resources</h3>
<P>Welcome to this class</P>
<P>Everyone will do an excellent job!</P>
</html>
```

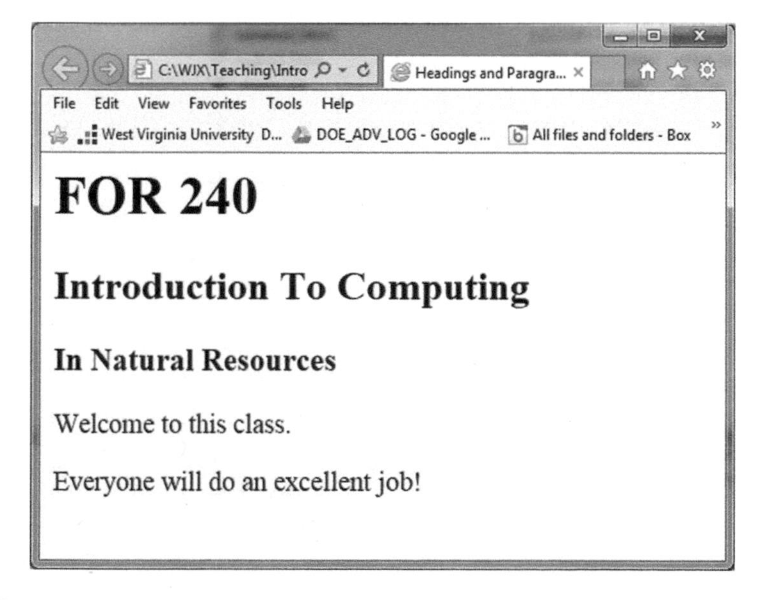

FIGURE 17.3
Headings and paragraphs.

17.3.7 Lists

HTML supports numbered and unnumbered lists. You can nest lists too, but use this feature sparingly because too many nested items can cause confusion.

A numbered list (also called an ordered list) uses `<ol>` tag. The items are tagged using the `<li>` tag.

```
<ol>
   <li>red oak
   <li>sugar maple
   <li>yellow poplar
</ol>
```

An unnumbered list uses `<ul>` to start the list, and its items are also tagged with `<li>`.

```
<ul>
   <li>red oak
   <li>sugar maple
   <li>yellow poplar
</ul>
```

Both numbered and unnumbered lists can be configured as nested lists.

```
<ul>
<li>red oak
<li>sugar maple
<li>yellow poplar
<li>pine
    <ul>
    <li>scotch pine
    <li>loblolly pine
    </ul>
</ul>
```

Here is an example to demonstrate how lists can be used:

```
<html>
<head>
<title>Using HTML lists<\title>
</head>
<body>
<p>Unnumbered list:</p>
      <ul>
      <li>red oak
      <li>pine
            <ul>
            <li>white pine
            <li>slash pine
            </ul>
```

```
        <li>sugar maple
        <li>yellow poplar
        </ul>
<p>Numbered list:</p>
        <ol>
        <li>red oak
        <li>sugar maple
        <li>yellow poplar
        </ol>
</body>
</html>
```

The output is Figure 17.4.

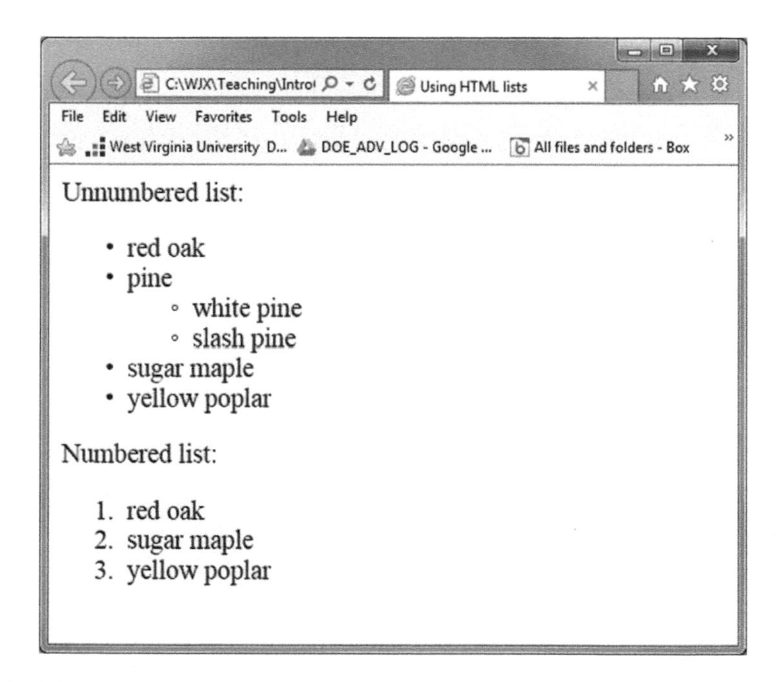

FIGURE 17.4
Output of HTML lists.

17.3.8 Forced Line Breaks

The `<br>` tag forces a line break in text with no extra (white) spaces between lines. Like `<p>` tag, `<br>` can be used to form a paragraph. However, the `<br>` tag can be efficient in text where the line breaks are significant.

```
355 Oakland Dr. <br> Morgantown, WV
```

will look like:

```
355 Oakland Dr.
Morgantown, WV
```

17.3.9 Tables and Images

HTML tables are useful to represent tabular data on a web page. The basic syntax for a table is:

```
<table>
<tr><td>column code</td></tr>
</table>
```

where
<table> defines the start of a table (you can align, color, and size here)
<tr> defines the table row (you can align, color, and size here)
<td> defines the table columns (you can align, color, size, and input data here)
<th> defines the table heading

For example, if we want to create a web page for our class and use a table to list three hyperlinks for the *syllabus*, *class notes*, and *project and lab assignments*, we'd use these codes:

```
<html>
<head>
<title>FOR 240 - Introduction to Computing ...</title>
</head>
<body>
<h2><center> FOR 240 </center>
<center>Introduction to Computing in</center>
<center>Natural Resources</center></h2>
<table ALIGN="center" BORDER="3" >
 <tr>
      <th width="100">No.</th>
      <th width="300">Contents</th>
 </tr>
 <tr>
      <td>1</td>
      <td><p><h3><A href="../courses/IntroCompuSyl2017.
        doc">Syllabus</a></h3>
      </td>
 </tr>
 <tr>
      <td>2</td>
      <td><p><h3><A href="../courses/notesFOR240.
        html">Class Notes</a></h3>
      </td>
```

```
</tr>
<tr>
    <td>3</td>
    <td><p><h3><A href="../courses/projectFOR240.
      html">Project and Lab
     Assignments</a></h3>
    </td>
</tr>
</table>
<br>
<h5 align="center"><font color="blue">This page was last
  modified on February 1, 2017.</font></h5>
</body>
</html>
```

The output page for this example is shown in Figure 17.5.

The <img> tag defines an image on an HTML web page. This tag has two required attributes: src and alt (W3Schools.com 2017). The src is to specify the source or URL of an image while the alt defines an alternate text of the image.

Here is another example of how to use a table to display both an image and texts. This is especially useful while designing a main web page.

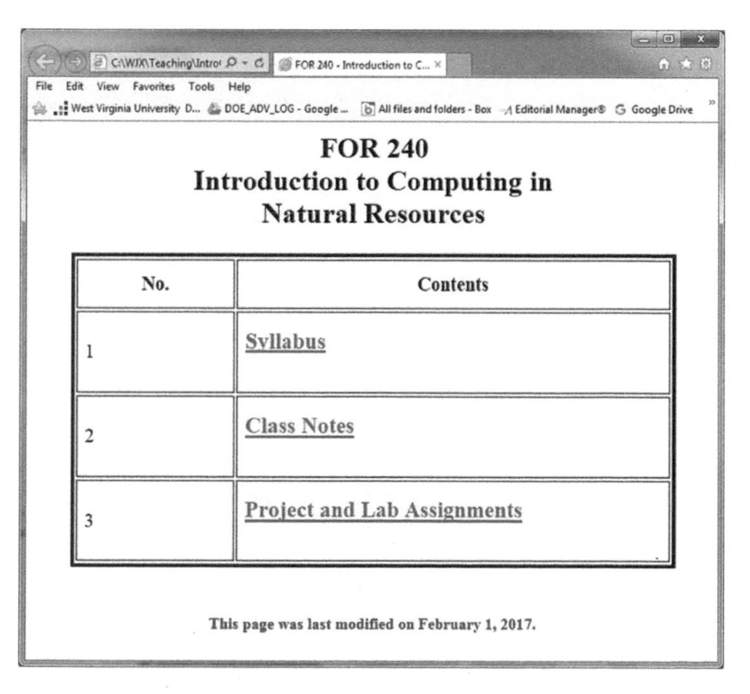

FIGURE 17.5
Output of our table example.

Open a Notepad, Expression Web, or other editor, type the following code, and save the file as "`image.html`."

```
<html>

<head>
<title>Timber Harvester</title>
</head/>

<body>
<table WIDTH="600">

<tr>
<td WIDTH="240" VALIGN="top" ALIGN="left"><img
  src="harvester.jpg" alt="Timber Harvester" border="0"
  height="154" width="164">
</td>
<td WIDTH="460" VALIGN="top" ALIGN="left" NOWRAP><br>
<p><h1><b>Timber Harvester</b></h1></p>
<p><h2><i>Harvesting Demonstration in Terrain</i></
  h2></p>
<p><b>Education</b>: Ph.D. (The University of Georgia)
  </p>
<p><b>Research Interest</b>: Forest Resources Management
  and Operations, Computer Simulations, System Modeling,
  Biomass and Bioenergy</p>
<p><b>Email Address</b>: <a href="mailto:jxwang@wvu.
  edu">jxwang@wvu.edu</a></p>
<p><b>Telephone</b>: (304) 293 - 2941 </p>
<p><b>Fax</b>: (304) 293 - 2441 </p>
<p><b>Office Room</b>: Percival Hall 317E </p>
<hr ALIGN="left" SIZE="20" WIDTH="95%" COLOR="#ff0000">
<p><font Size="2">This web page was last modified on
  January 10, 2017.</font></p>
</td>
</tr>

</table>
</body>

</html>
```

When you browse the file you just created, you will have a page like Figure 17.6. Again, in this example, we assigned an image (`harvester.jpg`) and an alternate (`Timber Harvester`) to the `src` and `alt` attributes, respectively,

```
<img src="harvester.jpg" alt="Timber Harvester">
```

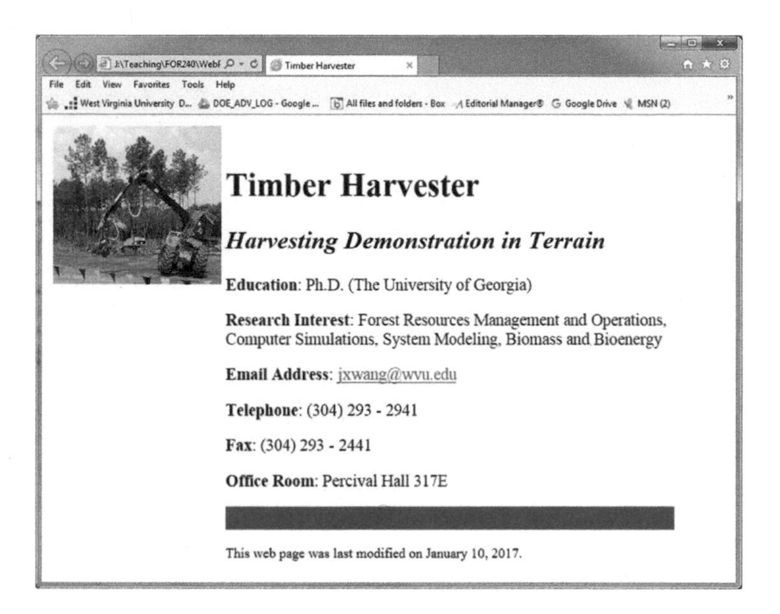

FIGURE 17.6
A page displaying image and texts in a tabular format.

The `alt` attribute is used to display text in case someone visits your page with a browser that can't show images, or in case they have turned off `image loading` so that pages will load more quickly.

The following example demonstrates how to enlarge a thumbnail-sized image (Figure 17.7). This could be useful when creating pages with tabular data containing thumbnail-sized images of, for example, animal, tree, or wood species and their explanations.

```
<html>
<title>Enlarge an Image</title>
<body>
<h3>Click the image to enlarge it!</h3>
<a href="image1.jpg" target="_top"><img src="image1.jpg"
  width="100" height="100" border="0"></a>
</body>
<html>
```

Once you click the image, it will be enlarged (Figure 17.8).

17.3.10 Forms

An HTML form is part of a web page that includes areas where readers can enter information to be sent back to the publisher of the web page. The following code details the basic form elements.

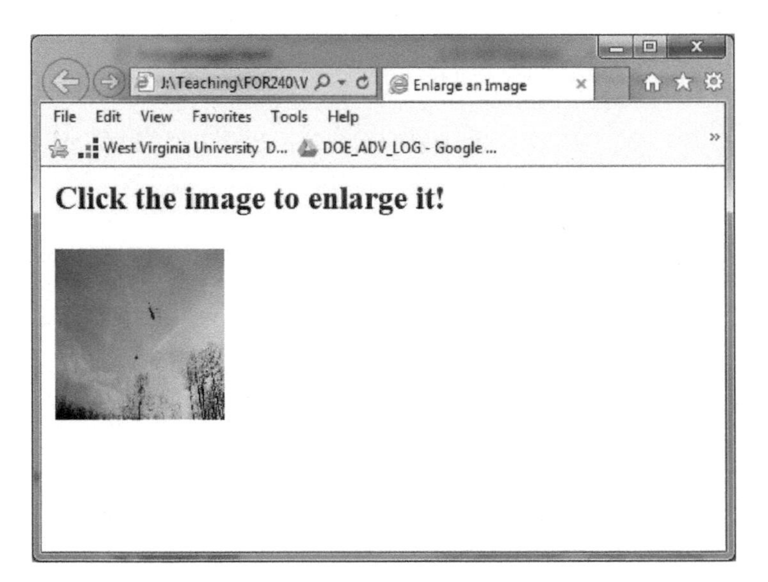

FIGURE 17.7
A thumbnail-sized image of helicopter logging.

FIGURE 17.8
An enlarged image of helicopter logging.

```
<form action="…">        (declares the form and tells what you want to do on submit)
<input type="text" NAME="subject"
  VALUE="title of form">                                  (defines a form field)
<input type="submit">                                (defines a submit button)
<input type="reset">                                  (defines a reset button)
</form>      ends the form.
```

There are numerous ways to receive input from readers of your web page. The following code specifies input types.

17.3.10.1 Types of Input

```
<input type="text">                    (text box)
<input type="radio">               (radio button)
<input type="checkbox">                (checkbox)
<input type="hidden">           (hidden text box)
<input type="password">       (password text box)
<input type="button">                    (button)
```

Let's use a simple example to demonstrate how each of the input types works on a web page. Here is the code listing.

```
<html>
<head><title>Form Example 1</title></head>
<body>
<h3>HTML Form Example 1 - Textbox, Checkbox, and Radio
  Button:</h3>
<Table>
     <form method="post" action="formtest22.asp">
          <tr>
          <td>Enter tree species: </td>
          <td><input type="text" name="v1" size="25" ></
            td></tr>
          <tr>
          <td>Select plot type: </td>
          <td><input type="radio" name="v2" value="DBH"
            size="15" ><label>DBH</label></td>
          <td><input type="radio" name="v2" value="Height"
            size="15"><label>Height</label></td>
          <tr>
          <td> Form harvesting system: </td>
          <td><input type="checkbox" name="v3" value="Feller-
            buncher" size="15"><label>Feller-buncher</
            label></td>
          <td><input type="checkbox" name="v3" value="Grapple
            skidder" size="15"><label>Grapple skidder</
            label></td>
          </tr>
          <tr>
```

FIGURE 17.9
Form for data entries—example 1.

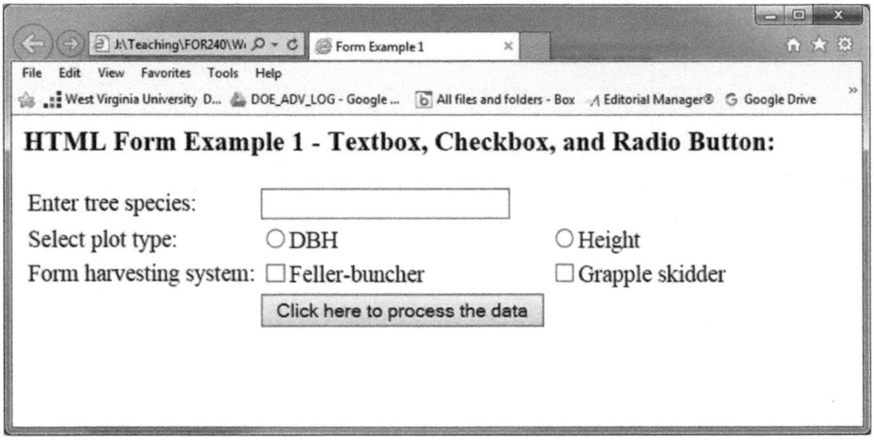

```
          <td></td>
          <td><input type="submit" value="Click here to
            process the data" border="0" align="top" /></td>
     </form>
     </tr>
</table>
</body></html>
```

The page should look like Figure 17.9.

Further info can be obtained from the reader via a drop-down menu, list-boxes, and text areas. Codes for these are in the following sections.

17.3.10.1.1 Drop-Down Menu

```
     <select name="…">
     <option>one</option>
     <option>two</option>
     <option>three</option>
     </select>
```

17.3.10.1.2 Listbox

```
     <select name="…" size=2 multiple>
     <option>one</option>
     <option>two</option>
     <option>three</option>
     </select>
```

17.3.10.1.3 Text Area

When you want to allow multiple lines of text in a single input item, use the `<textarea>` and `</textarea>` tags. For example,

```
<textarea id="comments" rows="4" cols="20">
Please send more information.
</textarea>
```

17.3.10.1.4 Getting Data from Forms

Forms collect data, but they are not used to process data. To process form data, you must include `method` and `action` in the `<form>` tag. `Method` can be set to `post` or `get`. The HTML specifications *technically* define the difference between `get` and `post` so that the former means that form data is to be encoded (by a browser) into a URL, while the latter means that the form data is to appear within a message body. As a simplification, we might say that `get` is basically for retrieving data, whereas `post` may involve anything, like storing or updating data, ordering a product, or sending an e-mail.

`Action` points to the address of the application that will process the form data. There are several ways to process the data collected by a form:

- E-mail the form data using "`mailto`."
- Use a server-based application to process the data.
- Process the form data with the client-based script.
- Send the form to a database.

Here is an example of using a drop-down menu, listbox, and text area on a web page to collect data. The code includes instructions for data to be processed.

```
<html><head><title>Form Example 2</title></head>
<body>
 <h4>HTML Form Example 2 - Dropdown menu, Listbox,
   and Text Area:</h4>
<Table>
<form method="post" action="dropdownlistbox.html">
<tr>
<td>Select tree species: </td>
<td>
<select name="dd" >
<option>Red Oak</option>
<option>Sugar Maple</option>
<option>Black Cherry</option>
```

```
</select>
</td></tr>
<tr>
<td># of trees selected: </td>
<td>
<select name="lm" size=2 multiple>
<option>One</option>
<option>Two</option>
<option>Three</option>
</select>
</td>
<tr>
<td>Field comments: </td>
<td><textarea name="ta" rows="4" cols="20">
Please enter more information here.
</textarea>
</td>
</tr>
<tr>
<td></td>
<td><input type="submit" value="Send data" border="0"
  align="top" /></td>
</form>
</tr>
</table>
</body></html>
```

If browsing this page, it should look like Figure 17.10.

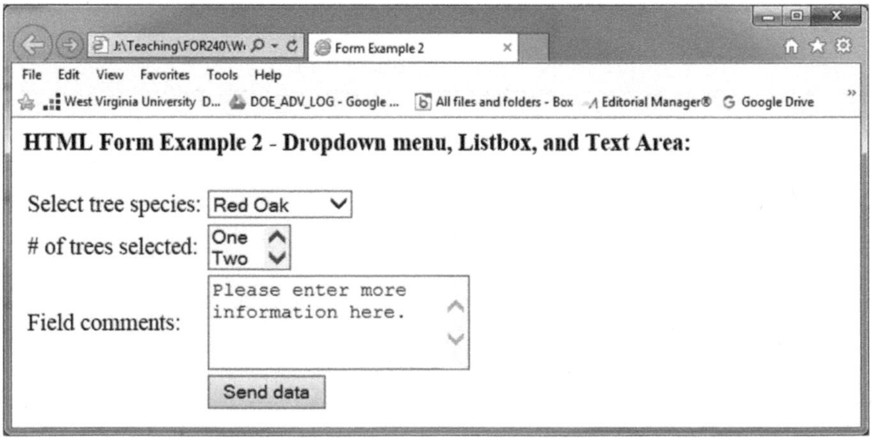

FIGURE 17.10
Form for data entries—example 2.

Class Exercises

1. Describe the structure of HTML code for a web page.
2. Why are tables and images so useful when developing web pages?
3. What are hypertext and hyper-references?
4. Create your own web pages.

   In this exercise, you need to use the HTML to create your web pages. While you are building the pages, the following items need to be considered:

   - A starting page (`index.html`) with your personal information
   - A second page about your coursework
   - A third page containing external links to other websites
   - A fourth page containing photos/images
   - At least one example of each of the following:
     - A table
     - A list
     - A change in color (font or background is acceptable)
     - A resizable picture/image

Reference

W3Schools.com. 2017. HTML <img> tag. https://www.w3schools.com/tags/tag_img.asp. Accessed on February 14, 2017.

18

Introduction to ASP.NET

Active Server Pages (ASP) is a Microsoft technology that allows us to develop hypertext markup language (HTML) pages using programming languages just before they are delivered to the browser (Buser et al. 1999). It is a great tool for creating dynamic web pages. ASP was introduced to the world by Microsoft in 1996. "It gained much wider recognition when it was bundled with version 3.0 of Microsoft's Internet Information Server (IIS) web server in 1997 and it has been gaining steadily in popularity since then" (Buser et al. 1999).

When Microsoft released ASP.NET 1.0 in 2000, many considered it a revolutionary leap forward in web application development. ASP.NET 4 continues to build on the foundation laid by the release of ASP.NET 1.0/2.0/3.5 (Evjen et al. 2010).

There are quite a few static web pages on the Internet. "A static page is a web page whose content consists of some HTML that was typed directly into a text editor such as Notepad and saved as an .htm or .html file" (Ullman et al. 2001). The static page's content is completely determined once it is developed. Alternatively, ASP.NET pages replace the hard-coded HTML code with a set of instructions that will be used to generate HTML for the page at the time the user requests the page. In other words, the page is generated dynamically on request and is called a dynamic web page.

18.1 ASP.NET Programming

When you develop an ASP.NET page, it is likely to be composed of a combination of three types of syntax, including some parts of ASP.NET, some parts of HTML tags, and some parts of pure text (Microsoft Corporation 2017). We save all these constituent parts of the ASP.NET page in a file with a .aspx extension.

To write and run ASP.NET pages requires the following programs:

- To write ASP.NET pages, we need a text editor or other web development tool. Notepad works fine for this purpose, but there are some other editors available, such as MS Visual InterDev or Expression Web or Visual Studio.

- In order to publish the pages, we'll need a web server that supports ASP.NET, such as Internet Information Server 7.0 or a later version. You can also run ASP.NET on a local machine with Personal Web Server installed.
- In order to view and test the pages, we'll need a web browser.

For example, if we use Microsoft Visual Studio to develop ASP.NET pages, we use the same integrated development environment (IDE) as we do for the VB.NET applications. To demonstrate the steps to start an IDE of ASP.NET, we'll use the following example in Visual Studio 2013 (or a later version):

a. Click *Start* and choose *All Programs* → *Visual Studio 2013* → *Visual Studio 2013.*
b. Select *Visual Basic Development Settings* as your default environment settings when you first use the Visual Studio. You can always change to other programming languages, such as Visual C++ or Visual C#.
c. Select *New Project…*, in the *New Project* dialog box, select *ASP.NET Web Application.*
d. You may rename the project and select a folder for the project, and click *OK.*
e. In the *Select a Template* dialog box, select *Web Forms* and click *OK*. You may save the Visual Studio project with as many ASP.NET pages as you want.
f. The ASP.NET IDE is displayed (Figure 18.1).

Like VB.NET IDE, the ASP.NET IDE consists of menu bar, toolbars, toolbox, search solution explorer window, properties window, windows for page design and coding, output window for debugging, and others.

Let's take a look at a simple ASP.NET page. Anything that falls between the `<%` and `%>` markers is ASP.NET script and will be processed on the web server by the ASP.NET script engine, after the ASP.NET page is requested and just before it is delivered to the browser.

Here is a simple example using ASP.NET (Figure 18.2). This page is saved as `Default.aspx`. The code listing is as follows:

```
<%@ Page Title="Home Page" Language="vb" MasterPageFile="~/
  Site.Master" AutoEventWireup="false"
    CodeBehind="Default.aspx.vb" Inherits="ASPNETapp1._
      Default" %>
<asp:Content ID="HeaderContent" runat="server" ContentPlaceHol
  derID="HeadContent">
</asp:Content>
<asp:Content ID="BodyContent" runat="server" ContentPlaceHolde
  rID="MainContent">
```

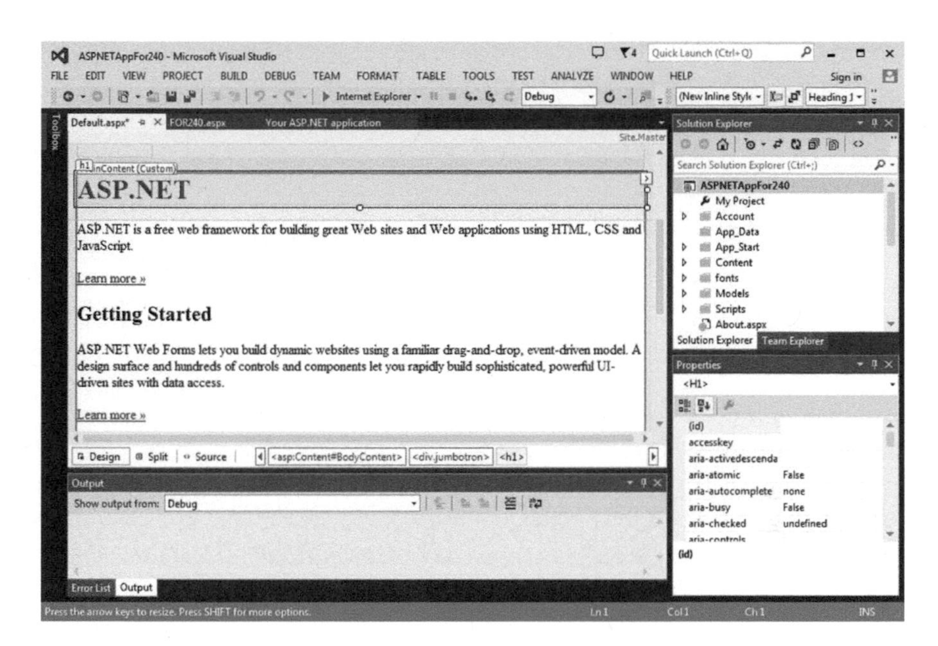

FIGURE 18.1
IDE of ASP.NET.

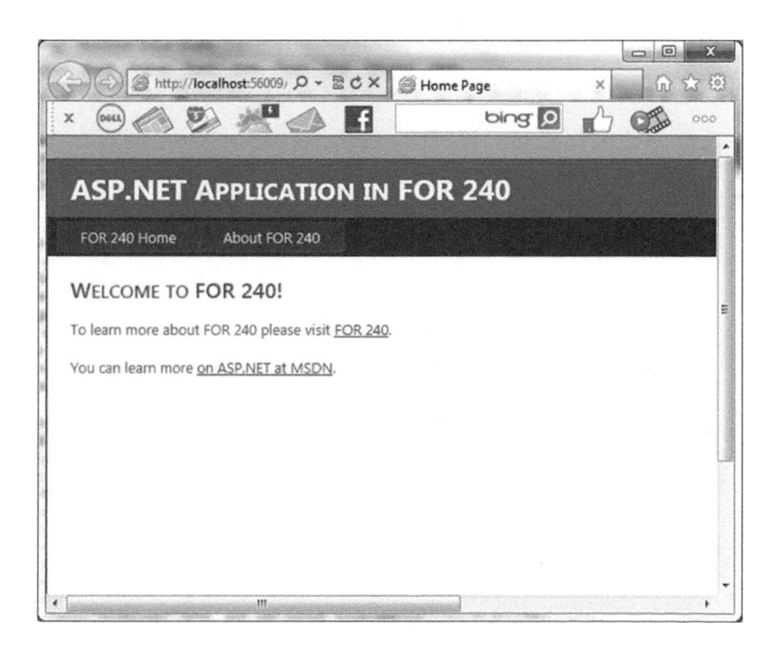

FIGURE 18.2
An ASP.NET page.

```
    <h2>
        Welcome to FOR 240!
    </h2>
    <p>To learn more about FOR 240 please visit
      <a href="http://www.wdscapps.caf.wvu.edu/jxwang/Courses/
      For240.html" title="ASP.NET Website">FOR 240</a>.
    </p>
    <p>You can learn more <a href="http://go.microsoft.com/fwl
      ink/?LinkID=152368&amp;clcid=0x409"
            title="MSDN ASP.NET Docs">on ASP.NET at MSDN</a>.
    </p>
</asp:Content>
```

The output is shown in Figure 18.2.

Another ASP.NET programming example is to use an array and a loop to calculate the sum and the average of an array of integers (Figure 18.3). The Response object is used here to send information directly to the client. The code listing is:

```
<%@ Page Language="vb" AutoEventWireup="false"
  CodeBehind="CalAvg.aspx.vb" Inherits="ASPNETapp1.WebForm1" %>

<html>
<head runat="server">
    <title></title>
</head>
<body>
    <form id="form1" runat="server">
    <div>
```

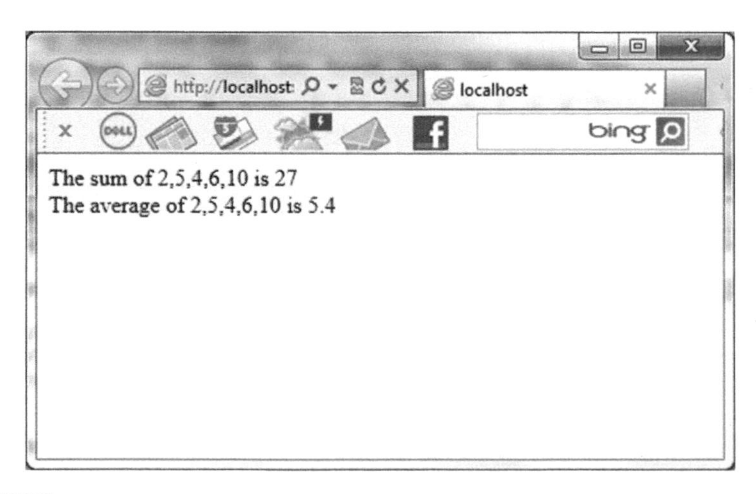

FIGURE 18.3
ASP.NET page for calculating the sum and the average of an array.

```
    <%  Dim i As Integer
        Dim intNum(5), sum, avg As Single
        intNum(0) = 2
        intNum(1) = 5
        intNum(2) = 4
        intNum(3) = 6
        intNum(4) = 10
        sum = 0
        For i = 0 To 4
            sum = sum + intNum(i)
        Next
        Response.Write("The sum of 2,5,4,6,10 is " & sum)
        avg = sum / i
        Response.Write("<br>")
        Response.Write("The average of 2,5,4,6,10 is " & avg)
        %>
    </div>
    </form>
</body>
</html>
```

The output is shown in Figure 18.3.

18.2 Basic ASP.NET Techniques

There are several intrinsic ASP.NET objects:

- Request
- Response
- Application
- Session
- Server

The following examples show you how to retrieve and process data.

18.2.1 Request and Response

This example demonstrates how to collect three integers from a user and then sum them and display the total on another page. The first file we need to create is an HTML file named "`WebForm1.aspx`."

```
<%@ Page Language="vb" AutoEventWireup="false"
  CodeBehind="WebForm1.aspx.vb" Inherits="ASPNETForm.
  WebForm1" %>
```

```
<html>
<head runat="server">
    <title></title>
    <style type="text/css">
        #form1
        {
            height: 20px;
        }
    </style>
</head>
<body>

    <div style="height: 36px">
    <h1>HTML Form Example - enter the data</h1>
    <Table>
    <form method="post" action="WebForm2.aspx">
    <tr>
    <td>Enter first integer here: </td>
    <td><input type="text" name="v1" size="25" /></td></tr>
    <tr>
    <td>Enter second integer value: </td>
    <td><input type="text" name="v2" size="25" /></td>
    <tr>
    <td>Enter third integer value: </td>
    <td><input type="text" name="v3" size="25" /></td>
    </tr>
    <p></p>
    <tr>
    <td></td>
    <td><input type="submit" value="Click here to process the
      data" border="0" align="top" /></td>
    </form>
    </tr>
    </table>

    </div>

</body>
</html>
```

The output of this form looks like what is shown in Figure 18.4.

In the code listing of `WebForm1.aspx`, the line `<form method="post" action="WebForm2.aspx">` tells that the current form data will be transferred to the address indicated by the ACTION item for processing, `WebForm2.aspx` in this case, once you enter the data and click the *Click here to process data* button. METHOD of data retrieval, post or get, is discussed in Section 17.3. The code listing of `WebForm2.aspx` is as follows:

```
<%@ Page Language="vb" AutoEventWireup="false"
  CodeBehind="WebForm2.aspx.vb" Inherits="ASPNETForm.WebForm2" %>
```

FIGURE 18.4
Interface of a form.

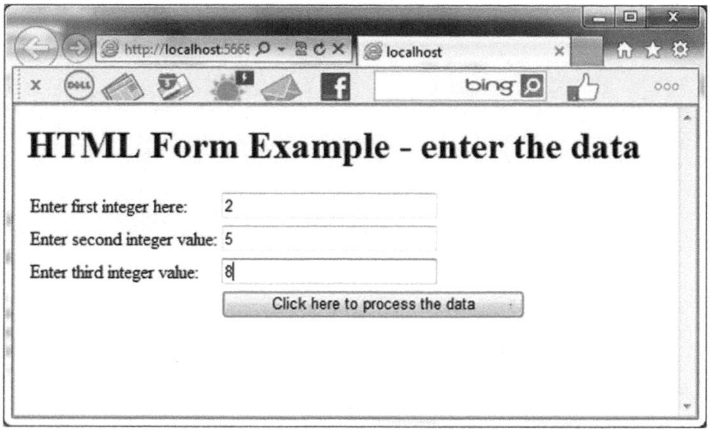

```
<html>
<head runat="server">
    <title></title>
</head>
<body>
    <form id="form1" runat="server">
    <div>
    <h1>HTML Form Example 1 - process the data</h1>
    <%
        Dim v1, v2, v3, sum As Single
        v1 = Request.Form("v1")
        v2 = Request.Form("v2")
        v3 = Request.Form("v3")
        sum = CSng(v1) + CSng(v2) + CSng(v3)
    %>
<Table>
<tr>
<td>The Sum of Three Values Entered = : </td>
<td><% Response.Write(sum)%></td></tr>
</table>

    </div>
    </form>
</body>
</html>
```

Once the data processing is done, the result will be displayed on another ASP.NET page (Figure 18.5).

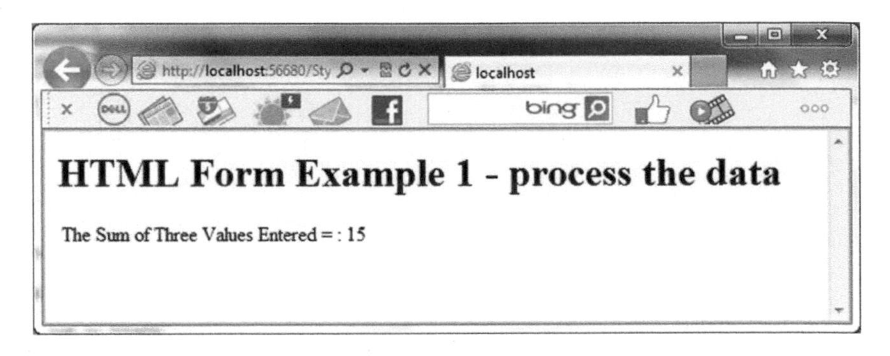

FIGURE 18.5
Result of the data processing.

18.2.2 QueryString

Another way to get the data from the user is by using the querystring. After the URL on some web pages you may notice there are some text strings. Here is an example: `www.wdsc.caf.wvu.edu?user=wdsc&pword=wvu`

From this URL, we get two text strings: `wdsc` retrieved for user name and wvu transferred for password.

Here is a simple example of how to get information from the user using querystring plus the input form. In this example, we simply want to send the species name entered in a textbox on one ASP.NET page (`WebForm1.aspx`) via querystring to another page (`getInfo.aspx`). The code listing of `WebForm1.aspx` is as follows:

```
<%@ Page Language="vb" AutoEventWireup="false"
  CodeBehind="WebForm1.aspx.vb" Inherits="ASPNETQueryString.
  WebForm1" %>


<html><head><title>Using Querystring</title></head>
<body>
<form name=gather method="get" action="getInfo.aspx">
<input name=Species>
<br>
<input type=submit value=Submit>
</form>
</body>
</html>
```

The input form of `WebForm1.aspx` should look like Figure 18.6.

FIGURE 18.6
Input form using querystring.

FIGURE 18.7
Output of using querystring.

The code for `getInfo.aspx` and the output of this querystring (Figure 18.7) are:

```
<%@ Page Language="vb" AutoEventWireup="false"
  CodeBehind="getinfo.aspx.vb" Inherits="ASPNETQueryString.
  WebForm2" %>

<html><head><title>Get Info Using Querystring</title></head>
<body>
<%
    Dim spp As String
    spp = Request.QueryString("Species")
    Response.Write("The species you entered is " & spp)
%>
</body>
</html>
```

Class Exercises

1. What is ASP.NET and how is it different from other web-based programming languages?
2. Are there any similarities between VB.NET and ASP.NET in terms of programming fundamentals?

References

Buser, D., J. Kauffman, J. Llibre, B. Francis, D. Sussman, C. Ullman, and J. Duckett. 1999. *Beginning Active Server Page 3.0*. Wrox Press Ltd., Birmingham, U.K., 1198pp.

Evjen, B., S. Hanselman, and D. Rader. 2010. *Professional ASP.NET 4 in C# and VB*. Wiley Publishing, Inc., Indianapolis, IN, 1477pp.

Microsoft Corporation. 2017. ASP.NET Overview. https://msdn.microsoft.com. Accessed on February 14, 2017.

Ullman, C., O. Cornes, J. Libre, and C. Goode. 2001. *Beginning ASP.NET Using VB.NET*. Wrox Press Ltd., Birmingham, U.K., 800pp.

19

ASP/ASP.NET Applications

We would like to conclude this textbook by offering a few examples of how the computing concepts of web applications have been addressed and applied in forest and natural resource management. We have developed several Active Server Pages (ASP) or ASP.NET applications in forest and natural resource management out of our previous research projects. Here are three examples: (1) a web-based data entry and retrieval system for forest health programs, (2) an online timber cruising program, and (3) a web-based decision support system (DSS) for timber harvesting cost analysis.

19.1 Web-Based Data Entry and Retrieval System for Forest Health Protection

The Forest Health Protection Program (FHP) is a granting mechanism by the USDA Forest Service. The FHP typically involves a lot of paperwork or proposal submission and project management. To simplify this process and improve work efficiency, we developed a web-based system for data entry, data review, and data reporting for the FHP (Grushecky et al. 2004). It was developed using ASP technology running on a Windows Internet Information Server. This system consists of a back end database (MS Access) that warehouses all new and old FHP proposal request data. The database resides on a server that receives requests from system users. All requests are processed on the server, which interacts with the database to deliver dynamic responses to the user based on the request (Figure 19.1). We programmed this application using a combination of ASP, Visual Basic Script, and Java Script and formatted it for Internet Explorer version 5 or a later version.

The major input/output forms of the application are shown in Figures 19.2 through 19.5. The first page is a login form that is specially programmed for the system security with required user name and password (Figure 19.2). It is a typical dynamic web page login form.

Figure 19.3 shows one of the data entry forms that allow users to enter or edit related data fields. You click the *Add Results* button to add the results to the related tables of the database. You can also go back to the main menu by clicking *Main Menu* button on that page.

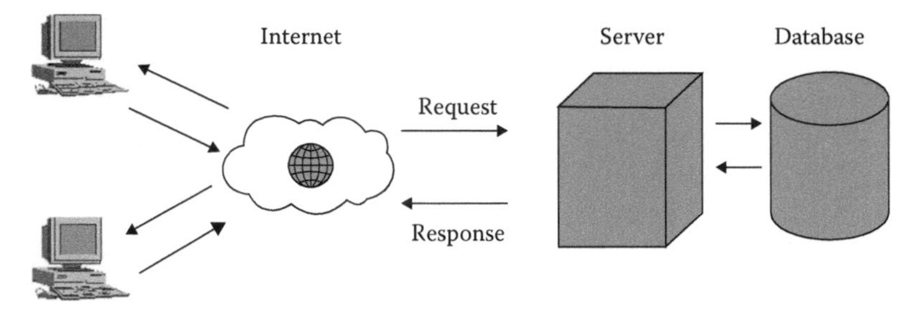

FIGURE 19.1
Schematic diagram of a web-based data storage and retrieval application. (From Grushecky, S. et al., Developing a web-based data entry and retrieval system for the forest health protection program, Final report to the USDA Forest Service, Newtown Square, PA, 2004.)

FIGURE 19.2
Login form.

ASP can also employ data editing sheets in an Excel format with multiple data types and business functions (Figure 19.4). This design can especially improve online data entry/editing efficiency.

Like MS Access, our system was also programmed to allow users to generate online reports depending on users' choices (Figure 19.5).

19.2 Online Timber Cruising System

In 2007, we began to develop an online cruising application using ASP—the Appalachian Hardwood Center (AHC) Timber Cruising Application. As discussed in Chapters 9 and 15, timber cruising first involves field measurements

FHP - Add proposal information User = Helen Butalla, logged in at 10:36:19 AM

Project Number	FWS-04-004	Project Type	Suppression
Agency	USDI F&WS	Ownership	Other Federal
Region	R-5	State	HI
Administrative Unit	Hakalau Forest NWR		
Fiscal Year	2004	Region Priority	4
Project Name	Blackberry control on Hakalau Forest NWR		
Primary Pest	Invasive Plants	Secondary Pest	
Primary Host	unknown	Secondary Host	
Acres Proposed for Treatment	1000	Treatment Method	Ground application

Project Activities	Units of Work (# of acres)	Original Fiscal Year Targets and Costs Total Cost ($)	Amount Funded by Job Codes
Pre-treatment surveys	0	0	SPFH: 84000
Treatment	1000	84000	SPCH: 0
Post-treatment monitoring	0	0	SPS4: 0
Other (NEPA or planning)		0	SPS5: 0
Direct Administrative Support		0	EPPFf: 0
Carryover		0	EPPFc: 0
Total Funding Requested		84000	

Project Accomplishments

Actual acres treated

Date completed (format mm/yyyy)

Status: completed

Amt carryover requested: 0

Comments

Select Counties (Hold Down Control Key to Select Multiple Counties)

AK — Aleutians East
AK — Anchorage
AK — Bethel
AK — Bristol Bay
AK — Dillingham
AK — Fairbanks North Star
AK — Haines

Add Results | Main Menu

FIGURE 19.3
Series of forms that allow regions to add or edit accomplishments.

and inventory data collection, and then follows with in-office data analysis using specifically designed software programs. These programs can be somewhat complicated for landowners and foresters to use. For user convenience and working effectiveness, we developed this online cruising program that allows the user to:

- Add a new cruise
- Add custom species
- Edit species
- Add custom grades
- Edit grades
- Add custom pricing

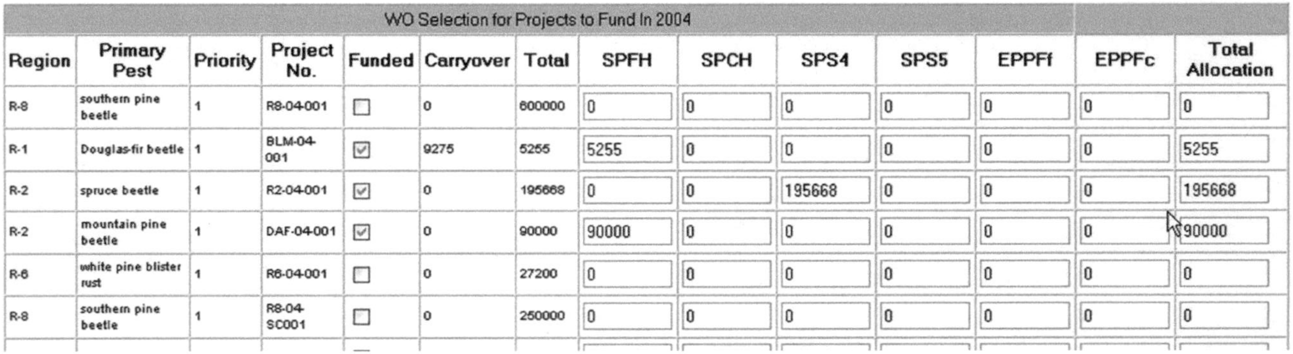

WO Selection for Projects to Fund In 2004													
Region	Primary Pest	Priority	Project No.	Funded	Carryover	Total	SPFH	SPCH	SPS4	SPS5	EPPFf	EPPFc	Total Allocation
R-8	southern pine beetle	1	R8-04-001	☐	0	600000	0	0	0	0	0	0	0
R-1	Douglas-fir beetle	1	BLM-04-001	☑	9275	5255	5255	0	0	0	0	0	5255
R-2	spruce beetle	1	R2-04-001	☑	0	195668	0	0	195668	0	0	0	195668
R-2	mountain pine beetle	1	DAF-04-001	☑	0	90000	90000	0	0	0	0	0	90000
R-6	white pine blister rust	1	R6-04-001	☐	0	27200	0	0	0	0	0	0	0
R-8	southern pine beetle	1	R8-04-SC001	☐	0	250000	0	0	0	0	0	0	0

FIGURE 19.4
Form that allows users to add and edit the funding requests.

Forest Health Proposals Funded by Primary Pest in 2004								
annosus root disease								
Prevention								
Forest Service								
Agency	**Region**	**State**	**Project No.**	**Treatment**	**Acres**	**Carryover**	**Allocations**	**Total**
USDA	R-5	CA	R5-04-004	Thinning	350	0	10000	10000
Armillaria root disease								
Suppression								
Forest Service								
Agency	**Region**	**State**	**Project No.**	**Treatment**	**Acres**	**Carryover**	**Allocations**	**Total**
USDA	R-2	CO	R2-04-037	Survey	153	0	40000	40000
USDA	R-2	CO	R2-04-047	Survey	31	0	16900	16900
USDA	R-2	CO	R2-04-024	Sanitation	20	0	13800	13800
black stain root disease								
Suppression								
Forest Service								
Agency	**Region**	**State**	**Project No.**	**Treatment**	**Acres**	**Carryover**	**Allocations**	**Total**
USDA	R-5	CA	R5-04-042	Thinning	68	0	25000	25000

FIGURE 19.5
One of the online reports generated by the program.

- Edit pricing
- Characterize stands
- Download data
- Report data

19.2.1 Main Menu

When you log onto the AHC Timber Cruise Application, you will be directed to the main menu (Figure 19.6). From this location, you can do the following: (1) add or edit cruise information (the first step in the program), (2) characterize previously entered plots into stands (you can base analyses on stand or tract level), (3) add or edit tree species (you can customize your tree list), (4) add or edit tree grades (you can have multiple grades for each tree species), (5) update timber prices (you can add prices for each species/grade combination), (6) report cruise data, and (7) download cruise data (download your data into MS Excel).

19.2.2 Add/Edit Data

When you have selected an option on the *Main Menu* form, additional forms with drop-down menus are generated to assist you in completing your

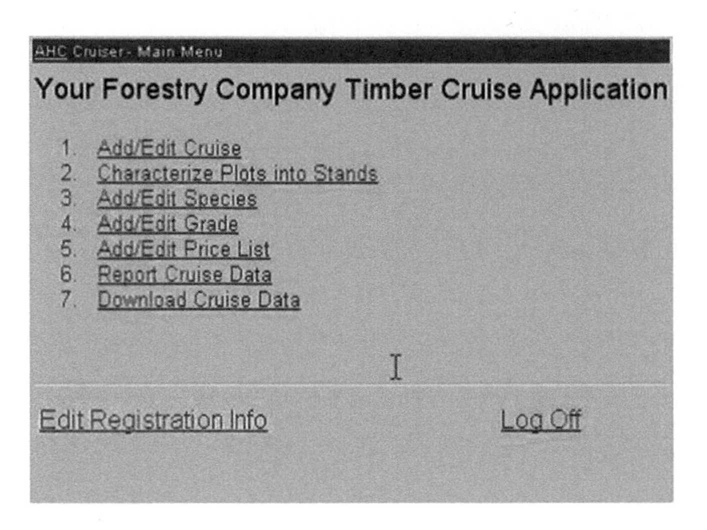

FIGURE 19.6
Main form of online cruiser.

data entry. For example, after adding species and grades for each species, you have the option to add current market prices for each species/grade combination. Simply select the species and grade from the drop-down menus (once a species is selected, the grade drop-down menu is filled with the grades for that species) and add the price in the text box. Click *Add Price* and the new market data will be visible at the bottom of the screen (Figure 19.7).

AHC Cruiser - Price List

Add Price

Species Name:

Grade:

Price ($/MBF):

Add Price | Reset | Main Menu

Price ID Click to Edit/Delete	Species	Grade	Price ($/MBF)
1	Red Oak	A	450
2	White Oak	B	125
3	Yellow-poplar	1	225
4	Yellow-poplar	2	150

FIGURE 19.7
Add/edit price.

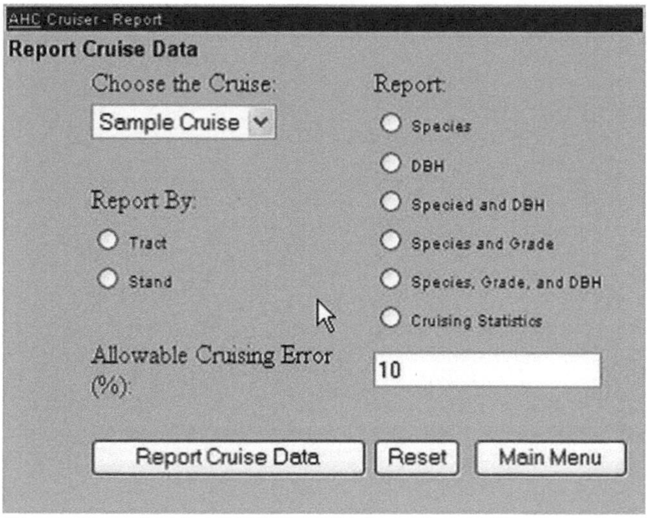

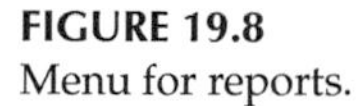

FIGURE 19.8
Menu for reports.

19.2.3 Reports

A total of 12 reports are currently available from the AHC Timber Cruising Application (Figure 19.8). To run a report, select the cruise you want to analyze from the drop-down menu, select how you want your report organized (by tract or by stand), note your allowable cruising error, and select a radio button that corresponds to the report you wish to view. Select *Stand* and *DBH* radio buttons and then click *Report Cruise Data* to run the report.

Figure 19.9 shows a sample report. This report was accessed by selecting the *Stand* and *DBH* radio buttons (Figure 19.8). Cruise information is in the upper left corner of the report. Tree information is found in the data grid. If you wish to further modify the parameters of the report or use the same format as MS Access, click *Report Design*.

19.3 Web-Based DSS for Analyzing Biomass and Timber Harvesting Costs and Productivity

An easy-to-use web-based DSS was developed to assist the analysis of timber harvesting costs and productivity (Wu et al. 2012). The system was designed using an ASP.NET platform via VB.NET language and an MS SQL server database. It consists of three web pages, including a main page for system configuration and simulation, a machine rate page for computing the

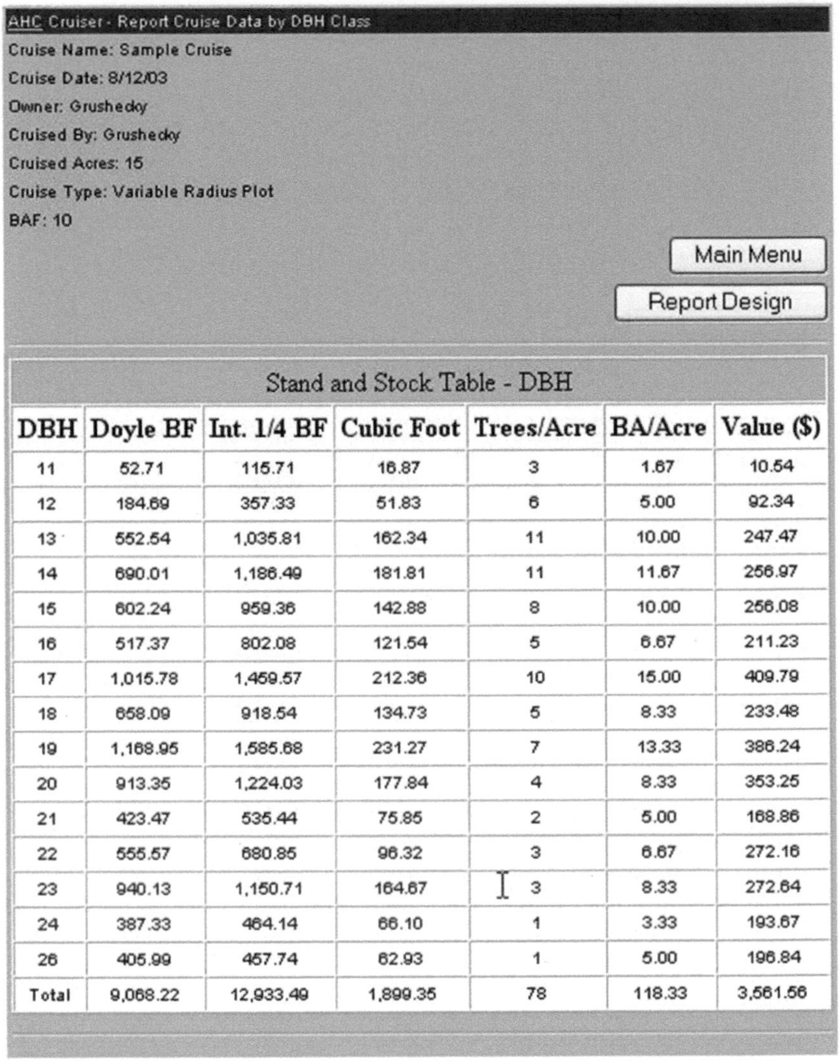

AHC Cruiser - Report Cruise Data by DBH Class

Cruise Name: Sample Cruise

Cruise Date: 8/12/03

Owner: Grushecky

Cruised By: Grushecky

Cruised Acres: 15

Cruise Type: Variable Radius Plot

BAF: 10

Main Menu

Report Design

Stand and Stock Table - DBH

DBH	Doyle BF	Int. 1/4 BF	Cubic Foot	Trees/Acre	BA/Acre	Value ($)
11	52.71	115.71	16.87	3	1.67	10.54
12	184.69	357.33	51.83	6	5.00	92.34
13	552.54	1,035.81	162.34	11	10.00	247.47
14	690.01	1,186.49	181.81	11	11.67	256.97
15	602.24	959.36	142.88	8	10.00	256.08
16	517.37	802.08	121.54	5	6.67	211.23
17	1,015.78	1,459.57	212.36	10	15.00	409.79
18	658.09	918.54	134.73	5	8.33	233.48
19	1,168.95	1,585.68	231.27	7	13.33	386.24
20	913.35	1,224.03	177.84	4	8.33	353.25
21	423.47	535.44	75.85	2	5.00	168.86
22	555.57	680.85	96.32	3	6.67	272.16
23	940.13	1,150.71	164.67	3	8.33	272.64
24	387.33	464.14	66.10	1	3.33	193.67
26	405.99	457.74	62.93	1	5.00	196.84
Total	9,068.22	12,933.49	1,899.35	78	118.33	3,561.56

FIGURE 19.9
A timber cruise report by DBH class.

handling machine costs requested by the main page, and a summary page for displaying all results. Stands, machines, and complete harvesting systems were modeled using previously published time and motion studies. A mathematical algorithm was designed to balance the harvesting system based on the machine productivity and utilization rates. The system can be used to analyze three harvesting systems used in the central Appalachian region including a chainsaw/cable skidder, a feller-buncher/grapple skidder, and a harvester/forwarder system.

19.3.1 System Design

Power (2002) defined a web-based DSS as a computerized system that delivers decision support information or decision support tools to a manager or business analyst using a "thin-client" web browser that is accessible through the Internet. Web-based DSS can be communications-driven, data-driven, document-driven, knowledge-driven, model-driven, or a hybrid of these methods. There are plenty of benefits associated with a web-based DSS, including a distributed infrastructure for information processing, timely information delivery, a user-friendly interface, and no restriction on time or geographic locations (Yao 2008).

Our web-based DSS consists of three major components: data inputs, data analysis, and data output. The basic architecture and data flow of the system is illustrated in Figure 15.11. The user interacts with the system through a specially designed interface that was modeled after the Auburn Harvesting Analyzer (Tufts et al. 1985, Greene and Lanford 1996) and the machine rate method for the cost estimates of harvesting machines (Miyata 1980). The graphical user interface (GUI) provides a user-friendly and comfortable environment to access stored information and to present and process information input by users.

The GUI includes three web pages: the *Main Page* for showing system configuration and simulation, the *Machine Rate Program* page for computing the hourly costs of the handling machines requested by the main page, and the *Summary* page (hidden by default) for showing the results from all the saved scenarios. The *Main Page* is organized into five sections similar to the Auburn Harvesting Analyzer (Tufts et al. 1985, Greene and Lanford 1996) with a number of additional functions. The *Machine Rate Program* page allows the user to calculate hourly machine costs for individual machines, which include general assumptions, fixed cost, variable cost, labor cost, and total cost in terms of $/PMH and $/scheduled machine hour (SMH). The *Summary* page provides a summary of the running results of each harvesting operation alternative.

19.3.2 Main Page

As discussed in Section 15.3 for Visual Basic for Applications with Excel worksheets, there are five sections in the *main page* of ASP.NET application: general information, machine, machine productivity, machine cost, and system.

19.3.2.1 General Information

The stand conditions are presented in a table and sourced from an MS SQL server database. Users are allowed to apply specific data to the table by clicking the *Edit* column. Most other information is also updatable, except for the automatically generated values shown in gray. If the user edits any of the information, the *Update* button on the top right corner should then be checked to update the whole section.

19.3.2.2 Machines

The felling operation could be accomplished by chainsaw, feller-buncher, or harvester; the extraction machines include rubber-tired cable skidder, rubber-tired grapple skidder, and forwarder; the knuckleboom loader is used in all the systems; the hauling can be accomplished by either a long log truck or a short log truck. This section provides the basis for the *Machine Productivity* and *Machine Cost* sections and should be addressed prior to continuing on to the other sections.

19.3.2.3 Machine Productivity

Machine productivity in terms of MBF (thousand board feet, 1 MBF = 4.59 m^3) per PMH (productive machine hour) is based on the selected machines and settings from this section. These settings include average distance between harvested trees, average extraction distance, average turn volume, load size, and others. Harvest time per tree (min/tree), harvest time per acre (hours/acre) (1 acre = 0.4 ha), and extraction time per turn are calculated from regression equations based on the selected machines, stand, and site data. Loading time and productivity are based on the selected product type (sawlogs, peeler logs, or pulp logs). In order to load the default values and compute the time and productivity for each type of equipment, users can click the *Get Values* button near the top right corner of this section. While the default values are provided by the system, users can adjust these values and recalculate the productivities based on their specific conditions.

19.3.2.4 Machine Cost

In this section, the machine hourly costs in each function are estimated and the system is balanced using a designed mathematical algorithm. A set of default Appalachian region costs is provided for the selected machines by the system. Users can click the *Default Costs* button to get the fixed costs, operating costs, and labor costs on an hourly basis. They can also opt to compute their own customized function costs by clicking a *Felling Cost*, *Extraction Cost*, or *Loading Cost* button. Since each individual machine in the system does not have the same productivity rate, there is a need to balance the harvesting system to minimize the cost per unit and maximize the system production rate (Wu et al. 2012). The following algorithm was used to accomplish system balancing:

$$\text{Number of machines in function } i = \left\lfloor \frac{\text{Max}(P_i \times U_i)}{P_i \times U_i} \right\rfloor,$$

where i is the function or operation stage, $i = (1, 2, 3)$, 1 = felling, 2 = extraction, 3 = loading. P_i is the single machine productivity in function i, U_i is the mechanical availability of the machine in function i. $\lfloor\,\rfloor$ is the integer floor function, which will return the least integer greater than or equal to $\text{Max}(P_i \times U_i)/P_i \times U_i$. By default, only one machine in each machine type is provided. Users have to click the *System Balance* button to balance the system in order to get the best system configuration. If loggers do not have the optimal number of machines available, they can manually input the actual number of machines they will use. The number of hauling trucks can be obtained by dividing the production rate of the balanced system (MBF/SMH) by the hauling productivity (MBF/SMH).

19.3.2.5 System

The machine productivity and unit cost for each harvesting function, system production rate, weekly production, onboard cost, and time required for harvesting a certain tract are reported in this section. The weekly production rate is provided in both volume per PMH and the number of truck loads. The total time required for harvesting the given tract is estimated and reported based on the user-supplied tract data and calculated system rate. When users click the *System Results* button, the results for the complete harvesting system are populated. After reviewing these specific results, users can click the *Save* button to save the system production rate/cost along with other summary information (stand conditions and machine rate/cost) into the MS SQL server database. Multiple simulations can be performed by changing parameters such as extraction distance, working schedule, and machine selection in order to make analyses and comparisons by system.

19.3.3 Machine Rate Program Page and Summary Page

The *Machine Rate Program* page is loaded when users click one of the *Felling Cost, Extraction Cost*, or *Loading Cost* buttons. This page is designed to provide an individual machine rate that includes fixed, operating, and labor costs that can be calculated based on the machine rate method (Miyata 1980). Default data from the Appalachian region are provided for purchase price, economic life, interest, insurance and taxes, fuel and lube consumption rate and prices, labor cost, and fringe benefits, as well as working schedule in order to assist the cost estimation process. Each of these variables can be adjusted to match the specific equipment and individual logger conditions. The calculated costs will be passed to the main page once the button *Return to Main Page* is triggered.

The Summary page displays the summary results from different running scenarios. After running all scenarios, users can hit the *Summary Report* button at the bottom of the *Main Page* to load the *Summary* page. Two charts are automatically generated and displayed on top of the page depicting the weekly system production and onboard cost. These graphs allow users to visually compare harvesting systems under different stand and harvest conditions. Three tables are pulled from the MS SQL server database and include stand summary, machine summary, and system summary. These tables provide all of the data and results' summaries of each system scenario. The database is cleaned out once the *Summary* page is unloaded to ensure that every user will only view his/her own analysis scenarios.

19.3.3.1 Implementation

When a user starts the web-based system, the default general information (such as site conditions and other support and road information) is loaded and displayed (Figure 19.10). Users can opt to click *Edit* in front of the tree table to update the tree information, and then click the *Update* button on the top right corner to get the tree volume in the designated tract. Users should also ensure the other default data are accurate for their scenario, or update those variables before moving on to the next section. In the *Machines* section, four drop-down boxes are used to select the machine type in each harvesting function. By clicking the *Get Values* and *Default Values* buttons in the sections of *Machine Productivity* and *Machine Cost*, the default productivities and costs

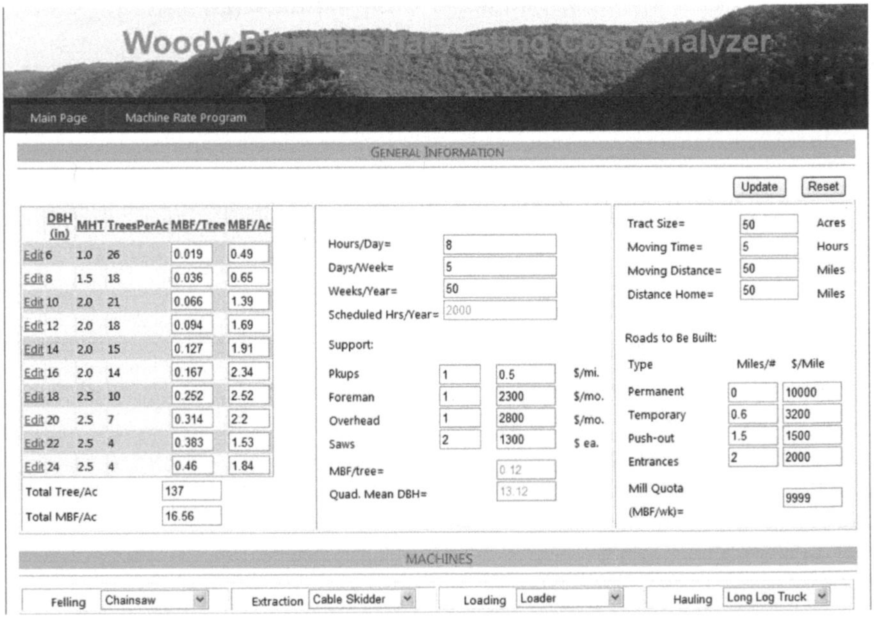

FIGURE 19.10
Partial view of the web-based DSS.

are retrieved based on the selected machines and knowledge base. The numbers of machines in each function are displayed as well. Users can click the button *System Balance* to get the optimal mix of machines in the harvesting system, or they can manually input the number of machines.

The default machine costs can also be updated by clicking the buttons for *Felling, Extraction*, or *Loading*. This action will lead users to the *Machine Rate Program* page (Figure 19.11). On this page, they will need to input specific information such as machine purchase price, economic life, interest, insurance and taxes, fuel consumption, mechanical availability, labor cost, and working schedule to get the machine cost and return to the main page. By clicking the *System Results* button at the bottom of the page, the system-related results, such as system rate, weekly production, onboard cost, and days required to cut the tract, are calculated and displayed in the *System* section. These results can be saved to the SQL database via the *Save* button. The whole process can be repeated for different harvesting systems, and each result can be saved to the same SQL database. To compare the different systems, users can click the *Summary Report* button at the bottom of the *Main Page*. In the resulting pop-up page, two charts related to harvesting system weekly production and onboard cost for each harvesting scenario, and three summary tables will be generated to provide detailed information for scenario comparisons.

Main Page Machine Rate Program

ESTIMATION OF HOURLY MACHINE COSTS

Purchase Price ($)	90000	Fuel	6.5	gal/PMH @ $	0.75	/gal
Salvage Value (%P)	25	Lube	4	qts/PMH @ $	1.16	/qt
Economic Life (years)	4	Repair and Maint. (% D)			100	
Interest (% AVI.)	12	Labor ($/SMH)			12	
Insurance (% AVI.)	5	Labor Fringe (% LR)			40	
Taxes (% AVI.)	3	Mechanical Availability (%)			65	
Weeks/year	50	SMH/Week			40	

FIXED COSTS:	$/SMH	$/PMH
Depreciation	8.44	12.98
Interest, Insurance and Taxes	6.47	9.95
Total Fixed Costs	14.91	22.93
VARIABLE COSTS:	$/SMH	$/PMH
Maintenance and Repair	8.44	12.98
Fuel and Lubrication	6.18	9.51
Total Variable Costs	14.62	22.49
LABOR COSTS:	$/SMH	$/PMH
Wages or Salaries	12	18.46
Fringe Benefits	4.8	7.38
Total Labor Costs	16.8	25.84
TOTAL HOURLY COSTS:	46.33	71.26

Return to Main Page | Calculate | Reset | Cancel

FIGURE 19.11
Machine Rate Program page.

19.3.3.2 Applications

This online program can be used by researchers and practitioners to analyze biomass or forest harvesting productivity and cost under a variety of site, stand, machine, and other operational conditions.

Class Exercises

1. What is the basic structure of a web-based, interactive application in forest and natural resource management?
2. Describe any other potential applications of web-based programming in forest and natural resources.

References

Greene, W.D. and B.L. Lanford. 1996. Logging cost analysis. Short course manual. Georgia Center for Continuing Education, Athens, GA, Vol. 78(119), p. 6.

Grushecky, S., J. Wang, and J. McNeel. 2004. Developing a web-based data entry and retrieval system for the forest health protection program. Final report to the USDA Forest Service, Newtown Square, PA.

Miyata, E.S. 1980. Determining fixed and operating costs of logging equipment. USDA Forest Service General Technical Report NC-55, St. Paul, MN.

Power, D.J. 2002. *Decision Support Systems: Concepts and Resources for Managers.* Greenwood Publishing Group, Westport, CT.

Tufts, R.A., B.L. Lanford, W.D. Greene, and J.O. Burrows. 1985. Auburn harvesting analyzer. *Compiler* 3(2): 14–15.

Wu, J., J. Wang, Y. Li, and B. Spong. 2012. A web-based decision support system for analyzing timber harvesting costs and productivity. *Northern Journal of Applied Forestry* 29(3): 141–149.

Yao, J. 2008. Web information fusion: A review of the state of the art. *Journal of Intelligent Systems* 17(1–3): 446.

Index

L

M

O

P

R

S

W